SUSTAINABLE BUILDINGS AND STRUCTURES: BUILDING A
SUSTAINABLE TOMORROW

PROCEEDINGS OF THE 2ND INTERNATIONAL CONFERENCE IN SUSTAINABLE BUILDINGS AND STRUCTURES, SUZHOU, CHINA, 25-27 OCTOBER 2019

Sustainable Buildings and Structures: Building a Sustainable Tomorrow

Editors

Konstantinos Papadikis, Chee S. Chin, Isaac Galobardes, Guobin Gong &

Fangyu Guo

Department of Civil Engineering, Xi'an Jiaotong – Liverpool University (XJTLU), Suzhou P.R. China

CRC Press
Taylor & Francis Group
Boca Raton London New York

CRC Press is an imprint of the
Taylor & Francis Group, an **informa** business

A BALKEMA BOOK

Published by:
CRC Press/Balkema
Schipholweg 107C, 2316 XC Leiden, The Netherlands

First issued in paperback 2023

© 2020 by Taylor & Francis Group, LLC
CRC Press/Balkema is an imprint of Taylor & Francis Group, an informa business

No claim to original U.S. Government works

ISBN-13: 978-1-03-257093-8 (pbk)
ISBN-13: 978-0-367-43019-1 (hbk)

DOI: https://doi.org/10.1201/9781003000716

Visit the Taylor & Francis Web site at
http://www.taylorandfrancis.com

and the CRC Press Web site at
http://www.crcpress.com

Typeset by Integra Software Services Pvt. Ltd., Pondicherry, India

Publisher's Note
The publisher has gone to great lengths to ensure the quality of this reprint but points out that some imperfections in the original copies may be apparent.

Table of Contents

Sustainable construction materials

Sustainable design in built environment

Green and low carbon buildings

Preface

The 2nd International Conference in Sustainable Buildings & Structures (ICSBS 2019) was held between the 25th and 27th of October 2019 in the South Campus of Xi'an Jiaotong – Liverpool University (XJTLU) in the historic city of Suzhou in China. The conference was organised by the Department of Civil Engineering in XJTLU, the Institute of Sustainable Materials and Environment (ISME) in XJTLU, the University of Hong Kong (HKU), and the Chinese Research Institute of Construction Management (CRIOCM).

Experts from various parts of the globe attended ICSBS 2019 and exchanged their innovative ideas and recent developments under the general theme 'Building a Sustainable Tomorrow' in the four distinct topical tracks of Sustainable Construction Materials, Sustainable Design in Built Environment, Green and Low Carbon Buildings, and Smart Construction and Construction Management. The conference aimed to bring together leaders in academia and industry experts in a forum that explored the future challenges of construction industry in the context of sustainable development. We were honoured to host 11 international keynote speakers from the academia and industry, 21 invited talks in selected areas of significance, and 51 oral presenters, graciously sharing their experiences and works among each other.

The efforts made by the international scientific committee in providing critical comments and improving the quality of the submitted works are gratefully appreciated. We are also very grateful to the Committee Secretaries, Sessions Chairs, student volunteers and all the professional services staff that unselfishly contributed to the successful organization of ICSBS – 2019. Last but not least, we would like to extend our thanks to all the authors and presenters of the conference. It is the quality of their work and their aspiration for a 'Sustainable Tomorrow' that made this conference a great success.

Dr. Konstantinos Papadikis
Organising Committee Chair, ICSBS 2019
Head of the Department of Civil Engineering,
Head of Design Cluster,
Xi'an Jiaotong – Liverpool University

Sustainable construction of buildings and structures is a core worldwide issue of concern that mankind has been trying to resolve due to the significance of resulting environmental impacts on earth and more importantly the depletion of raw resources. It would not be possible to identify solutions as we wish to achieve without gathering experts in the relevant areas to engage and discuss the way forward. The main aim of the Second International Conference on Sustainable Buildings and Structures (ICSBS 2019) is to continue its mission in fostering research interaction between academics and practitioners from various communities to focus on sustainability in built environment and infrastructure system. The themes of the ICSBS 2019 centre primarily on the state-of-the-art research in sustainable design and construction of infrastructures as well as development of smart and green buildings. The ICSBS 2019 publication has included 53 high quality peer reviewed papers and acknowledgements should go to all the contributors to this second volume. I would also like to thank Department of Civil Engineering and Institute for Sustainable Materials and Environment at Xi'an Jiaotong-Liverpool

University for their financial supports as well as the Organising and Technical Review Committees for all the efforts paid behind the scene to make this publication a successful endeavour.

Prof. Chee S. Chin

Scientific Committee Chair, ICSBS 2019

Dean of Learning and Teaching,

Director of the Institute of Sustainable Materials and Environment,

Xi'an Jiaotong – Liverpool University

Organisation

This volume contains the Proceedings of the 2nd International Conference on Sustainable Buildings & Structures – ICSBS 2019 held in Suzhou (China), from 25th to 27th of October 2019. ICSBS 2019 was organised by the Department of Civil Engineering at Xi'an Jiaotong – Liverpool University (XJTLU), the Institute of Sustainable Materials and Environment (ISME) at XJTLU, the University of Hong Kong (HKU), and the Chinese Research Institute of Construction Management (CRIOCM).

ORGANISING COMMITTEE

Chair: Dr. Konstantinos Papadikis, XJTLU, China

Co-Chair: Dr. Wilson Lu, HKU, China

Secretaries:
Dr. Guobin Gong, XJTLU, China
Dr. Fangyu Guo, XJTLU, China

Organising Committee:

Dr. Cheng Zhang	XJTLU, China
Dr. Hyungjoon Seo	XJTLU, China
Dr. Jian Li Hao	XJTLU, China
Dr. Jun Xia	XJTLU, China
Ms. Tingjia Wu	XJTLU, China
Ms. Xiaoyi Wu	XJTLU, China
Ms. Xiuran Xu	XJTLU, China
Ms. Yan Zhu	XJTLU, China

SCIENTIFIC COMMITTEE

Chair: Prof. Chee S. Chin, XJTLU, China

Secretary: Dr. Isaac Galobardes, XJTLU, China

External Scientific Committee:

Dr. Albert de la Fuente	Universitat Politecnica de Catalunya, Spain
Dr. Antonio Figueiredo	Universidade de Sao Paulo, Brazil
Dr. Arnold Yuan	Ryerson University, Canada
Dr. Berenice Martins	Universidade Federal de Londrina, Brazil
Dr. Byung Kang	University of Nottingham Ningbo, China
Dr. Chamila Gunasekara	RMIT University, Australia
Dr. David Law	RMIT University, Australia
Dr. Eshmaiel Ganjian	Coventry University, United Kingdom
Dr. Faisal Arain	Niagara College, Canada
Dr. Fan Xue	Hong Kong University, China
Dr. Guangming Chen	South China University of Technology, China

Dr. Hongping Yuan	Guangzhou University, China
Dr. Hongtao Wang	Sichuan University, China
Dr. Jian Yang	Shanghai Jiao Tong University, China
Dr. Kangkang Tang	University of Wolverhampton, United Kingdom
Dr. Karol Sikora	University of Wollongong in Dubai, United Arab Emirates
Dr. Ke Chen	Huazhong University of Science and Technology, China
Dr. Kincho Law	Stanford University, United States
Dr. Konstantinos Moustakas	National Technical University of Athens, Greece
Dr. Lan Kang	South China University of Technology, China
Dr. Lan Lin	Concordia University, Canada
Dr. Leo Gu Li	Guangdong University of Technology, China
Dr. Lin Liao	Taiyuan University of Technology, China
Dr. Luis Segura	Universidad de la Republica, Uruguay
Dr. Meng Ye	Hong Kong University, China
Dr. Ming-Xiang Xiong	Guangzhou University, China
Dr. Ricardo Pielarissi	Universidade Federal do Parana, Brazil
Dr. Ruoyu Jin	University of Brighton, United Kingdom
Dr. Sergio Cavalaro	Loughborough University, United Kingdom
Dr. Shang Gao	University of Melbourne, Australia
Dr. Shang Zhang	Suzhou University of Science and Technology, China
Dr. Stephen Wilkinson	University of Wollongong in Dubai, United Arab Emirates
Dr. Vivian Tam	Western Sydney University, Australia
Dr. Xiaobo Zhao	Central Queensland University, Australia
Dr. Xiaoling Zhang	City University of Hong Kong, Hong Kong
Dr. Yelda Turkan	Oregon State University, United States
Dr. Yi Peng	Zhejiang University of Finance & Economics, China
Dr. Yongtao Tan	Royal Melbourne Institute of Technology, Australia
Dr. You Dong	Hong Kong Polytechnic University, China
Dr. Yuanfeng Wang	Beijing Jiaotong University, China
Dr. Yujie Lu	Tongji University, China
Dr. Wilson Lu	Hong Kong University, China
Dr. Zhongfan Chen	Southeast University, China

Internal Scientific Committee:

Dr. Athanasios Makrodimopoulos	XJTLU, China
Dr. Bing Chen	XJTLU, China
Dr. Changhyun Jun	XJTLU, China
Dr. Charles Loo	XJTLU, China
Dr. Chee S. Chin	XJTLU, China
Dr. Cheng Zhang	XJTLU, China
Dr. Christiane Herr	XJTLU, China
Dr. Davide Lombardi	XJTLU, China
Dr. Fangyu Guo	XJTLU, China
Dr. Guobin Gong	XJTLU, China
Dr. Hyungjoon Seo	XJTLU, China
Dr. Isaac Galobardes	XJTLU, China
Dr. Jian Li Hao	XJTLU, China
Dr. Jun Xia	XJTLU, China
Dr. Konstantinos Papadikis	XJTLU, China

Dr. Lei Fan	XJTLU, China
Dr. Lin Lin	XJTLU, China
Dr. Moon Keun Kim	XJTLU, China
Dr. Ominda Nanayakkara	XJTLU, China
Dr. Theofanis Krevaikas	XJTLU, China
Dr. Xiaonan Tang	XJTLU, China

List of keynote and invited speakers

Keynote Speakers:

Dr. Anne Kerr	Mott MacDonald Group
Prof. Guang-Ming Chen	South China University of Technology, China
Prof. Jian Yang	Shanghai Jiao Tong University, China
Prof. Jian-Fei Chen	Queen's University Belfast, UK
Prof. Kincho H. Law	Stanford University, USA
Prof. Liyin Shen	Chongqing University, China
Prof. Miroslaw J. Skibniewski	University of Maryland, USA
Dr. Pingbo Tang	Arizona State University, USA
Ms. Ada Fung	Hong Kong Housing Authority
Prof. Xila Liu	Tsing University & Shanghai Jiao Tong University, China
Prof. Yan Xiao	Zhejiang University – University of Illinois at Urbana Champaign Institute, China

Invited Speakers:

Dr. Baoshan Han	TUS-Design Group,China
Mr. Hongkang Zhao	TUS-Design Group, China
Prof. Hongping Yuan	Guangzhou University, China
Dr. Hongtao Wang	Sichuan University, China
Mr. Jianglong Lu	ARTS Group, China
Dr. Lan Kang	South China University of Technology, China
Dr. Leo Gu Li	Guangdong University of Technology, China
Mr. Liang Qian	Bosch (China) Group
Prof. Linzhu Sun	Wenzhou University, China
Dr. Ming-Xiang Xiong	Guangzhou University, China
Mr. Pei Song	Shanghai Jundao Residential Industry Co., Ltd., China
Dr. Sergio Torres	Westlake University, China
Dr. Shang Zhang	Suzhou University of Science and Technology, China
Mr. Xin Yuan	XinY Structural Consultants, China
Mr. Xuezhao Zhan	China Design Group Co., Ltd., China
Dr. Yelda Turkan	Oregon State University, China
Prof. Yuanfeng Wang	Beijing Jiaotong University, China
Mr. Yuhua Liu	Anhui Huasheng Construction Group Co., Ltd., China
Dr. Yujie Lu	Tongji University, China
Prof. Yun Zou	Jiangnan University, China
Prof. Zhongfan Chen	Southeast University, China

Acknowledgements

The Local Organising Committee wish to express their sincere gratitude for the financial assistance from the following organisations: Department of Civil Engineering at Xi'an Jiaotong-Liverpool University (XJTLU), the Institute of Sustainable Materials and Environment (ISME) at XJTLU, Shanghai Jundao Residential Industry Co., Ltd. and Hangzhou Popwil Instrument Co., Ltd.. The Committee would like to extend their gratitude to the following organisations for their great help with the organisation: the University of Hong Kong (HKU) and the Chinese Research Institute of Construction Management (CRIOCM).

Acknowledgements are due to all the Scientific Committee members, who provided constructive feedback for improvement and/or criticism on all the submitted papers during the peer review process.

Special thanks are also due to the individuals listed below:

Ms. Xiaoyi Wu & Ms. Xiuran Xu, both from Academic Services Office at XJTLU, for their kind and patient assistance with the organisation of this conference.

Ms. Tingjia Wu & Ms. Yan Zhu, secretaries of Department of Civil Engineering and of Design Cluster at XJTLU respectively, for their professional services provided.

Mr. Balal Khan and Ms. Junyao Gao, both from University Marketing and Communications at XJTLU, for their continuous help with the setup and updates of the conference website and flyer.

Mr. Shiva Prashanth Kumar Kodicherla and Mr. Yida Guo, both PhD students at XJTLU, for their assistance with the format checking of all the manuscripts.

Mr. Zhikang Bao, PhD student at University of Hong Kong (HKU), for his help with coordination between XJTLU and HKU.

Prof. Jian Yang and Prof. Guang-Ming Chen, both Keynote Speakers, from Shanghai Jiao Tong University and South China University of Technology respectively, for their help with the promotion of this conference.

Leon Bijnsdorp and Lukas Goosen, both from Taylor & Francis Group, for their patience and timely help in resolving the publication issues.

Finally, we would like to thank all the conference presenter and participants as well as the student volunteers, who have contributed to the success of ICSBS 2019.

ICSBS 2019 Organising Committee

Keynote lectures

Corporate social responsibilities for sustainable design and construction

M.J. Skibniewski

Deptment of Civil and Environmental Engineering, University of Maryland, Maryland, USA

ABSTRACT: Sustainable construction is vital for the embodiment of sustainable development, and corporate social responsibility practice is critical for construction firms to realize sustainable construction. However, the link between sustainable construction and corporate social responsibility is still in its infancy, and previous studies have not developed comprehensive methods for evaluating sustainable construction performance. We first established a conceptual corporate social responsibility framework based on the five aspects of sustainable construction: economic, environmental, social, stakeholders, and health and safety. Then, a set of criteria assessing sustainable construction performance was identified through a comprehensive literature review and interviews with industry practitioners. A questionnaire survey was conducted to collect primary data and identify the significance of the proposed criteria. Partial least squares structural equation modeling technique was used to statistically validate the conceptual model and to identify key factors relevant for sustainable construction. The obtained results reveal that workers' well-being and environmental protection are the key aspects of concern, while site inspection and audits, providing safe working environment, effective emergency management procedures and safety supervisions, compliance with environmental laws and regulations, reducing pollution and waste, and corporate environmental management are critical to achieving sustainable construction to a larger extent than other sustainability factors. Our study underlines the link between sustainable construction and corporate social responsibility, and it provides new guidelines for practical applications of corporate governance tools in achieving sustainability goals.

A software platform for civil infrastructure health monitoring

K.H. Law
Stanford University, Stanford, California, USA

ABSTRACT: The past couple decades have witnessed remarkable advances in sensor technologies and their deployments in all types of physical systems, including civil infrastructures. The increasing deployment of sensors on civil infrastructures has enhanced the monitoring and management of the physical structures. However, effective management and utilization of the massive amount and the wide variety of data collected pose what generally known as the big data problem. This presentation discusses some of recent technology developments for dealing with the data issues and how they may be deployed for civil infrastructure monitoring. The discussion includes: (1) information modeling and database management to enable information interoperability and to support structural modeling, engineering analyses and sensing and monitoring (Jeong et al. 2016, Jeong et al. 2017); (2) the deployment of a distributed cloud-based database environment for efficient storage and retrieval of data (Jeong & Law 2018, Jeong et al. 2019a); and (3) a cyberinfrastructure framework utilizing cloud computing platforms that allows big data analytic for structural monitoring applications (Jeong et al. 2019a, Jeong et al. 2019b, Hou et al. 2019). The scalable cyberinfrastructure framework is demonstrated for the health monitoring of bridge structures.

REFERENCES

Hou, R. Jeong, S. Law, K. H. & Lynch, J. P. 2019.Reidentification of trucks in highway corridors using convolutional neural networks to link truck weights to bridge responses. *Proceedings of the SPIE Smart Structures/NDE Conference. Denver, CO, USA* March 4- 7.

Jeong, S. Hou, R. Lynch, J. P. Sohn, H. & Law, K. H. 2017. An information modeling framework for bridge monitoring. *Advances in engineering software* 114:11-31.

Jeong, S. & Law, K. H. 2018. An IoT platform for Civil Infrastructure Monitoring Application. *IT in Practice (ITiP) Symposium, IEEE COMPSAC (IEEE Computer Society Conference on Computers, Software and Applications)*, Tokyo, Japan July 23- 27.

Jeong, S. Ferguson, M. & Law, K.H. 2019b. Sensor Data Reconstruction and Anomaly Detection using Bidirectional Recurrent Neural Network. *Proceedings of the SPIE Smart Structures/NDE Conference. Denver, CO, USA* March 4- 7.

Jeong, S. Hou, R. Lynch, J. P. Sohn, H. & Law, K. H. 2019a. A Scalable Cloud-based Cyberinfrastructure Platform for Bridge Monitoring. *Structure and Infrastructure Engineering – Maintenance, Management and Life-Cycle Design and Performance, Taylor & Francis* 15(1):82-102.

Jeong, S. Zhang, Y. O'Connor, S. Lynch, J. P. Sohn, H. & Law, K. H. 2016. A NoSQL Data Management Infrastructure for Bridge Monitoring. *Smart structures and systems* 17(4):669-690.

Invited lectures

Sustainability assessment of bridge structures based on big data and carbon emission reduction policies

Y.F. Wang

School of Civil Engineering, Beijing Jiaotong University, Beijing, PR China

ABSTRACT: Since the concept of sustainable development was proposed, it has been gradually integrated into various industries. However, the concept and method of sustainability has not been implemented and applied in bridge engineering. Also, with the development of informatization and digitalization, the traditional bridge design can no longer meet the requirements of sustainability and digital design thinking based on big data can bring new potentials and opportunities for traditional bridge design. As the country with the largest bridge construction, the scale of new medium and small span bridges in China is increasing every year. While promoting urbanization, the bridge engineering plays an important role in achieving the United Nation's sustainable development goals (UN SDGs).

Therefore, this study comprehensively assessed the status of the bridges achieving the carbon emission reduction target in the General Atlas of Highway and Bridge Design (GAHBD), so as to provide quantifiable and feasible suggestions ultimately for sustainable design of highway bridges. In the study, a simply supported T-beam bridge with a span of 30 meters in the GAHBD was selected as a specific design case. The research contents are as follows: (1) simulating the bridge design schemes (100,000 magnitude) by changing the geometric and physical parameters of the bridge, analyzing all possibilities of designed bridges comprehensively and quantifying the CO_2 emissions of all schemes; (2) using the big data-based analysis method the Maximal Information Coefficient (MIC) to operate correlation analysis of the geometric and physical parameters with the reliability, life cycle assessment (LCA) and life cycle cost (LCC) of the simulated bridges; (3) establishing the limit state of carbon emissions per unit GDP for the bridges under different scenarios for the next 20 years which considers relevent policies on the implementation of climate change action and the GDP growth scenarios under different shared socioeconomic pathways in China. Bridges that meet the carbon emission limit state was selected after analyzing the failure status of CO_2 emission reduction in the bridge design stage; (4) calculating the reliability, LCA and LCC of the bridges that meet the carbon emission limit state, establishing a comprehensive sustainable model and obtaining the final sustainable design.

Here we show, in the current policy scenario, the bridge design schemes based on the GAHBD could meet the carbon reduction target of 2030; in the new policy scenario, the bridge design schemes only meet the carbon reduction target of 2025; in the sustainable development scenario, however, none of the bridge design schemes meet the carbon emission reduction targets in any period of time.

Sustainable Buildings and Structures: Building a Sustainable Tomorrow – Papadikis et al. (Eds)
© 2020 Taylor & Francis Group, London, ISBN 978-0-367-43019-1

Study on compressive performance of prefabricated composite column with steel angles

J. Ding, Y. Zou, C. Wang & T. Li
School of Environmental and Civil Engineering, Jiangnan University, Wuxi, China

M. Chen
Open-Steel Structure Co., Ltd, Shanghai, China

ABSTRACT: This paper proposed experimental and numerical studies on a novel prefabricated composite column using steel angles and infilled reinforced concrete (LSRC) subjected to axial compression. Steel angles, which were connected by batten plates, were placed at the corners of cross section. The reinforcement cage was made up of longitudinal bars, hoops and cross ties, and it was used outside steel angles. It can be seen as a reinforced concrete (RC) column with embedded steel angles. Four specimens were designed and tested to investigate the difference in mechanical properties of LSRC columns and concrete-filled steel tube encased cover (CFTEC) column. The tests also discuss the effects of different batten plate spacing on failure modes, bearing capacity, strength-to-weight ratio and mid-span strain. The test results indicated that the embedded reinforcement cage effectively prevent the cracking in the cover concrete from propagating quickly, so that LSRC specimens have higher compressive strength and strength-to-weight ratio than CFTEC specimen. For LSRC columns, bearing capacity and strength-to-weight ratio were significantly enhanced by appropriately increasing the spacing of batten plates. Compared to CFTEC column, LSRC columns with large batten plate spacing made better use of mid-span steel material at the peak point.

In addition, three-dimension finite element (FE) models were established using ABAQUS, in which the nonlinear material behavior and material modelling of confined concrete were adopted. By comparing the load-deformation curves from the test results, the rationality and reliability of the FE models has been verified. Numerical simulation was modeled to clarify the failure mechanism of specimens under axial compression and analyze the effect of the steel angle thickness and length, cross sectional area of longitudinal bars and steel ratio of cross section on mechanical performance of LSRC columns. Furthermore, the FE analysis confirmed that confinement effect of batten plates gradually increases with increased axial loading.

On the basis of test and FE results, the practical formula is proposed to predict the ultimate bearing capacity of new precast composite column. It was found that the predicted results agree reasonably well with the experimental and FE values.

Ductile fracture of welded steel structures

L. Kang
School of Civil Engineering and Transportation, South China University of Technology, Guangzhou, Guangdong Province, People's Republic of China

H.B. Ge
Department of Civil Engineering, Meijo University, Nagoya, Japan

ABSTRACT: Ductile fracture of steel structures is one important failure model of thick-walled steel structures under strong earthquakes. In the past eight years, the authors experimentally and numerically investigated the ductile fracture of thick-walled steel structures. On one hand, at the material level, four important issues were studied in detail, including the ductile fracture of welded steels, the ductile fracture of steels at complex stress states, the ductile fracture of steels under cyclic loadings, the ductile fracture of steels after high temperatures, and so on. First of all, for welded steels, the ductile fracture behavior of the base metal, weld and HAZ was investigated based on the smooth flat bar, U-notch flat bars, and V-notch flat bars, and a three-stage and two-parameter ductile fracture model was developed. Therefore, for different and complex stress states, the ductile fracture behavior of steels was tested, including negative triaxiality, low stress triaxiality, medium and high stress triaxiality. One side U-notch and V-notch steel flat bars and shear steel specimens were tested in order to model different stress states. One novel ductile fracture model including the effect of complex stress states was developed. Later, the experiments and FEA results of steel specimens under cyclic loadings were reported. Besides, the post fire mechanical properties and ductile fracture behavior of Q460 and Q690 steels were obtained based on our experimental results and other researchers' test results. On the other hand, the ductile fracture behavior of steel bridge piers, steel beam-to-columns, steel braces and other steel members were experimentally and numerically investigated in detail at the member level. The ductile crack initiation, propagation and final failure of steel members were obtained during tests. And the mesh size sensitivity during numerical study has been solved through a nonlocal model. Furthermore, a simplified method using fiber beam element model was developed to facilitate engineering application. This study provides abundant basis data and important method to evaluate the ductile fracture of steel structures.

Life cycle assessment and management of sustainable buildings

H.T. Wang
College of Architecture and Environment, Sichuan University, Chengdu, China

L. Zhang & Y.W. Liang
IKE Environmental Technology Co. Ltd., Chengdu, China

Y.Y. Liao
College of Architecture and Environment, Sichuan University, Chengdu, China

ABSTRACT: With the increasing concern on sustainability of buildings, life cycle assessment (LCA) has been more widely applied in the field of building materials and construction. For example, ISO 21930 (2017) proposes to adopt LCA of building material supply chain to analyze life cycle environmental performance of buildings. European Union established a set of LCA standards for building materials and construction, e.g. EN15804, EN 15978. And on this basis, almost all the EU building materials suppliers provide LCA reports, i.e. Environmental Product Declaration, to building developers. Moreover, LEED, BREEAM, DGNB and other green building rating systems set LCA-based scores of building materials and construction. In China, national standard for building carbon emission calculation (GB/T 51366-2019) based on LCA has been published and will come into effect soon.

In this presentation, we will introduce the Chinese LCA core database (CLCD) and the included building materials database, as well as the method for establishing the database and data quality assessment. We have done the first LEED-approved LCA study of a building in China, based on an online LCA platform for building material and construction developed by IKE. More on-going LCA studies of buildings and building materials will be introduced. Furthermore, based on the online platform, it is possible to establish a low-carbon and green management system for the full life cycle and full supply chain of buildings.

REFERENCES

EN 15804. 2012. *Sustainability of construction works – Environmental product declarations – Core rules for the product category of construction product.*

EN 15978. 2011. *Sustainability of construction works - Assessment of environmental performance of buildings - Calculation method.*

ISO 21930. 2017. *Sustainability in buildings and civil engineering works – Core rules for Environmental Product Declarations of construction products and services.*

Sustainable Buildings and Structures: Building a Sustainable Tomorrow – Papadikis et al. (Eds)
© 2020 Taylor & Francis Group, London, ISBN 978-0-367-43019-1

Deep semantic segmentation for 3D as-is bridge model generation

Y. Turkan & Y. Xu

School of Civil Engineering & Construction Engineering, Oregon State University, Corvallis, Oregon, USA

ABSTRACT: The mobility of people and goods is highly dependent on the health of a nation's transportation system. Timely inspection and effective maintenance of bridges is crucial to avoid any issues that may have a negative impact on public mobility. However, current bridge inspection practice inhibits the collection and analysis of information regarding the status of bridges in an efficient and timely manner. A Bridge Information Model (BrIM) is an object-oriented database that enables storing all bridge data, including its 2D drawings and 3D models, material specifications, inspection notes, images and maintenance information. Recent research efforts have focused on implementing BrIM for bridge structural condition assessment, and concluded that it is a suitable concept and technology that can be used to improve the current bridge inspection and management processes. However, there are several challenges that needs to be overcome for wide adoption of BrIM for bridge inspections and management tasks. These challenges include the following: 1) manual development of 3D BrIM for existing bridges from their 2D drawings and specifications is labor-intensive and time-consuming; 2) 3D BrIM that are built based on the original 2D drawings may be inaccurate representations of the current status and geometrical information of the bridges; and 3) transportation agencies do not have the resources or personnel time to develop 3D BrIM for all the bridges they maintain and operate. There is, therefore, an urgent need to establish an automated and cost-effective method for developing 3D BrIM for existing bridges. In order to overcome these challenges associated with the development of 3D BrIMs, this study presents a novel data collection and analysis framework that enables rapid collection of 3D geometrical information from existing bridges in the form of 3D dense point clouds, and converts them into 3D BrIM in an automated and efficient manner. 3D point clouds are obtained by applying Structure from Motion (SfM) algorithms to the images collected using an Unmanned Aerial System (UAS). In the next step, these 3D point clouds are automatically segmented and classified into different structural components using deep learning algorithms. In the final step, the labeled point clouds are automatically converted to 3D BrIM. In summary, this study develops a framework that will make it more convenient and faster to develop and implement 3D BrIM, which can improve the current bridge inspection and management practice significantly in terms of efficiency and safety, thus help improve public mobility.

Sustainable Buildings and Structures: Building a Sustainable Tomorrow – Papadikis et al. (Eds)
© 2020 Taylor & Francis Group, London, ISBN 978-0-367-43019-1

A brief on cost controls of highway construction projects

Y. Liu
Anhui Huasheng Construction Group Co., Ltd, P.R. China

G. Gong
Department of Civil Engineering, Xi'an Jiaotong – Liverpool University, Suzhou, P.R. China

X. Zhan
Institute of Urban Construction Planning & Design, China Design Group Co., Ltd, P.R. China

ABSTRACT: Cost control of construction projects is to minimize all kinds of consumptions and achieve the predetermined cost target using necessary technology and management measures without affecting the predetermined construction pe-riod and quality. The control measures include organizational measures, tech-nical measures and economic measures. The control process includes construc-tion project cost estimation, cost planning and implementation, cost accounting, cost analysis etc. The control methods include target control, dynamic control, comprehensive control and key control. This talk discusses the cost control of highway construction projects.

New achievements and new opportunities and challenges in tunnel engineering in China

X. Zhan
Institute of Urban Construction Planning & Design, China Design Group Co., Ltd, P.R. China

G. Gong
Department of Civil Engineering, Xi'an Jiaotong – Liverpool University, Suzhou, P.R. China

Y. Liu
Anhui Huasheng Construction Group Co., Ltd, P.R. China

ABSTRACT: In recent years, there have been all sorts of tunnel engineering projects in China. With the sustained, healthy and rapid development of China's national economy and the further enhancement of its comprehensive national strength, there have been significant typical tunnel projects in the construction fields of high-speed railway, highway, urban rail transit, diversion and utility tunnels. Considerable progress has been made in tunnel construction technology as well as in operation and maintenance management.

On the basis of the existing cases investigated on tunnel construction, management and operation and maintenance in China, this talk puts forward new challenges in the field of tunnel construction in China by systematically sorting out and thoroughly summarizing the open-cut method (cut and cover method), cover-excavation method, pipe jacking method, shield tunneling method, mining method and immersed pipe method, and points out the future challenges. The talk is expected to promote the new development of tunnel engineering technology.

Sustainable construction materials

Sustainable Buildings and Structures: Building a Sustainable Tomorrow – Papadikis et al. (Eds)
© 2020 Taylor & Francis Group, London, ISBN 978-0-367-43019-1

Roles of nano-silica and micro-silica in durability of concrete

L.G. Li
Department of Civil Engineering, Guangdong University of Technology, Guangzhou, China

J.Y. Zheng
Guangzhou Institute of Construction Industry, Guangzhou, China

A.K.H. Kwan
Department of Civil Engineering, The University of Hong Kong, Hong Kong, China

ABSTRACT: It has been proven by various studies that adding micro-silica (MS) alone or nano-silica (NS) alone can provide significant beneficial effects on the durability of concrete. In theory, MS and NS may also be added together, but up to now, such possible combined usage of MS and NS for improving the durability of concrete has not been deeply explored. In this paper, the individual and combined roles of MS and NS in the chloride and carbonation resistance of concrete were investigated by producing a series of concrete mixes containing varying MS, NS and water contents for rapid chloride permeability test and carbonation test. It was found that the single addition of MS or NS can significantly improve the chloride and carbonation resistance, but combined addition of MS and NS can further enhance the sulphate and chloride resistance.

1   INTRODUCTION

Along with increasingly stringent strength and durability requirements of civil infrastructures as well as advances in concrete technology, high-strength concrete and high-performance concrete (HPC) are becoming popular. To produce these concretes, supplementary cementitious material (SCM) is typically an essential ingredient. As a byproduct from the production of ferrosilicon alloys and silicon crystals, micro-silica (MS), also called silica fume (SF), has been used widely in HPC due to its high fineness and good pozzolanic reactivity. Apart from offering the beneficial effect of strength enhancement (Mazloom et al. 2004), SF can also effectively improve the durability (Khan & Siddique 2011, Rostami & Behfarnia 2017).

Along with rapid advancement of nanotechnology, many nano-materials have been developed for producing HPC. Amongst these, nano-silica (NS), with its ultrafine particle size and high pozzolanic reactivity, has attracted lots of attention and is quite possibly the one having the greatest potential for practical applications. Being finer than SF, NS can fill into very small voids to enhance the imperviousness and durability of concrete (Du et al. 2017, Ghafari et al. 2015).

In contrast with the use of SCM alone or nano-material alone, the combined usage of SCM and nano-material may offer greater positive effects on the durability of concrete. For instance, Supit & Shaikh (2015) found that 38% fly ash plus 2% NS is the optimum combination for durability improvement of high-volume fly ash concrete. Mohseni & Tsavdaridis (2016) demonstrated that the combined use of 25% FA and 1% to 5% nano-Al_2O_3 in mortar can result in denser microstructure and lower water and chloride permeability. However, besides these explorative studies, further in-depth and more systematic studies are still needed to evaluate the combined effects of SCM and nano-material and to enable better understanding of the mechanism involved.

In this study, with the aim to evaluate the effects of combined usage of SF and NS on the durability of concrete, a series of concrete mixes containing different water, SF and NS contents were made for testing.

2 EXPERIMENTAL DETAILS

2.1 Materials

Three binder materials were used. They are ordinary portland cement (OPC), silica fume (SF) and nano-silica (NS). The OPC has strength of class 52.5 and specific gravity of 3.11. The SF was a condensed silica fume having particle size of about 0.1 μm and specific gravity of 2.20. The NS was in the form of white colour powder and has particle size falling in the range of 5 to 20 nm and specific gravity of 1.94. Both coarse and fine aggregates were employed in the concrete mixes. The coarse aggregate employed was crushed granitic rock with maximum aggregate size of 10 mm, specific gravity of 2.68, water absorption of 1.04% and moisture content of 0.11%. The fine aggregate employed was river sand with maximum aggregate size of 5 mm, specific gravity of 2.64, water absorption of 1.10% and moisture content of 0.11%. An aqueous form superplasticizer (SP) was dosed to each of the concrete mixes to attain sufficient workability. The SP was of the polycarboxylate-based type. It had a specific gravity of 1.03 and a solid mass content of 20%.

2.2 Mix proportions

The testing programme encompassed 24 trial concrete mixes. Four mix parameters, namely the water/cementitious materials (W/CM) ratio, SF content, NS content and SP dosage, were varied in the following manner. The W/CM ratio by mass was varied from 0.30 to 0.45 in increments of 0.05, the SF content was varied among 0% and 5% of total binder by mass, and the NS content was varied among 0%, 0.5% and 1% of total binder by mass. The choice of SF content and NS content was based on practical considerations of workability and economy of the concrete mixes. To impart adequate workability to the concrete mixes, SP was dosed to each mix to achieve 150 ± 50 mm slump, and the corresponding SP dosage (as percentage of total binder by mass) was first determined by trial mixing. During trial mixing, the SP was added in small fractions until the slump reached the target range. Then, the SP dosage so determined was adopted during the formal production of concrete mixes for undergoing testing. The other two mix parameters: the paste volume and fine/total aggregate ratio, were fixed. The paste volume was set at 30% of the total concrete volume. As a result, the aggregate volume was set constant at 70%. The fine/total aggregate ratio by mass was set at 0.4. Each concrete mix was annotated by a mix number in the format of A-B-C, which corresponds to: (W/CM ratio)-(SF content in percentage)-(NS content in percentage), as shown in the first column of Table 1.

2.3 Testing methods

The workability of the concrete mixes was evaluated from the slump cone test. During the test, the drop in height of the concrete mix after lifting of the slump cone to form a patty was taken as the slump value. The chloride resistance of the concrete mixes was evaluated from the rapid chloride permeability test (RCPT) in accordance with the relevant Chinese Standard GB/T 50082-2009. The test procedures were very similar to those in American Standard ASTM C1202–19. The carbonation test stipulated in Chinese Standard GB/T 50082-2009 was carried out to evaluate the carbonation resistance of the hardened concrete.

3 TEST RESULTS

3.1 *SP dosage and slump*

The SP dosages applied in the actual production of the concrete mixes are tabulated in the second column of Table 1, and its variation with the W/CM ratio is plotted in Figure 1. Since the SP served the purpose to adjust the workability to within the target range, the SP dosage reflected the influence of various mix parameters on the workability. For example, regardless of the SF and NS contents, the higher was the W/CM ratio, the lower was the SP dosage. Such phenomenon is expected as a result of increasing the W/CM ratio. Conversely, at given W/CM ratio, the SP dosage significantly increased when the SF and/or NS contents increased. This was due to the very large specific surface areas of the ultrafine SF and NS particles, both of which demanded much more SP to disperse than the OPC particles.

Although the SP dosage was adjusted such that the workability was within the target range, the measured slump varied slightly due to the addition of the SP in increments. For record purpose, the slump results are summarised in the third column of Table 1, where it can be seen that the slump varied from 110 to 188 mm, all within the target range of 150 ± 50 mm.

3.2 *RCPT total charge*

The RCPT total electrical charge passed (in Coulomb) of each concrete mix was calculated, as presented in the fourth column of Table 1, and its variation with the W/CM ratio is plotted in Figure 2. From these results, it is found that at given SF and/or NS contents, the total charge passed increased with the W/CM ratio, indicating that the chloride resistance was lower for concrete with higher W/CM ratio.

Table 1. Test results.

Mix number	SP dosage (%)	Slump (mm)	RCPT total charge (Coulomb)	Carbonation depth (mm)
0.30-0-0	4.50	140	1056	4.2
0.30-0-0.5	5.00	158	913	3.7
0.30-0-1	5.40	130	799	3.5
0.30-5-0	5.20	140	585	3.7
0.30-5-0.5	5.58	122	435	3.4
0.30-5-1	5.90	110	266	3.1
0.35-0-0	2.40	157	1732	6.7
0.35-0-0.5	2.60	160	1455	5.2
0.35-0-1	3.00	150	1375	4.2
0.35-5-0	2.75	128	786	5.3
0.35-5-0.5	3.15	145	579	3.9
0.35-5-1	3.35	129	395	3.5
0.40-0-0	0.70	181	2600	10.6
0.40-0-0.5	0.85	175	2080	8.2
0.40-0-1	1.18	169	1970	6.5
0.40-5-0	0.98	181	883	8.8
0.40-5-0.5	1.35	188	649	5.8
0.40-5-1	1.50	187	483	4.4
0.45-0-0	0.30	125	3433	17.9
0.45-0-0.5	0.50	120	2952	12.6
0.45-0-1	0.83	110	2584	9.7
0.45-5-0	0.64	115	1083	12.8
0.45-5-0.5	0.98	169	883	9.1
0.45-5-1	1.15	166	665	6.3

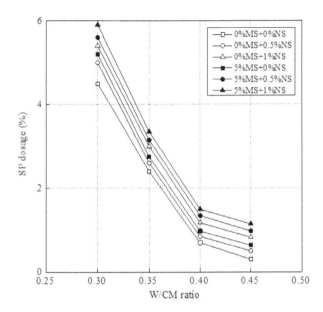

Figure 1. SP dosage versus W/CM ratio.

From the figure, it is also noted that the RCPT total charge passed-W/CM ratio curves shift downwards following the sequence of (0% SF + 0% NS), (0% SF + 0.5% NS), (0% SF + 1% NS), (5% SF + 0% NS), (5% SF + 0.5% NS), and (5% SF + 1% NS). Such downward shifting of the total charge passed-W/CM ratio curves as the SF and/or NS content increases reveals that adding SF and/or NS can enhance the chloride resistance of concrete.

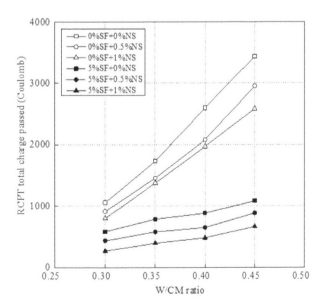

Figure 2. RCPT total charge versus W/CM ratio.

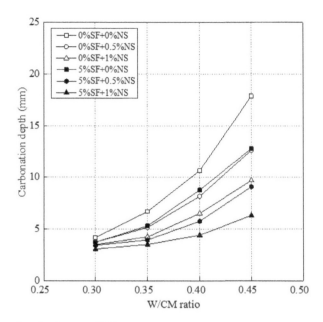

Figure 3.   Carbonation depth versus W/CM ratio.

It is noteworthy that for concrete with SF and NS added together, the total charge passed was always lower than that of concrete with SF added alone or with NS added alone, showing that the usage of SF and NS together is a better way of enhancing the chloride resistance.

3.3   *Carbonation depth*

The 28-day carbonation depth results are respectively listed in the last column of Table 1, and plotted against the W/CM ratio for different SF + NS contents in Figure 3. From the figure, it is obvious that at given SF and NS contents, the carbonation depth decreased with decreasing W/CM ratio. More importantly, at given W/CM ratio, both the carbonation depths significantly decreased with increasing SF and/or NS contents, indicating that the addition of SF and/or NS had effectively improve the carbonation resistance.

Evidently, the carbonation depths with both SF and NS added were always smaller than those with only SF added or only NS added. For instance, at carbonation time of 56 days and W/CM ratio of 0.45, with 0% SF and 1% NS added, the carbonation depth was decreased to 15.1 mm, with 5% SF and 0% NS added, the carbonation depth was decreased to 18.3 mm, but with 5% SF and 0.5% NS added, the carbonation depth was decreased to 12.5 mm, and with 5% SF and 1% NS added, the carbonation depth was further decreased to 8.6 mm. Hence, the combined usage of SF and NS had enabled further enhancement of the carbonation resistance to higher than possible with only SF added or only NS added.

4   COMBINED EFFECTS

4.1   *Combined effects on chloride resistance*

To quantify the combined effects of SF and NS addition on chloride resistance, the corresponding percentages decreases in RCPT total charge are listed in Table 2. These results indicate that at a given W/CM ratio, the percentage decrease in total charge passed by adding 5% SF alone was always larger than that by adding 1% NS alone, whereas the percentage decrease in total charge passed by adding SF and NS together was always larger than that by adding the SF alone or the NS alone.

Table 2.   Effects of SF and NS on chloride resistance.

Resulted from addition of	Percentage decrease in RCPT total charge passed			
	W/CM ratio = 0.30	W/CM ratio = 0.35	W/CM ratio = 0.40	W/CM ratio = 0.45
0.5% NS only	13.5	16.0	20.0	14.0
1% NS only	24.3	20.6	24.2	24.7
5% SF only	44.6	54.6	66.0	68.5
5% SF + 0.5% NS	58.8	66.6	75.0	74.3
5% SF + 1% NS	74.8	77.2	81.4	80.6

Meanwhile, it is observed that the percentage decrease in total charge passed resulted from addition of both SF and NS was only occasionally greater than the sum of the individual percentage decreases in total charge passed resulted from single addition of the SF and from single addition of the NS. When the W/CM ratio was equal to 0.30, the decrease in total charge passed by adding 5% SF was 44.6%, and the decrease in total charge passed by adding 1% NS was 24.3%, while the decrease in total charge passed resulted from combined addition of 5% SF and 1% NS was 74.8%, which was larger than the sum of 44.6% and 24.3% (i.e. 68.9%). This reveals that in some cases, the usage of SF and NS together could provide certain synergistic effect on the chloride resistance.

4.2   Combined effects on carbonation resistance

To quantify the effects of SF and NS on carbonation resistance, the percentage decreases in carbonation depth are computed and listed in Table 3. From this table, it is evident that at given SF and/or NS contents, the percentage decreases in carbonation depth were generally larger at higher W/CM ratio. Particularly, at a W/C ratio of 0.30, the addition of 5% SF and 1% NS decreased the 28-day carbonation depth by 26.8% whereas at a W/C ratio of 0.45, the addition of 5% SF and 1% NS decreased the 28-day carbonation depth by 64.6%. Therefore, the effectiveness of adding SF and/or NS to mitigate carbonation was higher when the W/CM ratio was higher.

On the other hand, at a given W/CM ratio, the percentage decrease in carbonation depth by adding 1% NS alone was always larger than that by adding 0.5% NS alone, and the percentage decrease by adding 0.5% NS alone was always larger than that by adding 5% SF alone. This demonstrates that the addition of 0.5% NS is more effective than the addition of 5% SF in improving the carbonation resistance. More importantly, the percentage decrease in carbonation depth by adding SF and NS together was always larger than that by adding SF alone or NS alone. Hence, the combined usage of SF and NS is always better than the single usage of SF alone or NS alone in improving the carbonation resistance.

Table 3.   Effects of SF and NS on carbonation resistance.

Due to addition of	Percentage decrease in carbonation depth			
	W/CM = 0.30	W/CM = 0.35	W/CM = 0.40	W/CM = 0.45
0.5% NS only	11.0	23.0	23.2	29.4
1% NS only	16.7	36.8	38.7	45.7
5% SF only	10.9	20.8	17.5	28.4
5% SF + 0.5% NS	18.1	41.3	45.9	49.1
5% SF + 1% NS	26.8	47.7	58.6	64.6

5 CONCLUSIONS

The effects of silica fume (SF) and nano-silica (NS), whether added alone or together, on the chloride and carbonation resistance of concrete have been investigated. The research findings are summarized hereunder:

(i) The use of SF and/or NS increased the SP demand required to achieve the target slump, enhanced the chloride resistance as demonstrated by reduced total charge passed during rapid chloride test, and improved the carbonation resistance as demonstrated by reduced carbonation depth.

(ii) With both SF and NS added together, the chloride resistance and carbonation resistance were further improved, compared to the cases with SF added alone or with NS added alone, indicating that the combined use of SF and NS have additive and positive effects on the sulphate and chloride resistance.

REFERENCES

Du, H. Du, S. & Liu, X. 2014. Durability performances of concrete with nano-silica. *Construction and Building Materials* 73: 705-712.

Ghafari, E. Arezoumandi, M. Costa, H. & Júlio, E. 2015. Influence of nano-silica addition on durability of UHPC. *Construction and Building Materials* 94: 181-188.

Khan, M.I. & Siddique, R. 2011. Utilization of silica fume in concrete: review of durability properties. *Resources, Conservation and Recycling* 57: 30-35.

Mazloom, M. Ramezanianpour, A.A. & Brooks, J.J. 2004. Effect of silica fume on mechanical properties of high-strength concrete. *Cement & Concrete Composites* 26(4): 347-357.

Mohseni E. & Tsavdaridis K.D. 2016. Effect of nano-alumina on pore structure and durability of class F fly ash self-compacting mortar. *American Journal of Engineering and Applied Sciences* 9(2): 323-333.

Rostami, M. & Behfarnia, K. 2017. The effect of silica fume on durability of alkali activated slag concrete. *Construction and Building Materials* 134: 262-268.

Supit, S.W.M. & Shaikh, F.U.A. 2015. Durability properties of high volume fly ash concrete containing nano-silica. *Materials and Structures* 48(8): 2431-2445.

Recycled aggregate from wet mix concrete

D.W. Law, C. Gunasekara & S. Setunge
Civil & Infrastructure Engineering, RMIT University, Melbourne, Australia

M. Hogan & D. Thomas
Cycrete Pty Ltd, Unit 9 Birubi Chambers, Hawker, Australia

ABSTRACT: Quarry aggregate reserves are depleting fast, particularly in some desert regions of the world. Worldwide Quarry aggregate production is about 4.5 billion tonnes, and Australia alone consumes about 130 million tonnes of aggregates annually. The process energy usage and greenhouse gas emissions from Quarry aggregate is around 7.4 to 8.0 kg CO_2-e per tonne. Aggregate demand is also increasing with the expansion of construction. In contrast, the annual production of premix concrete in Australia is about 30 million cubic meters, and 3–5 % of concrete delivered to site remains unused and ends up in the landfills or crushing plants. Thus, manufacturing coarse aggregates using waste concrete is a sustainable approach in terms of environmental and economic aspects. This paper presents the mechanical performance results of concrete produce using a novel manufactured coarse aggregate recycled directly from premix concrete. A series of mixes of were produced using this manufactured coarse aggregate with the replacement by weight of quarry aggregate by 25%, 50%, 75% and 100%. The mixes investigated were 100% Ordinary Portland cement and 25% replacement with Fly Ash for a 25 MPa and a 32 MPa concrete. All mixes were tested at 7 and 28 days for the compressive strength, and then compared with the concrete produced with quarry aggregate. Test results demonstrated that concrete produced with the recycled aggregate achieved a mean 28 day compressive strength equivalent to that of natural granite aggregate for the 25 MPa mix and slightly lower (1-3 MPa) for the 32 MPa mix. No effect on compressive strength was observed based on the level of replacement of recycled aggregate added.

1 INTRODUCTION

Concrete is the most commonly used construction material in the world with coarse aggregates, forming up to 65% of the volume of the concrete. Annual worldwide aggregate usage in concrete is about 4.5 billion tons (Alexander and Mindess, 2010), of which Australia alone contributes about 130 million tons (Australia, 2016). If current consumption rates are maintained then by 2050 the Australian industry will need to produce about 210 million tonne per year, which is a 60% increase in production. This will create many challenging issues for the next decades, concerning environmental impacts due to aggregate production and loss of resources. Thus, a major challenge for the aggregate and construction industries is finding alternative aggregate sources.

A number of replacement materials for aggregates in concrete have been suggested including quarry dust, silt, agriculture waste (sugarcane bagasse ash, rice husk waste), glass, crumbed rubber and demolition waste, with a number of these now used in the construction industry. One of these is concrete demolition waste which has been widely adopted as an excellent source of recycled coarse aggregate (RCA) for new concrete production. Studies show that concrete making using RCA can be as strong as traditional concrete (Evangelista and de Brito, 2007). Hence, using RCA can be extremely beneficial to the preservation of our environment. However, RCA have disadvantages in that they require significant energy to crush and grade the aggregate and also as it has higher cement content on the aggregate surface. Therefore, when using RCA additional water has to be added to the concrete, which can affect the strength and durability.

One possible alternative source of the RCA is that of the waste concrete that is unused in construction projects. At present this is discharged from the mixer, allowed to set and then crushed to produce RCA. This process requires both time and energy in detouring to discharge the unused concrete, the setting time and then the crushing and grading. An alternative approach has been developed which directly treats the wet mix. Using a patented hygroscopic chemical formula, a powder is added to the wet concrete. Due to the hygroscopic property it absorbs virtually all of the water and moisture in the mix. Once the mix is dry it allows the fine and the coarse aggregate to be separated and stored for later use. This process does not require the concrete to harden as in most concrete recycling techniques, and therefore significantly reduces the time needed in traditional concrete recycling. In addition, the procedure does not require heavy machinery to crush down any concrete, nor does it require transportation to storage sites, hence conserving energy and fossil fuel usage. Furthermore, due to the highly hygroscopic nature of the powder, when used in sufficient amounts it will absorb the majority of the water and moisture in the mixer, leaving the interior almost dry with no sticky residue. This makes it significantly easier to clean the trucks potentially using only one tenth of the usual amount of water to clean the mixing equipment compared to a mixer used for traditional concrete.

In order to assess the viability of this RCA an investigation project has been undertaken to determine the effect of replacing standard coarse granite aggregate with the RCA at 0, 25, 50, 75 and 100%. To investigate the strength that could be achieved by the RCA two different binders and three different standard concrete design strengths were employed. These corresponded to a 100% Ordinary Portland Cement (OPC) and a 25% Fly ash (FA) mix with specified compressive strengths of 25 MPa and 32 MPa, corresponding to the two lowest concrete mix designs in the Australian concrete code AS 3600 (AS3600, 2018). This paper reports the 7 and 28 day compressive strength of the concrete produced and the effect of the replacement proportion of the RCA on performance.

2 METHODOLOGY

The RCA used was provided by Cycrete Pty Ltd in accordance with their patented technology. The key steps in which are;

1. Estimate the quantity of excess wet concrete left over in the mixer
2. Slump and concrete properties are entered in the programmable logic controller (PLC)
3. PLC calculates the volume of additive required
4. Compressor and rotary valve activated
5. Additive added into agitator which mixes the material for a period time determined by PLC (usually 2-5 minutes)
6. Processed material is fed through a trammel which separates the coarse and fine aggregates

Commercially available, crushed granite aggregate and Type 1 Portland cement (PC) conforming to ASTM C150 standard (ASTM, 2016), were used. The RCA coarse aggregate was sieved to match the grading curve of the commercially coarse aggregate used.

A series of mix designs were investigated using, 0, 25, 50, 75 and 100% RCA replacement by weight of aggregate. The composition of the mix is summarized in Table 1. The ratio of components was calculated using the absolute volume method (Neville, 1996). The fine aggregate used was river sand in uncrushed form with a specific gravity of 2.5 and a fineness modulus of 3.0. Both types of coarse aggregates used in concrete were in saturated surface dry condition to prevent the water absorption from the concrete mix. The quantity of total aggregates in both concretes was kept to 60% in the entire mixture by volume. Consequently, the total weight of cement and water was changed in order to maintain the material volume and water/cement ratio (0.42) constant.

Table 1. Mix design (kg/m^3).

Concrete Mix	Cement	Fly ash	Sand	Aggregates	Water
25 MPa OPC mix	270	0	910	1015	165
32 MPa OPC mix	320	0	855	1015	170
25 MPa FA mix	225	70	845	1015	170
32 MPa FA mix	255	85	800	1015	175

A 120L planetary concrete mixer was used to produce concrete. Initially, the fine aggregate and coarse aggregates (dry materials) were mixed for two minutes. The cement was then added and mixed for a further minute. The water was then mixed into dry material and mixed for another four minutes. A slump test was conducted using Australian standards, AS 1012.3.1 (AS, 2014) to ensure concretes achieved the required workability range of 150 to 175 mm. The concrete mix was then poured into moulds and vibrated using a vibration table for 2 minutes to remove air bubbles.

The samples were demoulded after 24 hours and then water cured until testing. Compressive strength tests were performed using an MTS machine (loading rate=20 MPa/min) according to Australian standards, AS 1012.9 (AS, 1999).

3 RESULTS

The strength results for the 25 MPa and 32 MPa 100% OPC concrete are shown in Figures 1 and 2 respectively.

The data shows that only the 0% RCA and 100% RCA specimens achieved the 25 MPa design strength and only the 50% RCA the 32 MPa mix design strength. However, based upon the control strengths the RCA mixes at 25 MPa and 32 MPa all achieved similar strengths to that of the 0% control mix, with the exception of the 75% RCA, 25 MPa mix. It is

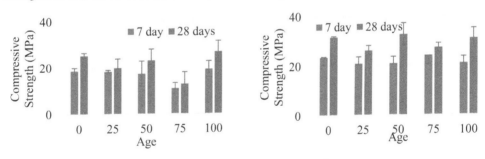

Figure 1. Compressive strength vs Age, 25 MPa OPC & 25% FA mix.

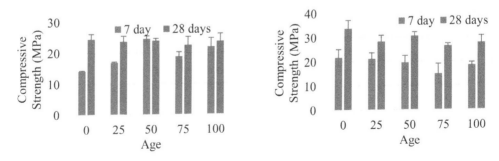

Figure 2. Compressive strength vs Age, 32 MPa OPC & 25% FA mix.

unclear why this mix did not achieve similar strengths to the other RCA mixes. Given that the 100% RCA, 25 MPa mix gave the highest compressive strengths at 7 and 28 days this cannot be attributed to the level of replacement with the RCA aggregate. Indeed, based upon the strength data there is no clear trend between strength and RCA level.

The RCA mixes achieved similar strengths for the 25 MPa and 32 MPa mixes, with the highest strength achieved by the 75% replacement mix, but this was only 5 MPa less than the strength achieved by the control mix at 28 days. The remaining three mixes ranged from 10 to 15 MPa lower than the control mix and were actually less than those achieved by the same replacement level for the 32 MPa mix designs. This would indicate that for the 32MPa concrete the RCA material does not achieve the strength of the control 100% OPC concrete.

As would be expected due to a slower pozzolanic reaction rate in fly ash, the FA concrete generally has a low compressive strength at 7 days [6]. The exceptions to this are the 50, 75 and 100% RCA 25 MPa mixes, which have a higher strength than the corresponding OPC mixes. It is hypothesized that this may be due to some residual cement particles on the RCA following the treatment. These particles are activated and undergo hydration during the mixing process and add extra early age strength to the FA mixes with a higher percentage of RCA replacement. While the 7 day strengths in the 32 MPa 25% FA mixes do not exceed those for the corresponding OPC mixes, those with a higher percentage of RCA added do generally display higher 7 day strengths than those with a lower level of RCA replacement.

At 28 days the 25 MPa and 32 MPa mixes for all FA specimens achieve similar strengths to those displayed by the OPC mixes. As for the OPC specimens there is no clear variation in the effect of RCA level on the 28 day strengths achieved. For the 25 MPa mix all of the compressive strengths, 0 – 100% RCA, are within 2 MPa indicating that the use of RCA has not any effect on the compressive strength of the concrete. For the 32 MPa mix designs all of the RCA mixes again have similar strength, ranging from 3 to 7 MPa lower than the control mix.

4 CONCLUSIONS

Overall the data indicates that at 25 MPa, the use of the RCA up to 100% has no evident impact on the 28 days strength of the concrete for the OPC or FA mix designs. At 32 MPa similar strengths are also observed for the OPC mix design up to 100% RCA. For the FA mixes a higher 7 day strength is observed in those mixes with 50% and above RCA content. However, at 28 days the RCA mixes all have slightly lower strength than the control mix with 100% natural aggregate. Again no effect of RCA level on strength is observed for the 32 MPa concretes.

The results indicate that the use of RCA recovered from the wet mix can achieve similar strengths to OPC and FA concretes using natural aggregates for concrete strengths 25 MPa mix designs. However, for 32 MPa mix designs while the OPC can again achieve similar strengths for the RCA, for 25% FA replacement the 28 day strengths are slightly lower.

REFERENCES

Alexander, M & Mindess, S. 2010. *Aggregates in concrete*, CRC Press.
AS 3600 (ed.) 2018. *Concrete Structures*, Sydney, Australia: Australian Standard.
AS 1999. Method of testing concrete, Method 9: Determination of the compressive strength of concrete specimens. *AS (Australian Standards)*. Australia: Standards Australia.
AS 2014. Determination of properties related to the consistency of concrete - Slump test. *AS (Australian Standards)*. Australia: Standards Australia.
ASTM. 2016. Standard Specification for Portland Cement. West Conshohocken: ASTM International.
Australia, C.C.A. 2016. Available: Available from: http://www.concrete.net.au/iMIS_Prod.
Evangelista, L. & De Brito, J. 2007. Mechanical behaviour of concrete made with fine recycled concrete aggregates. *Cement & Concrete Composites* 29(5): 397-401.
Neville, A. M. 1996. *Properties of Concrete*. Harlow, Pearson Education Limited.

Overview of the influence of internal curing in recycled aggregate concrete

B.F.N. Rahmasari & Y. Yu
Shanghai Jiao Tong University, Shanghai, China

J. Yang
Shanghai Jiao Tong University, Shanghai, China
University of Birmingham, Birmingham, UK

ABSTRACT: The use of recycled concrete aggregates retrieved from construction and demolition wastes in new concrete mixes is becoming a popular application. Recycled aggregates are composed of original aggregates and adhered mortar. Recycled aggregates concrete has several drawbacks that cannot be overcome by external curing. Insufficient curing might significantly reduce the expected performance from the concrete, therefore internal curing is needed. Internal curing requires some sort of water store that can supply water to the cement paste during cementitious reaction. General requirements for internal curing of cementitious materials are explained in this paper. Several materials are proved to have the beneficial effect of supplying curing water internally. Internal curing can enhance concrete properties such as compressive strength, splitting tensile strength, autogenous shrinkage.

1 INTRODUCTION

Recycled aggregates are composed of adhered mortar and original aggregates (Etxeberria et al. 2007). Previous research (Etxeberria et al. 2007) has confirmed that the weak adhered mortar in recycled concrete aggregate (RCA) has some negative characteristics such as lower density, higher porosity, and increased water absorption compared to natural aggregate. The increased absorption characteristics of recycled aggregate mean that the concrete made with natural sand and recycled coarse aggregate natural sand typically needs 5% more water than conventional concrete in order to obtain the same workability (Sri Ravindrarajah & Tam 1985). RAC (recycled aggregate concrete) research results have shown that RAC has lower durability than natural aggregate concrete. This problem also leads to self-desiccation and autogenous shrinkage.

Autogenous shrinkage is the dominant volume change for recycled aggregate concrete (RAC), characterized by the high risk of early-age cracking of RAC due to the large magnitude of early-age autogenous shrinkage caused by the drying of the internal part of the concrete during the reaction of binder hydration (Paillere et al. 1989). In order to avoid the early-age cracking of RAC, as the hydration process of cement takes place, it is crucial to prevent the decrease of internal relative humidity. To mitigate the drawbacks of RAC, numerous researchers have explored various approaches and techniques to reduce or partially compensate the detrimental effects of RCA or to improve the characteristics of the resultant concrete (Shaban et al. 2019). External curing is standard for limiting shrinkage of concrete (Lura et al. 2014). Conventional external curing methods perform to keep concrete warm and moist, so the hydration of cement continues until its total strength gain (Taylor 2013). The example of those methods, such as put in the water pond or water and covering with wet burlap, are difficult to have significant effects because of the relatively low surface porosity of cementitious material.

However, once the capillary pores are depercolate, it will be more challenging to provide adequate external water curing. External curing procedures are not sufficiently effective to address the problems in RAC, external curing may eliminate the autogenous shrinkage in small cross-sections but not in sections bigger than about 50 mm (Bentur 2000). Insufficient curing might significantly reduce the expected performance from the specified water/cement (w/c) ratio and cement content. Conventional external curing might not be sufficient to prevent the self-desiccation at the center part of thick concrete elements, especially in low permeability concrete. This "curing" problem was recognized nearly 30 years ago by another study, who suggested incorporating saturated lightweight fine aggregate (LWFA) into the concrete mixture in order to provide the internal water source necessary to replace water that consumed during the hydration process. Pre-soaked lightweight aggregate in the mix act as an internal water reservoir preventing the reduction of relative humidity (Weber & Reinhardt 1997). In ACI Committee 308, Curing Concrete, "internal curing refers to the hydration process of cement occurs due to the availability of additional internal water that is not part of the mixing water (Bentz et al. 2005).

Internal curing is introducing internal water reservoirs into the concrete mixture (Jensen et al., 2006). Previous research reported that high absorption lightweight aggregate might have the beneficial effect of supplying curing water internally. Cells and gels are also good examples of water containers found in nature. As the cement hydration process progresses, the humidity in the cement paste decreases due to the consumption of capillary pore water which leads to the formation of menisci in the capillaries (Suzuki et al. 2009). The radius of the menisci at the interface between air and pore fluid progressively decreases as the largest pores empty, and porosity fills up with hydration products (Chen et al. 2013). This process is known as the primary mechanism that leads to autogenous shrinkage development of the cementitious paste.

In order to reduce self-desiccation and support cement hydration, additional water (curing water) might be provided. In concrete mixtures that internally cured, the internal curing material acts as a water reservoir and the pore network of the cement paste absorbs the water that contained in internal curing material by capillary suction to maintain water-filled capillary pores and to avoid menisci formation and internal capillary tension development. Immediately after casting, the system is in a fluid state, and the aggregate and cement mixture tend to settle due to gravity forcing pore water to the surface. Adequate water is supplied after concrete setting to promote hydration of the cement matrix for internal curing in order to reduce shrinkage.

2 INTERNAL CURING MATERIALS

Internal water curing of concrete requires some water store that can supply water to the cement paste during the cementitious reactions (Jensen et al. 2006). General requirements for internal curing materials of cement-based materials include: 1) thermodynamic availability requires the water to have an activity close to one, in other words an equilibrium RH close to 100%; and 2) kinetic availability refers to the transport of water from the reservoir to all parts of the self-desiccating cementitious material (Jensen et al. 2006). The example of internal curing material that has investigated extensively is the use of lightweight aggregate (LWA) (Weber & Reinhardt 1997), while the other is the use of water-absorbing polymers known as a superabsorbent polymer (SAP) (Jensen et al. 2006). The use of rice husk ash also has been extensively investigated (Rößler et al. 2014).

2.1 *Substances containing physically adsorbed water*

Superabsorbent polymer (SAP) and bentonite clay are examples of substances containing physically adsorbed water. The particle size of SAP usually employed less than 63 μm in the dry state (Justs et al. 2015).

2.2 Lightweight aggregate (LWA)

Since 1960, the study already stated that some degree of internal curing could be achieved through the use of saturated porous aggregate such as certain types of lightweight aggregates. Lightweight aggregate can be in the form of fine and coarse aggregate.

2.3 Porous superfine powders

Porous superfine powders main characteristics are small particle size materials with large specific surface area, mesoporous structure and absorbing ability of the aqueous phase (Liu et al. 2017) that enable the supply of water for further cement hydration. Porous superfine powders have nm-sized pores (Rößler et al. 2013), such as RHA, coal bottom ash, and wood-derived powder. The mean particle size of porous superfine powders generally ranges from 5 μm to 10 μm.

3 EFFECT OF INTERNAL CURING ON MECHANICAL PERFORMANCE OF RECYCLED AGGREGATE CONCRETE

3.1 Compressive strength

One of the essential properties in concrete is the compressive strength of concrete (Koushkba-ghi et al. 2019). As stated in (Suwan et al. 2017), recycled aggregate concrete (RAC) has compressive strength at least 76% and modulus of elasticity 60% to 100% of the natural aggregate concrete. Effect of the internal curing on the RAC depends on the type and amount of material used for internal curing. The use of waste porous crushed ceramic aggregate (PCCA) in (Suzuki et al. 2009) showed an insignificant impact towards the compressive strength at the early ages. However, the compressive stress or recycled concrete with PCCA after seven days starts to overcome the concrete without PCCA.

Padhi et al. (2018) incorporating the use of rice husk ash and 100% RCA and the 7-day compressive strength show a significant decrease compared to normal aggregate concrete (Figure 1). RHA also improved the compressive strength (Koushkbaghi et al. 2019) due to the release of absorbed trapped mixture water into the pores, which improves the hydration.

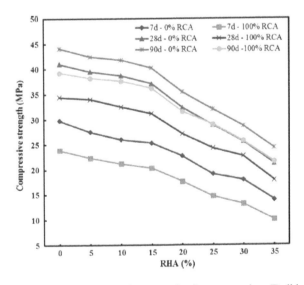

Figure 1.   Effect of RHA (%) on the compressive strength of concrete mixes (Padhi et al. 2018).

3.2 *Splitting tensile strength (Justs et al. 2015)*

Although the compressive strength of concrete is the most important and primarily studied, the tensile strength is critical for concrete subjected to tensile forces such as those induced by the development of autogenous shrinkage (Suzuki et al. 2009). The splitting tensile test is usually performed due to its simplicity (Deb et al. 2014). In RAC, around 10-15% reduction was detected compared to normal aggregate concrete (Etxeberria et al. 2007). Premature tensile failure may occur in a cementitious material that experienced autogenous shrinkage like RAC (Igarashi et al. 1999). Figure 2 showed the influence of the porous crushed ceramic aggregate (PCCA) as internal curing material to splitting tensile strength of RCA.

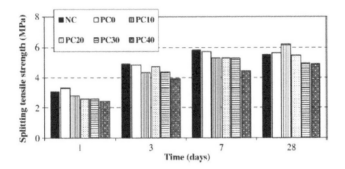

Figure 2. Influence of the incorporation of the PCCA on splitting tensile strength development (Suzuki et al. 2009).

3.3 *Autogenous shrinkage*

Autogenous shrinkage developed more rapidly within the first day. The internally entrained water through internal curing helps to alleviate the adverse effects of concrete drying by replacing any evaporated water before self-desiccation of the concrete that may cause the shrinkage cracks (Bentz & Weiss 2011).

Figure 3 showed that certain amount of internal curing material might eliminate autogenous shrinkage effectively depends mainly on the water transport characteristics within the cement matrix which are intimately related to internal curing materials parameters including proportions, pore size, and grain size distribution. From the picture, it can be observed that when more internal curing material was added, the autogenous shrinkage of RAC was significantly reduced. The grain size distribution might influence the water reservoirs distribution in the cement matrix, which means water transport may become limited to the distance surrounding the internal curing materials.

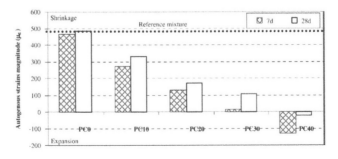

Figure 3. The contribution of the internal curing using PCCA in autogenous shrinkage reduction (Suzuki et al. 2009).

4 DISCUSSION

4.1 *Comparison of the internal curing material*

Based on the previous studies, internal curing materials could be classified into two types, that are substances containing physically adsorb water and porous material. Substances containing physically adsorb water consists of material that can retain water inside its body without dissolving, such as SAP and bentonite clay. The advantage of using substances containing physically adsorb water is that it can retain a large amount of water by swelling because of the osmotic pressure that is proportional to the concentration of ions in the aqueous solution.

Porous material could be classified to lightweight aggregate and porous superfine powder. LWA is the most common method used as a water reservoir. The advantage of using lightweight aggregate is that the LWA-internally cured concrete can gain compressive strength improved up to 20% compared to concrete without LWA, which means even when the water in LWA has been absorbed during hydration, it stays in its original size so does not leave pores that may decrease the compressive strength. The small pores in LWA allow for a steady moisture release to all parts of the concrete. The disadvantage of LWA is several studies stated that concrete containing LWA takes longer drying time.

The most common porous superfine powder used for internal curing material is rice husk ash (Lura et al. 2014) because besides it can retain and supply water later, using RHA is a smart attempt to recycle valueless waste materials into renewable materials. Porous superfine powders can delay and slow down the decrease of RH (self-desiccation) of cementitious materials, leading to mitigate autogenous shrinkage significantly.

4.2 *Effect of internal curing towards recycled aggregate concrete (RAC)*

Replacement of natural aggregate with RCA has a significant effect on the mechanical and durability properties of concrete. The unit weight decreased with increasing RCA content.

Based on the previous studies, the possibility of reducing autogenous shrinkage by the internal curing materials, which can retain water and release it later during the hydration process, are examined. Compared to the RAC without internal curing, the evolution of autogenous shrinkage of RAC with internal curing sample is different. The addition of internal curing materials contributes to the increase of hydration, the absence of large empty pores, the mitigation of autogenous shrinkage, and the enhancement of later age compressive strength. The slump values are affected by the amount of water content with the increase of extra water provided by the internal curing material.

5 SUMMARY AND CONCLUSIONS

This paper presents an overview of internal curing materials and the effect of internal curing to RAC. Several studies proved that 100% saturated RCA mixtures appear to gain compressive strength at a higher rate for longer during early-age strength gain and reduce autogenous shrinkage, compared to unsaturated or not fully saturated RCA. Internal curing proved to decrease autogenous shrinkage happened to RAC.

Addition of internal curing materials mentioned in this paper is a quick and easy method to address the problems presented by eco-efficient concrete with recycled aggregates. The recommended material used is the lightweight aggregate, for example, porous ceramic and pumice, followed by porous superfine powders, such as RHA and wood-derived powders. The lightweight aggregate and porous superfine powders considered to be the most effective as internal curing materials in terms of material cost and sustainable development. SAP as internal curing material considered high-cost material even though it has great potential to enhance the properties of RAC. Most of the internal curing materials addition to RAC is currently being examined on a small scale under laboratory conditions.

REFERENCES

Bentur, A. 2000. Early age shrinkage and cracking in cementitious systems. In *International RILEM Workshop on Shrinkage of Concrete* (1-20). RILEM Publications SARL.

Bentz, D.P. Lura, P. & Roberts, J.W. 2005. Mixture proportioning for internal curing. *Concrete international* 27(2): 35-40.

Bentz, D.P. & Weiss, W.J. 2011. *Internal curing: a 2010 state-of-the-art review*. Gaithersburg, Maryland: US Department of Commerce, National Institute of Standards and Technology.

Chen, H. Wyrzykowski, M. Scrivener, K. & Lura, P. 2013. Prediction of self-desiccation in low water-to-cement ratio pastes based on pore structure evolution. *Cement and Concrete Research* 49: 38-47.

Choi, H. Choi, H. Lim, M. Inoue, M. Kitagaki, R. & Noguchi, T. 2016. Evaluation on the mechanical performance of low-quality recycled aggregate through interface enhancement between cement matrix and coarse aggregate by surface modification technology. *International Journal of Concrete Structures and Materials* 10(1): 87-97.

Deb, P.S. Nath, P. & Sarker, P.K. 2014. The effects of ground granulated blast-furnace slag blending with fly ash and activator content on the workability and strength properties of geopolymer concrete cured at ambient temperature. *Materials & Design* 62:32-39.

Etxeberria, M. Vazquez, E. Mari, A. & Barra, M. 2007. Influence of amount of recycled coarse aggregates and production process on properties of recycled aggregate concrete. *Cement and Concrete Research* 37(5): 735-742.

Igarashi, S. Bentur, A. & Kovler, K. 1999. Stresses and creep relaxation induced in restrained autogenous shrinkage of high-strength pastes and concretes. *Advances in Cement Research* 11(4): 169-177.

Jensen, O. M. & Lura, P. 2006. Techniques and materials for internal water curing of concrete. *Materials and Structures* 39(9): 817-825.

Justs, J. Wyrzykowski, M. Bajare, D. & Lura, P. 2015. Internal curing by superabsorbent polymers in ultra-high performance concrete. *Cement and Concrete Research* 76: 82-90.

Koushkbaghi, M. Kazemi, M.J. Mosavi, H. & Mohseni, E. 2019. Acid resistance and durability properties of steel fiber-reinforced concrete incorporating rice husk ash and recycled aggregate. *Construction and Building Materials* 202: 266-275.

Liu, J. Shi, C. Ma, X. Khayat, K.H. Zhang, J. & Wang, D. 2017. An overview on the effect of internal curing on shrinkage of high performance cement-based materials. *Construction and Building Materials*, 146: 702-712.

Lura, P. Wyrzykowski, M. Tang, C. & Lehmann, E. 2014. Internal curing with lightweight aggregate produced from biomass-derived waste. *Cement and concrete research* 59: 24-33.

Maruyama, I. & Sato, R. 2005. A trial of reducing autogenous shrinkage by recycled aggregate. In *Proceedings of 4th International Seminar on Self-Desiccation and its Importance in Concrete Technology* (264-270).

Padhi, R.S. Patra, R.K. Mukharjee, B.B. & Dey, T. 2018. Influence of incorporation of rice husk ash and coarse recycled concrete aggregates on properties of concrete. *Construction and Building Materials* 173:289-297.

Paillere, A. Buil, M. & Serrano, J.J. 1989. Effect of fiber addition on the autogenous shrinkage of silica fume. *Materials Journal* 86(2):139-144.

Roßler, C. Bui, D.D. & Ludwig, H.M. 2014. Rice husk ash as both pozzolanic admixture and internal curing agent in ultra-high performance concrete. *Cement and Concrete Composites* 53: 270-278.

Roßler, C. Bui, D.D. & Ludwig, H.M. 2013. Mesoporous structure and pozzolanic reactivity of rice husk ash in cementitious system. *Construction and Building Materials* 43:208-216.

Shaban, W.M. Yang, J. Su, H. Mo, K.H. Li, L. & Xie, J. 2019. Quality Improvement Techniques for Recycled Concrete Aggregate: A review. *Journal of Advanced Concrete Technology* 17(4):151-167.

Sri Ravindrarajah, R. & Tam, C.T. 1985. Properties of concrete made with crushed concrete as coarse aggregate. *Magazine of concrete research* 37(130):29-38.

Suwan, T. & Wattanachai, P. 2017. Properties and internal curing of concrete containing recycled autoclaved aerated lightweight concrete as aggregate. *Advances in Materials Science and Engineering* 2394641.

Suzuki, M. Meddah, M.S.. & Sato, R. 2009. Use of porous ceramic waste aggregates for internal curing of high-performance concrete. *Cement and Concrete Research* 39(5): 373-381.

Taylor, P.C. 2013. *Curing concrete*. CRC Press.

Weber, S. & Reinhardt, H.W. 1997. A new generation of high performance concrete: concrete with autogenous curing. *Advanced Cement Based Materials* 6(2):59-68.

Sustainable Buildings and Structures: Building a Sustainable Tomorrow – Papadikis et al. (Eds)
© 2020 Taylor & Francis Group, London, ISBN 978-0-367-43019-1

Sustainability analysis of functionally graded concrete produced with fibres and recycled aggregates

R. Chan, T. Hu, X. Liu, I. Galobardes, Charles K.S. Moy & J.L. Hao
Department of Civil Engineering, Xi'an Jiaotong-Liverpool University, Suzhou, China

K. Krabbenhoft
School of Engineering, University of Liverpool, Liverpool, UK

ABSTRACT: Fibre reinforced concrete is widely used in the pavement construction despite presenting some economic and environmental issues. To address these matters, the concept of functionally graded material (FGM) concept with materials that present changeable properties over its volume is adopted on concrete, generating functionally graded concrete (FGC). To understand these new composite materials more studies are needed. In this study, eight mixes of FGC concretes were produced to assess the effect of the recycled aggregate and different depths of the fibre reinforced layer on the flexural performance. Besides, a simplified sustainability analysis was performed to assess the goodness of these special concretes. The experimental results indicate that the FGC studied could be used in structural applications requiring lower loading capacity. The sustainability analysis indicates that a cut in the total volume of steel fibres and the use of recycled aggregates would enhance the benefits of the FGC.

1 INTRODUCTION

Conventional reinforced concrete (RC) is widely used in construction. However, this traditional material presents drawbacks regarding technical issues and high carbon footprint (Liu et al. 2018). In that sense, fibre reinforced concrete (FRC) may improve RC drawback in terms of controlling crack propagation (Liu et al. 2018). Current studies present FRC as optimisation of RC to produce certain types of applications such as pavements or tunnel linings (Chan et al. 2019). However, the homogenous distribution of fibres in the whole volume of concrete makes it a non-efficient economic material for applications under bending (Liu et al. 2018). On the other hand, to reduce the RC environmental drawbacks, recycled aggregate is used to substitute natural aggregate in concrete (Chan et al. 2019). Despite their low mechanical properties, recycled aggregate concrete (RAC) leads to a reduction of landfill use and quarry exploitation (Liu et al. 2018).

The concept of functionally graded material (FGM) to create composite materials, which present changeable properties over its volume, can be adopted to combine the benefits of FRC and RAC generating a functionally graded concrete (FGC) (Liu et al. 2018, Rio et al. 2015). In that sense, this paper aims to present an investigation on the flexural performance of FGC and an analysis of its sustainability regarding the materials used to produce them. Additionally, natural and recycled aggregates and fibres were considered to fabricate the FGC.

2 EXPERIMENTAL PROGRAM

2.1 *Materials*

Plain cement CEM I 42.5N, was used. Mixes were produced with tap water controlled at room temperature (20°C). Two types of aggregate, natural and recycled, were used. Limestone

was used as natural aggregate, whereas the recycled aggregate was composed by local demolition waste (Suzhou, China). The particle size distributions and the main properties of the aggregates, such as saturated surface dry density, water absorption, uniformity coefficient and coefficient of curvature, are the ones presented by Liu et al. (2018).

Finally, hooked-end steel fibres were used in the study. The fibres presented 60 mm length and aspect ratio of 80, being recommended to produced fibre reinforced concrete (Liu et al. 2018, Chan et al. 2019). Only one content of fibre (cf) equal to 40 kg/m^3 (approximately 0.50% in volume fraction).

2.2 *Types of concrete*

In this study, four concrete groups are considered, as shown in Figure 1. Two groups were produced with one concrete mix through their entire volume: FRC and FRRAC (see Figure 1a and Figure 1b), while the FGM concept was applied in the other two groups. In that sense, two-layered systems were adopted being the top layer produced with plain cement concrete (PCC) and the bottom layer with FRC (see Figure 1c) and FRRAC (see Figure 1d), respectively.

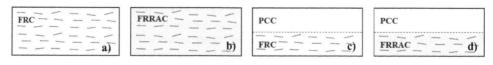

Figure 1.   Concrete mixes considered: (a) FRC (b) FRRAC (c) PCC+FRC and (d) PCC+FRRAC.

All FGC mixes are designed to absorb tension in the bottom layer since the paper aims to study the flexural behaviour of the mixes (Liu et al. 2018). In order to study the effects of the layer reinforced with fibres different relationships between reinforced layer height and total height (h/H): 0.25, 0.50, and 0.75 are considered (see Figure 2). The following codification was used to identify the mixes: First number + Letter + Second number. The first number stands for the relation h/H, and the letter is N for natural aggregates and R is for recycled aggregates, and the second number stands for the content of fibre in the reinforced layer (cf). For simplification, all concrete mixes produced with only natural aggregates will be referred to as N-concretes and concretes containing recycled aggregates, R-concretes.

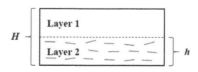

Figure 2.   FGC samples: bottom layer thickness/total thickness relation (h/H).

2.3 *Reference mix design and mix production*

Mixes were produced with 475 kg/m^3 and a water-cement ratio (w/c) of 0.45. Notice that the actual amount of water used was adjusted with consideration of the water absorption results presented by Liu et al. (2018). The amount of these for coarse and fine aggregates was 790 and 890 kg/m^3 and 805 and 715 kg/m^3 for natural and recycled aggregate mixes, respectively.

The casting procedure described in ASTM C 192/C 192M - 02 was followed to produce homogeneous concretes (ASTM, 2018). Regarding the functionally graded concretes, they were produced according to the following method: (1) the concrete of the bottom layer was mixed and placed in the mould; (2) then it was compacted for 20 s by mechanical vibration, and its height was verified afterwards, and (3) the upper layer was cast 20 min after finishing the previous layer, being vibrated for half the time spent for the bottom layer (10 s) in order to avoid mixing the layers to each other. This production method was used based on previous studies to assure bonding between layers (Liu et al. 2018, Rio et al. 2015). Finally, all samples were demoulded 24 hours after casting and cured in water at a temperature of approximately 20 °C.

2.4 *Bending test*

The 4-point bending test described in the ASTM standard C 1609/C 1609M-06 was used in the study (ASTM 2007). Four prismatic samples per mix were tested at the age of 28 days. In this test, the load (F) is controlled, and the net deflection (δ) is measured using a pair of LVDTs sustained by a steel frame, obtaining the F–δ curve. These results are used to determine the flexural strength at the first crack (f_1), and the residual flexural strengths are corresponding to a net deflection equal to $L/600$ and $L/150$ ($f_{0.75}$ and $f_{3.0}$, respectively), being L the span length of the test.

3 EXPERIMENTAL RESULTS AND ANALYSIS

The average F–δ curves obtained from the bending test for each type of concrete are presented in Figure 3. In general, it is observed that an increase in h/H results in a higher post-cracking response, being translated in softening post-cracking behaviour in concretes with lower values of h/H ($h/H \leq 0.50$), and hardening behaviour for higher values of h/H ($h/H \geq 0.75$). This may happen because the post-cracking response depends on the size of the cracking area containing fibres, which increases with h/H (Figueiredo et al. 2000, Liu et al. 2018).

An exception occurs when comparing the curves between different values of h/H for R-concretes. In this case, when h/H is equal to 1.00, which refers to homogeneous concrete, even

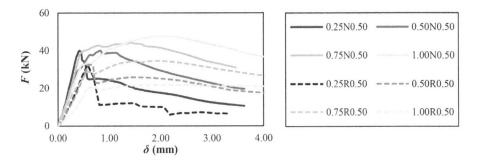

Figure 3. F – δ curves of concretes with different h/H for cf equal to 0.50%.

containing fibres in its whole volume, the post-cracking response falls back to a similar level when h/H is equal to 0.50. This fact highlights the influence of the type of aggregate in the post-cracking behaviour. Considering that the content of fibre has no significant influence in the modulus of elasticity (Carneiro et al. 2014), the higher modulus of elasticity of the non-cracked plain cement concrete, in comparison to the recycled aggregate concrete (McNeil & Kang 2013), reduces the stresses transferred to the bottom layers of the sample. Thus, in the case of homogeneous FRRAC, which presents a lower modulus of elasticity, the concrete cracks under lower loads. Furthermore, the hardening post-cracking behaviour is presented in N-concretes, suggesting that this phenomenon is linked to the compressive strength of concrete (Carneiro et al. 2014). In this sense, the lower compressive strength of R-concretes may decrease the pull-out strength, diminishing the post-cracking response (Isla et al. 2015).

Table 1 presents the results of flexural strength at the first crack (f_1) and residual flexural strengths corresponding to a net deflection equal to 0.75 mm and 3.0 mm ($f_{0.75}$ and $f_{3.0}$, respectively). The average results are expressed in MPa, and the coefficient of variation is added in parenthesis.

Table 1. Results obtained using the bending test.

Concrete	f_1 (MPa)	$f_{0.75}$ (MPa)	$f_{3.0}$ (MPa)
0.25N0.50	5.36 (6.86%)	3.35 (11.48%)	1.66 (16.73%)
0.50N0.50	5.44 (10.63%)	5.21 (7.88%)	3.03 (21.94%)
0.75N0.50	5.60 (6.77%)	5.51 (6.76%)	4.57 (10.56%)
1.00N0.50	4.73 (3.02%)	4.74 (7.41%)	5.75 (21.79%)
0.25R0.50	5.01 (9.15%)	1.47 (8.54%)	0.91 (10.92%)
0.50R0.50	4.34 (2.04%)	3.05 (9.41%)	2.79 (11.00%)
0.75R0.50	3.60 (15.46%)	3.74 (14.97%)	4.10 (12.15%)
1.00R0.50	3.02 (8.80%)	2.54 (8.36%)	3.15 (11.87%)

Flexural strength (f_1) results tend to present a lower coefficient of variation than the results for residual flexural strengths ($f_{0.75}$ and $f_{3.0}$), with average values of 7.84%, 9.35% and 14.62%, respectively. This tendency was expected (Figueiredo et al. 2000). Regarding the type of aggregate, in general, the results for R-concretes are systematically lower than those for N-concretes, when the same values of h/H are considered. Therefore, the FGC studied could be used in structural applications requiring lower loading capacity, such as low traffic roads, car parks, cycling lanes, or pedestrian pavements.

The f_1 results for the N-concretes present no significant change with h/H. However, R-concretes suggest decreasing f_1 values with the increase of h/H. This may be explained by the dependence of flexural strength on the properties of the concrete matrix (Figueiredo et al. 2000).

In the case of $f_{0.75}$, the parameter rises until h/H reaches the value of 0.75; after this value, the results drop. The increase of $f_{0.75}$ with h/H is expected, since the area containing fibres is increased with h/H, enhancing the post-cracking response (Figueiredo et al. 2000, Liu et al. 2018). On the other hand, the decrease of $f_{0.75}$ after reaching a particular value of h/H may be explained by the random distribution of fibres throughout the reinforced layer. In that sense, since the cracking area is concentrated in the bottom of the sample when the net deflection is 0.75 mm, the parameter $f_{0.75}$ is conditioned by the number of fibres in the cracked area, which can fluctuate randomly (Figueiredo et al. 2000).

Regarding the values of $f_{3.0}$, N-concretes show a growing tendency for this parameter with increasing reinforced layer thickness. This is expected since, as presented previously, the area containing fibres is augmented with h/H and enhances the post-cracking response (Figueiredo et al. 2000, Liu et al. 2018), especially for larger net deflections, such as 3.0 mm, where the cracked area advances towards the top layer of the specimen. For R-concretes, a growing tendency with a maximum point within the h/H range of 0.75 to 1.00 was observed.

In the case showing a growing tendency with a maximum point, the response may be due to the influence of the modulus of elasticity. As explained previously, homogeneous FRRAC cracks under lower loads than FGC with FRRAC because it presents a lower modulus of elasticity than FRC (Carneiro et al. 2014). Furthermore, an optimal h/H ratio is expected within the range of 0.75 to 1.00, in which a maximum $f_{3.0}$ would be reached, showing the benefits of using FGC with FRRAC instead of homogeneous FRRAC. This optimised value of h/H has to be assessed in future studies.

4 SIMPLIFIED SUSTAINABILITY ANALYSIS

A sustainability analysis was performed to compare the embodied CO_2 and costs between the concrete mixes produced and assessed in this study. In this comparative analysis, the different concrete mixes are considered to be used for the same applications and building processes. Therefore, the simplified methodology used by Caceres et al. (2014) was adopted, which focus on the impact of CO_2 emissions and costs of the production of materials, rather their application. In that sense, several processes are not taken into account, such as construction, maintenance, and demolishing. Also, since the quantity of cement used for each concrete mix is the same, only the quantities of natural aggregates, recycled aggregates and steel fibres were considered.

Table 2 presents the embodied CO_2 (*EC*) and the cost, per ton of material, for natural fine and coarse aggregates (*NFA* and *NCA*, respectively), recycled fine and coarse aggregates (*RFA* and *RCA*, respectively) and steel fibres (*SF*). The cost values are the same values used in the analysis presented by Liu et al. (2018), which were given by the suppliers. Meanwhile, the embodied CO_2 values for aggregates and steel fibres were obtained from studies published by Hossain et al. (2016) and Stengel & Schießl (2008), respectively.

Table 2. Unit embodied CO_2 and cost of materials.

Parameter	NFA	NCA	RFA	RCA	SF
EC ($kgCO_{2-eq}$/ton)	23.0	32.0	12.0	11.0	2680
Cost (USD/ton)	28.57	38.10	6.15	7.99	1714.25

The total *EC* and costs (in brackets) values to produce one cubic meter of each concrete mix, along with the values for each material, are presented in Table 3. Besides, Figure 4 shows the relationship between *EC* and costs and the residual strength $f_{3.0}$ to consider the performance of the mixes as sustainability analysis.

From Table 3, it can be observed that, for the same *h/H*, the total *EC* and costs values for R-concretes mixes are always lower than the values for N-concretes. This is expected since the only variable changed is the type of aggregate and, as presented in Table 2, recycled aggregates

Table 3. Embodied CO_2 ($kgCO_{2-eq}$) and costs (in brackets) (USD) to produce one m³ of concrete.

Concrete	NFA	NCA	RFA	RCA	SF	Total
0.25N0.50	18.2 (22.57)	28.5 (33.91)	- (-)	- (-)	26.1 (16.71)	72.8 (73.19)
0.50N0.50	18.2 (22.57)	28.5 (33.91)	- (-)	- (-)	52.3 (33.43)	98.9 (89.91)
0.75N0.50	18.2 (22.57)	28.5 (33.91)	- (-)	- (-)	78.4 (50.14)	125 (106.62)
1.00N0.50	18.2 (22.57)	28.5 (33.91)	- (-)	- (-)	104.5 (66.86)	151 (123.34)
0.25R0.50	13.6 (16.93)	21.4 (25.43)	2.42 (1.24)	1.97 (1.43)	26.1 (16.71)	65.5 (61.74)
0.50R0.50	9.09 (11.28)	14.2 (16.96)	4.83 (2.47)	3.93 (2.85)	52.3 (33.43)	84.4 (67.00)
0.75R0.50	4.54 (5.64)	7.12 (8.48)	7.25 (3.71)	5.90 (4.28)	78.4 (50.14)	103 (72.26)
1.00R0.50	- (-)	- (-)	9.66 (4.95)	7.87 (5.71)	105 (66.86)	122 (77.52)

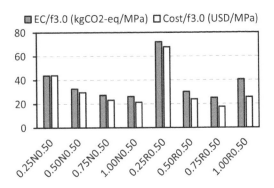

Figure 4. Relationship between EC and costs and $f_{3.0}$.

are less polluting and also cheaper than natural aggregates. Furthermore, the difference in both total EC and costs values between R-concretes and N-concretes increases with increasing h/H. This behaviour indicates that the influence of the total amount of steel fibres, in terms of embodied CO_2 and costs, is more significant than the replacement ratio of natural aggregates for recycled aggregates.

When homogenous fibre-reinforced concrete mixes are considered, the effect of the steel fibres in the total EC can be as high as 69% and 86%, for N-concretes and R-concretes, respectively. Regarding the total costs, steel fibres can contribute as much as 54% and 86%, for N-concretes and R-concretes, respectively. Therefore, a cut in the total volume of steel fibres, due to a reduction of h/H, is substantially significant to minimise the total EC and costs of concrete. Although the replacement ratio of natural aggregates for recycled aggregates does not affect so much as the content of steel fibres, the role played by the recycled aggregates cannot be neglected, specifically for higher values of h/H, when the reduction in EC and cost can be as high as 19% and 37%, respectively. So, regarding both, the use of recycled aggregates and the reduction of h, for h/H equal to 0.50 or 0.75, the FGC with FRRAC present advantages in terms of EC and cost (See Figure 4). This highlights the benefits of FGC produced with fibres and recycled aggregates.

5 CONCLUSIONS

An investigation on the flexural performance of FGC incorporating fibres, natural and recycled aggregates where the mechanical properties were evaluated, and a simplified sustainability analysis was presented in this paper. The experimental results indicated that the FGC studied could be used in structural applications requiring lower loading capacity, such as low traffic roads, car parks, cycling lanes, or pedestrian pavements. The sustainability analysis indicated that a cut in the total volume of steel fibres, due to a reduction of h/H, was substantially significant to minimise the total EC and cost of concrete. Also, although the replacement ratio of natural aggregates for recycled aggregates did not affect so much as the content of steel fibres, in terms of EC and cost, the role played by recycled aggregates cannot be neglected, specifically for higher values of h/H. Overall, this paper highlighted the benefits of FGC produced with fibres and recycled aggregates.

ACKNOWLEDGMENT

The authors would like to thank XJTLU Research Development Fund for the financial support received from the project with reference RDF-16-02-42.

REFERENCES

ASTM 2007. *ASTM C1609 / C1609M - 12 Standard Test Method for Flexural Performance of Fiber-Reinforced Concrete (Using Beam With Third-Point Loading)*. American Society for Testing and Materials.

ASTM 2018. *ASTM C 192/C 192M – 02 Standard Practice for Making and Curing Concrete Test Specimens in the Laboratory1*. American Society for Testing and Materials.

Cáceres, A. John, V. & Figueiredo, A. 2014. *Comparação entre pavimentos de concreto armado e de concreto reforçado com fibras com relação às emissões globais de CO2*. Anais Do 56o Congresso Brasileiro Do Concreto, IBRACON.

Carneiro, J.A. Lima, P.R.L. Leite, M.B. & Filho, R. D. T. 2014. Compressive stress–strain behavior of steel fiber reinforced-recycled aggregate concrete. *Cement & Concrete Composites* 46:65-72.

Chan, R. Santana, M.A. Oda, A.M. Paniguel, R.C. Vieira, L.D.B.P. De figueiredo, A.D. & Galobardes, I. 2019. Analysis of potential use of fibre reinforced recycled aggregate concrete for sustainable pavements. *Journal of Cleaner Production* 218:183-191.

Figueiredo, A.D. Nunes, N.L. & Tanesi, J. 2000. *Mix design analysis on steel fiber reinforced concrete. Fifth International RILEM Symposium on Fibre-Reinforced Concrete (FRC)*. RILEM Publications SARL 103-118.

Hossain, M.U. Poon, C.S. Lo, I.M.C. & Cheng, J.C.P. 2016. Comparative environmental evaluation of aggregate production from recycled waste materials and virgin sources by LCA. *Resources Conservation and Recycling* 109: 67-77.

Isla, F. Ruano, G. & Luccioni, B. 2015. Analysis of steel fibers pull-out. Experimental study. *Construction and Building Materials* 100:183-193.

Liu, X. Yan, M. Galobardes, I. & Sikora, K. 2018. Assessing the potential of functionally graded concrete using fibre reinforced and recycled aggregate concrete. *Construction and Building Materials* 171:793-801.

Mcneil, K. & Kang, T.H.K. 2013. Recycled Concrete Aggregates: A Review. *International Journal of Concrete Structures and Materials* 7:61-69.

Rio, O. Nguyen, V.D. & Nguyen, K. 2015. Exploring the potential of the functionally graded SCCC for developing sustainable concrete solutions. *Journal of Advanced Concrete Technology* 13:193-204.

Stengel, T. & Schießl, P. 2008. Sustainable construction with UHPC–from life cycle inventory data collection to environmental impact assessment. *Proceedings of the 2nd international symposium on ultra high performance concrete. Kassel University Press, Kassel* 461-468.

Sustainable Buildings and Structures: Building a Sustainable Tomorrow – Papadikis et al. (Eds)
© 2020 Taylor & Francis Group, London, ISBN 978-0-367-43019-1

Effects of particle morphology on the macroscopic behaviour of ellipsoids: A discrete element investigation

S.P.K. Kodicherla, G. Gong, L. Fan & Charles K.S. Moy
Department of Civil Engineering, Xi'an Jiaotong – Liverpool University, Suzhou, P.R. China

ABSTRACT: The use of granular materials are ubiquitous in the infrastructural projects. Due to the dramatic degradation of natural resources and rapid development of urbanization, new challenges in sustainable development have led the civil engineering community to consider a breakthrough approach based on a fundamental understanding of natural scale of grains and their interactions. In this study, a three-dimensional DEM based investigation of particle morphology on the macroscopic behaviour of ellipsoids is carried out under conventional drained triaxial test conditions. Six numerical simulations are performed by controlling two parameters i.e., 3ξ and 3β, where ξ is the ratio of the smallest to the largest sphere and β is the maximum sphere-sphere intersection angle. It has been found that irrespective of ξ and β, all stress-strain curves exhibit strain-softening behaviour and a critical state is achieved at large axial strains. The shear strength of ellipsoids at both peak and critical states is found to increase with decreasing β. Moreover, the peak state friction angles are found to increase with increasing ξ, whereas critical state friction angles decrease with increasing ξ. These findings highlight the significance of particle morphology on the macroscopic response of granular materials.

1 INTRODUCTION

The interest of particle morphology of granular materials has called the attention of geotechnical and geological researchers for many decades. Many experimental studies evidenced that the mechanical properties of granular materials, such as compressibility, shear strength, dilation and crushability, are highly influenced by the morphological features of the constitutive particles (Guo & Su 2007, Tsomokos & Georgiannou 2010, Altuhafi & Coop 2011, Yang & Luo 2015). As an alternative to the laboratory experimentation on investigating the fundamental soil behaviors, the discrete element method (DEM) has made substantial contributions towards elucidating the micromechanics of the particle morphology affecting the mechanical properties of granular soils (Wang & Gutierrez 2010, Zhou et al. 2013, Fu et al. 2017). To date, many numerical simulations have been performed by employing various irregular shapes, such as clumped discs (Maeda et al. 2010, Saint-Cyr et al. 2011), polygons (Tillemans et al. 1995, Hosseininia 2012), spheropolygons (Alonso-Marroquin 2009) in two-dimensions, and polyhedrons (Zhao et al. 2015), spherocylinders (Pournin et al. 2005), superellipsoids (Wellmann et al. 2008, Cleary 2010) in three dimensions. The particle shape underpins various facets of the mechanical responses for granular materials and should be thoroughly accounted for in their modelling (Shin & Santamarina 2012, Payan et al. 2016, Alshibli & Cil 2017, Xiao et al. 2018, Shi et al. 2018). Moreover, the surface textures will also play a key role in affecting the mechanical behaviour of granular materials (Kozicki et al. 2012, Kozicki et al. 2014).

Ellipsoids are able to capture many realistic features of granular material and represent a wide range of particle shapes that exist in nature (Lin & Ng 1997, Ng 2004, Bagherzadeh-Khalkhali et al. 2009, Zhou et al. 2013). In this study, a commercial DEM code, particle flow code (PFC^{3D}) (Itasca 2018) was used to explore the effects of particle morphology on the macroscopic behaviour of ellipsoids. The morphological features of a particle is controlled by two parameters i.e., the ratio of the smallest to the largest sphere, ξ and the maximum sphere-sphere intersection angle, β. One direct explanation is that ξ will control the particle shape whereas β controls the surface texture of the particle. For a detailed understanding of the macroscopic response of ellipsoids, a brief discussion is presented in the following sections.

2 METHODOLOGY

In PFC^{3D}, one may use clump to represent irregular particle shape by joining and overlapping different spheres. Given an ellipsoid surface enclosing a volume, a Delaunay tetrahedralization is initially constructed and then for each tetrahedron, the centre and radius of its circumscribed sphere are recorded as the balls of a clump. This approach can be implemented in the Kubrix automatic mesh generation software package (Taghavi 2000, Simulation Works 2009) as an option called 'BubblePack' which automatically generates clump templates for PFC^{3D}. The 'BubblePack' outputs a file containing the description of clump template which is compatible to read in PFC^{3D} (see Figure 1).

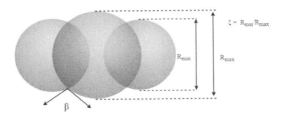

Figure 1.   A clump representing ellipsoid with three spheres.

Due to computational limitations, we, focus on the following cases: β =150° with ξ = 0.4, 0.6 and 0.8; and ξ = 0.4 with β = 100°, 130° and 160°. To optimize these two parameters, the development of clump generation with different parameters for a typical ellipsoid particle is shown in Figure 2. It is clear that with the increasing β and decreasing ξ, the shape features of the clump and the number of filling spheres dramatically increases.

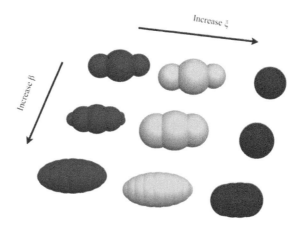

Figure 2.   The effects of varying β and ξ for ellipsoidal clumps.

3 DEM SIMULATIONS OF DRAINED TRIAXIAL TESTS

The conventional drained triaxial specimen comprised of six independent rigid walls formed as a cubic box. Taking the computational limitations into account, each assembly is restricted to a cloud of 2,845 ellipsoids which means that in a total of 25,605 individual rigid spheres (pebbles) in the assembly. The interactions between clump-clump and clump-wall are set to obey a general linear force-displacement contact law, as used by many researchers (Abedi & Mirghasemi 2011, Gu et al. 2014, Stahl & Konietzky 2011, Yan 2009, Yang et al. 2016). The friction coefficient between clump-clump is set to 0.5, whereas for the clump-wall is set zero. The input DEM parameters are: the stiffness ratio is 4/3; the notional particle density is 2700 (kg/m^3); effective modulus E is 100 MPa; friction between the clump-clump is 0.5; friction between clump-wall is 0.0 and damping constant is 0.7. It should be noted that all specimens are subjected to isotropic compression of 100 kPa with a low strain rate over a large number of small timesteps. The isotropic assemblies are subjected to compression by permitting top and bottom walls toward (compression) at a constant loading strain rate of 0.004 m/s (which is sufficiently small to ensure quasi-static shear), whereas the remaining walls hold the same confining pressure using a stress-controlled servo-mechanism.

4 RESULTS AND DISCUSSION

4.1 *Macroscopic response*

In the conventional drained triaxial test, the minor principal stress $\sigma_3(\sigma_2 = \sigma_3)$ is maintained constant and failure results from the increase in the major principal stress (axial stress, σ_1) under full drainage conditions. The rate of loading or deformation is so arranged that negligible excess pore pressure is generated in the specimen at any time during the application of axial stress and particularly at failure. Stress tensor is adopted to quantify the macroscopic response of granular material during shearing. It can be calculated from the discrete data with the proposed definition is given as (Christoffersen et al. 1981):

$$\sigma_{ij} = \frac{1}{V} \sum_{c\partial V} f_i^c l_j^c \tag{1}$$

where V is the total volume of the assembly, f^c is the contact forces at contact c, and l^c is the branch vector connecting the centres of the two ellipsoidal particles in contact. From Eq. (1), the deviator stress q can be calculated as:

$$q = \sqrt{3J_2} \tag{2}$$

where $I_1 = \sigma_{ii}$ is the first invariant of the stress tensor σ_{ij}, and $J_2 = \sigma'_{ij}\sigma'_{ij}/2$ is the second invariant of the deviatoric stress tensor σ_{ij}. The axial strain ε_a and volumetric strain ε_v are estimated from the displacements of the rigid walls (boundaries) as:

$$\varepsilon_1 = -\int_{H_0}^{H} \frac{dh}{h} = \ln\frac{H_0}{H} \tag{3}$$

$$\varepsilon_v = \varepsilon_1 + \varepsilon_2 + \varepsilon_3 = -\int_{V_0}^{V} \frac{dv}{v} = \ln\frac{V_0}{V} \tag{4}$$

where H and V are the height and volume of the specimen during simulation, respectively, and H_0 and H are their initial values of the boundaries before shear.

47

Figures 3-4 show the stress-strain response of ellipsoidal assemblies for $\xi = 0.4$ with different β values and $\beta = 150°$ with different ξ, respectively. In all simulations, irrespective of ξ and β, all curves exhibit peak stress, i.e., the so-called strain-softening behaviour, which is quite obvious in dense assembly. However, at large axial strain (i.e., greater than 40%), the deviator stress flattens off and tends to reach a constant value which can be identified as a critical state. Overall, despite the initial peaks, all samples exhibit very similar behaviour and show significant dilation up to the critical state, while minor contractive behaviour noticed at the early stages of shearing.

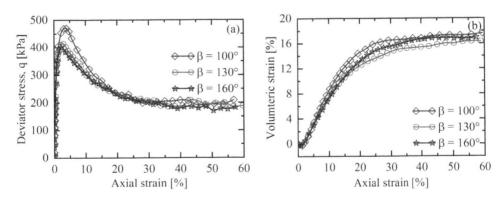

Figure 3.   Deviator stress (a) and volumetric strain (b) against axial strain for $\xi = 0.4$ with different β.

4.2   Friction angles

The peak state is considered at which the deviator stress shows a peak value against axial strain, whereas the critical state is considered to be the continuous deformation of particles at constant volume and constant stress. Referring to Figures 3(a) and 4(a), it appears that all specimens reach critical state at an axial strain ranging between 40% and 60%. The internal friction angle (or angle of internal friction) ϕ, which represents the shear strength of the granular materials, is given as:

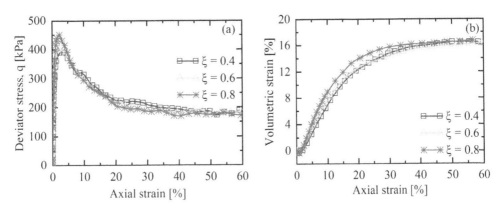

Figure 4.   Deviator stress (a) and volumetric strain (b) against axial strain for $\beta = 150°$ with different ξ.

$$\phi = \sin^{-1}\left(\frac{\sigma_1 - \sigma_3}{\sigma_1 + \sigma_3}\right) \tag{5}$$

where σ_1 and σ_3 are the major and minor principal stresses, respectively. Figure 5 shows the influence of β and ξ on the friction angles. Error bars represent the standard deviation of friction angles at critical state. It can be seen from Figure 5(a) that both peak and critical state friction angles decrease with increasing β which indicates that rougher the surface texture, higher the shear strength. From Figure 5(b), it can be noted that peak state friction angles increase with increasing ξ, whereas critical state friction angles decrease with increasing ξ. This variation may be due to the transformation in particle shape.

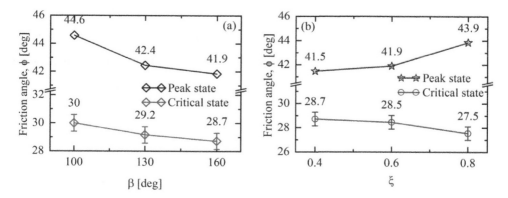

Figure 5.   The influence of β and ξ on the friction angles.

5   CONCLUSIONS

This paper reports the effects of particle morphological features on the macroscopic behaviour of ellipsoids. The two parameters i.e., the ratio of the smallest to the largest sphere, ξ and the maximum sphere-sphere intersection angle, β controls the shape and surface texture of the particle. It is found that irrespective of ξ and β, all curves exhibit strain-softening behaviour and reaches critical state at large axial strains. The shear strength of ellipsoids at both peak and critical states are found to increase with decreasing β. The peak state friction angles increase with increasing ξ, whereas critical state friction angles decrease with increasing ξ. The future study will focus on the evolution of microscopic response of ellipsoids.

ACKNOWLEDGEMENTS

The authors acknowledge the financial support from National Natural Science Foundation of China (Grant No: NSFC 51578499 and 51825803) and from Xi'an Jiaotong - Liverpool University (RDF-15-01-38, RDF-14-02-44 and RDF-18-01-23). Also, the funding supported by Key Program Special Fund in XJTLU (Grant No: KSF-E-19) and Natural Science Foundation of Jiangsu Province (Grant number BK20160393) is greatly appreciated.

REFERENCES

Abedi, S. & Mirghasemi, A.A. 2011. Particle shape consideration in numerical simulation of assemblies of irregularly shaped particles. *Particuology* 9(4): 387-397.

Alonso-Marroquin, F. & Wang, Y. 2009. An efficient algorithm for granular dynamics simulations with complex-shaped objects. *Granular Matter* 11(5): 317-329.

Alshibli, K.A. & Cil, M.B. 2017. Influence of particle morphology on the friction and dilatancy of sand. *Journal of Geotechnical and Geoenvironmental Engineering* 144(3): 04017118.

Altuhafi, F. & Coop, M. 2011. Changes to particle characteristics associated with the compression of sands. *Geotechnique* 61(6):459-471.

Bagherzadeh-Khalkhali, A. & Mirghasemi, A. 2009. Numerical and experimental direct shear tests for coarse-grained soils. *Particuology* 7: 83-91.

Christoffersen, J, Mehrabadi, M. & Nemat-Nasser, S. 1981. A micromechanical description of granular material behavior. *Journal of Applied Mechanics* 48(2): 339-344.

Cleary, P.W. 2010. DEM prediction of industrial and geophysical particle flows. *Particuology* 8: 106-118.

Fu, R. Hu, X. & Zhou, B. 2017. Discrete element modeling of crushable sands considering realistic particle shape effect. *Computers and Geotechnics* 91:179–191.

Gu, X.Q. Huang, M.S. & Qian, J.G. 2014. DEM investigation on the evolution of microstructure in granular soils under shearing. *Granular Matter* 16(1): 91-106.

Guo, P. & Su, X. 2007. Shear strength, interparticle locking, and dilatancy of granular materials. *Canadian Geotechnical Journal* 44(5):579–591.

Hosseininia, E.S. 2012. Investigating the micromechanical evolutions with inherently anisotropic granular materials using discrete element method. *Granular Matter* 14: 483-503.

Itasca Consulting Group. 2018. *Particle flow code in three dimensions (PFC^{3D})*, Minneapolis.

Kozicki, J. Tejchman, J. & Mroz, Z. 2012. Effect of grain roughness on strength, volume changes, elastic and dissipated energies during quasi-static homogeneous triaxial compression using DEM. *Granular Matter* 14: 457-468.

Kozicki, J. Tejchman, J. & Muhlhaus, H.B. 2014. Discrete simulations of a triaxial compression test for sand by DEM. *International Journal for Numerical and Analytical Methods in Geomechanics* 38(18):1923-1952.

Lin, X. & Ng. T.T. 1997. A three dimensional element model using arrays of ellipsoids. *Geotechnique* 47(2): 319-329.

Maeda, K. Sakai, H. Kondo, A. Yamaguchi, T. Fukuma, M. & Nukudani, E. 2010. Stress-chain based micromechanics of sand with grain shape effect. *Granular Matter* 12: 499-505.

Payan, M. Khoshghalb, A. Senetakis, K. & Khalili, N. 2016. Effect of particle shape and validity of G_{max} models for sand: A critical review and a new expression. *Computers and Geotechnics* 72: 28-41.

Pournin, L. Weber, M. Tsukahara, M. Ferrez, J.A. Ramaioli, M. & Liebling, T.M. 2005. Three-dimensional distinct element simulation of spherocylinder crystallization. *Granular Matter* 7: 119-126.

Saint-Cyr, B. Delenne, J.Y. Voivret, C. Radjai, F. & Sornay, P. 2011. Rheology of granular materials composed of nonconvex particles. *Physical Review. E* 84: 0141302.

Shi, X. Herle, I. & Muir Wood, D. 2018. A consolidation model for lumpy composite soils in open-pit mining. *Geotechnique* 68(3): 189–204.

Shin, H. & Santamarina, J. 2012. Role of particle angularity on the mechanical behavior of granular mixtures. *Journal of Geotechnical and Geoenvironmental Engineering* 139(2): 353-355.

Simulation Works, Inc. 2009. *Kubrix, Version 11.6 Manual*. St. Paul: Simulation Works.

Stahl, M. & Konietzky, H. 2011. Discrete element simulation of ballast and gravel under special consideration of grain-shape, grain-size and relative density. *Granular Matter* 13(4): 417-428.

Taghavi, R. 2000. Automatic Block Decomposition Using Fuzzy Logic Analysis. *Proceedings of the 9th International Meshing Roundtable*. New Orleans: Sandia National Laboratories 187-192.

Tillemans, H.J. & Herrmann, H.J. 1995. Simulating deformations of granular solids under shear. *Physica A: Statistical mechanics and its applications* 217(3-4):261–288.

Tsomokos, A. & Georgiannou, V. 2010. Effect of grain shape and angularity on the undrained response of fine sands. *Canadian Geotechnical Journal* 47(5):539–551.

Wellmann, C. Lillie, C. & Wriggers, P. 2008. A contact detection algorithm for superellipsoids based on the common-normal concept. *Engineering Computations* 25(5):432-442.

Xiao, Y. Long, L. Evans, T.M. Zhou, H. Liu, H. & Stuedlein, A.W. 2018. Effect of particle shape on Stress-Dilatancy responses of Medium-Dense sands. *Journal of Geotechnical and Geoenvironmental Engineering* 145(2): 04018105.

Yan, W.M. 2009. Fabric evolution in a numerical direct shear test. *Computers and Geotechnics* 36(4):597-603.

Yang, J. & Luo, X. 2015. Exploring the relationship between critical state and particle shape for granular materials. *Journal of Mechanics and Physics of Solids* 84:196–213.

Yang, Y. Wang, J.F. & Cheng, Y.M. 2016. Quantified evaluation of particle shape effects from micro-to-macroscales for non-convex grains. *Particuology* 25: 23-35.

Zhao, S. Zhou, X. Liu, W. & Lai, C. 2015. Random packing of tetrahedral particles using the polyhedral discrete element method. *Particuology* 23: 109-117.

Zhou, B. Huang, R. Wang, H. & Wang, J. 2013. DEM investigation of particle anti-rotation effects on the micromechanical response of granular materials. *Granular Matter* 15(3): 315-326.

Sustainable Buildings and Structures: Building a Sustainable Tomorrow – Papadikis et al. (Eds)
© 2020 Taylor & Francis Group, London, ISBN 978-0-367-43019-1

Electrochemical investigation of steel bar in slag-cement paste under coupled chloride and sulphate attack

H. Davoudi & X.B. Zuo
Department of Civil Engineering, School of Science, Nanjing University of Science and Technology, Nanjing, P.R. China

ABSTRACT: The current paper reports on corrosion behavior of steel bar embedded in slag-cement paste under coupled chloride and sulphate attack. Electrochemical study, particularly Tafel Polarization method, was considered to investigate the corrosion resistance of steel bar in hardened slag-cement immersed in *3.5%NaCl+5%Na2SO4* in a static corrosion environment. Three types of specimens with 0%, 20% and 40% slag content as a supplementary cementitious material (SCM) were used. The experiment was executed within 270 days in an ambient temperature. The results showed that the sample S2-LB3 with 20% slag has the most corrosion resistance after 270 days under chloride and sulphate attack.

1 INTRODUCTION

The phenomena of steel reinforcement corrosion in concrete has been one of the main reasons for its service life reduction (Gartner & Macphee 2011). Concrete generally provides reinforcing steel with strong corrosion protection (Hope et al. 1986) and Portland cement produces a strong alkaline environment which is able to passivate the steel and keep it safe from corrosion (Pradhan & Bhattacharjee 2007). However, a relation via chloride ion in concrete and corrosion of steel in concrete is now stoutly determinate. The existence of chloride ion can lead to depassivation of passive film on the surface of steel reinforcement speeding up the corrosion process. Both free and bound chloride are present in concrete. Free chloride is responsible for depassivation of steel and bound chloride either reacts to tricalcium aluminate (C_3A) chemically or bounds to C-S-H gel. There are some reasons which lead to establishment of bound chloride such as the amount of C_3A in cement or the usage of additive in the mix (for instance slag) (Pradhan & Bhattacharjee 2007). For example, there are many highways in Canada where the material in concrete presents a high amount of chloride, nonetheless the steel bars in these structures do not show remarkable corrosion after several years of service scope (Okafor et al. 2015). Consequently, reduction in porosity and expanding the chloride binding capacity of concrete led to a durable concrete. Additives such as slag are proven to have excellent chloride binding capacity. The slag has conspicuous performance on strength, durability, and long-term corrosion resistance of the concrete in chloide-contaminated environment (Sanusi et al. 2018).

The research on the corrosion mechanism of reinforcing bar in concrete during the process of chloride and the mineral admixture such as slag can improve the performance of concrete, and is of great significance for the durability evaluation and service-life prediction of concrete structures in marine or coastal environment. Electrochemistry is one of the effective means to study the corrosion mechanism of steel bar in corrosive environment (Lee 2006). In this research, the influence of partial replacement of slag as a supplementary cementitious material on corrosion behavior of steel bar embedded in slag-cement paste under chloride and sulphate attack was investigated.

2 EXPERIMENT

2.1 *Materials*

The cement and the supplementary cementitious material (SCM) used in this experiment are 52.5 grade Ordinary Portland Cement (OPC) and S95 grade blast furnace slag powder. The cement and slag densities and specific surface areas are 3100 kg/m^3, 350 m^2/kg and 2900 kg/m^3, 435 m^2/kg respectively. The chemical composition of OPC and slag are shown in Table 1. Ammonium chloride and sodium chloride reagents with solubility 37.2 g and 36.0 g per 100 ml, respectively, are used in an ambient temperature to make an accelerated corrosion media. Tap water is used to design mixing proportion and deionized water is utilized as corrosion solution in which their proportion is listed in Table 2. HPB235 ordinary steel is used presenting tensile strength and elastic modulus of 235 MPa and 210 GPa, respectively. Its chemical composition is shown in Table 3.

3 FABRICATION OF SPECIMEN

First of all, the steel is fixed in a drilling machine. The drilling machine has a needle on top of it used to drill a hole of 1 cm in the steel. The steel is then taken out of the machine and a 10 cm-copper wire in inserted. The wire is then soldered into the steel. Secondly, the PVC is cut into 10 cm apart vertically along the Y-axis. The PVC is then wounded around with a tape in order not to permit the leakage of paste injected into the PVC with the aid of a syringe. Thirdly, it is located in the center of the PVC, followed by the pouring of the cement paste using a syringe into the mold. The steel is then fixed into the mold using a tape to prevent it from moving. Finally, after 24 h, the paste is hardened and the specimen is demolded. Just after that the specimen is placed in a water container and maintained in a curing machine for 90 days. Notice that the top and bottom of the steel is covered with epoxy. The schematic diagram of ultimate specimen for EIS measurement is shown in Figure 1.

Table 1. Chemical composition of cement and slag.

Components	SiO$_2$	Al$_2$O$_3$	CaO	MgO	SO$_3$	Fe$_2$O$_3$	K$_2$O	Na$_2$O	TiO$_2$
Cement	21.1	5.56	62.48	1.76	3.59	3.98	0.94	0.20	0.14
Slag	33.4	12.21	42.70	6.70	0.30	0.43	0.56	1.27	1.07

Table 2. Specimen and its soaking solution.

Sample number	Thickness of protective layer (mm)	Content of each component of cementitious material		Corrosion solution
		Cement (%)	Slag (%)	
C-LB1/2/3	6	100	0	3.5%NaCl+5%Na$_2$SO$_4$
S2-LB1/2/3	6	80	20	3.5%NaCl+5%Na$_2$SO$_4$
S4-LB1/2/3	6	60	40	3.5%NaCl+5%Na$_2$SO$_4$

Table 3. Chemical composition of HPB235 steel (wt. %).

Elements	C	S	P	Mn	Si	Fe
Cement	0.20	0.039	0.021	1.42	0.56	Other

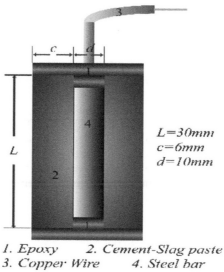

$L=30mm$
$c=6mm$
$d=10mm$

1. Epoxy   2. Cement-Slag paste
3. Copper Wire   4. Steel bar

Figure 1.   Schematic diagram of specimen structure.

4   CORROSION SOLUTION

In order to study the electrochemical corrosion process of slag-cement paste specimens under the coupling attack of chloride and sulphate, the electrochemical test was carried out within $3.5\%NaCl+5\%Na_2SO_4$ as corrosion media. During the experiment, a continuous immersion corrosion mechanism was used to immerse the cement paste reinforced bar specimens (the working electrode specimen) in the corrosion solution as shown in Table 2. The sides of the steel were extensively studied for the corrosion purposes. The corrosion area of the cylinder steel bar is 9.42 cm^2. Among them, C indicates no mineral admixtures, L means 6 mm for protection layer, B indicates 3.5% sodium chloride and +5.0% sodium sulphate, 1/2/3 represents three samples, and also S2 and S4 for 20% and 40% slag. The three specimens with the same number were immersed in the same container.

5   ELECTROCHEMICAL MEASUREMENT

The electrochemical measurement was executed at an ambient temperature (25 ± 2 °C) in a conventional three-electrode system. The commercial saturated calomel electrode (SCE) and platinum electrode were utilized as working electrode, reference electrode and counter electrode, respectively. At different immersion time, the electrochemical specimens were electrochemically measured. Every 30 days, the data from Tafel Polarization curves were drawn using the software ORIGIN. The open circuit potential need to be steady before execution of these tests. The measurement system used was a CHI660E electrochemical workstation. The EIS measurement was performed in the frequency range between 0.01 Hz and 106 Hz, and sinusoidal voltage of ±10 mV was supplied. Five data were recorded at every frequency order of magnitude, and the data were fitted using the software ZSimp- Win. The Tafel polarization (TP) measurement was conducted by arranging a potential ranging from -100 mV to +150 mV for the corrosion potential (Ecorr) with a scan rate of 0.1 mV.

6 RESULT AND DISCUSSION

6.1 *Tafel polarization*

Figure 2 illustrates the time-varying curves of Tafel polarization and corrosion current density for the electrochemical specimen immersed within *3.5%NaCl+5%Na₂SO₄* coupling solution. In order to study the influence of slag on the corrosion of steel reinforced slag-cement paste under the coupled corrosion of chloride and sulphate attack, polarization curve is extracted. Figure 4 shows the Tafel polarization curves of C-LB1, S2-LB3 and S4-LB2 specimens.

In order to further the analysis on corrosion behavior of steel in slag-cement paste under coupled *3.5%NaCl+5%Na₂SO₄* attack, Table 4 extracted from Figure 2 in which, β_a and β_c, represent the slope of the anode and cathode in Tafel curves, respectively. $E_{corr\ (SCE)}$ is the corrosion potential of the samples, I_{corr} indicates the corrosion current density and R_p is the polarization resistance. Equation 1 is used to calculate the polarization resistance (Li et al. 2016).

$$R_p = \frac{\beta_a \beta_c}{2.3031\, I_{corr}(\beta_a + \beta_c)} \tag{1}$$

According to the extraction obtained from the above polarization curves, firstly, the corrosion potential tends to shift to negative value by increasing the immersion time. Secondly, according to Table 4, the corrosion potential value of C-LB1 is changed from -0.06 v to -0.70 v during the 1 to 120 days. Besides S2-LB3 and S4-LB2 corrosion potential is changed from -0.12 v to -0.50 v and -0.13 v to -0.50 v, respectively, after 257 days The shift of the corrosion potential to a negative value shows the thermodynamic tendency to corrosion (Lee 2006, Sanusi et al. 2018). Therefore, it can be concluded that by increasing the immersion time, thermodynamic tendency of the system to corrosion is increased or in other words, the system is more active. The increment of activity in C-LB1 is more than other samples which approximately show the same activity. The corrosion potential in all samples has a noticeable increase

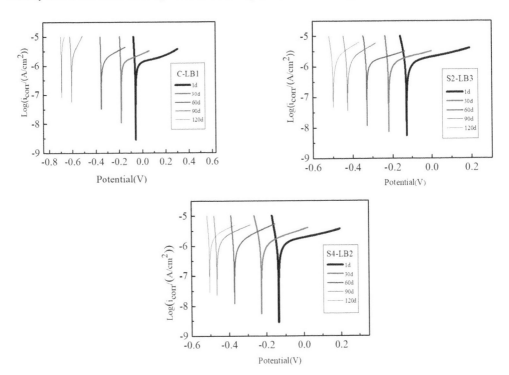

Figure 2.   Polarization curve of C-LB1, S2-LB3 and. S4-LB2 specimen after 1, 30, 90, 150 and 270 d.

Table 4. Results of polarization test of (a) C-LB1, S2-LB3 and S4-LB2 specimens.

Sample	Immersion time (d)	β_a(v.dec^{-1})	$-\beta_c$(v.dec^{-1})	E_{corr} SCE (V)	i_{corr} (μA/cm^2)	R_p (Kohm.cm^2)
C-LB1	1	0.961	0.013	-0.06	0.37	15.05
	30	0.557	0.008	-0.18	0.78	4.39
	60	0.445	0.005	-0.35	0.85	2.52
	90	0.125	0.019	-0.60	3.01	2.38
	120	0.071	0.014	-0.70	12.55	0.41
S2-LB3	1	0.626	0.033	-0.12	0.64	21.27
	30	0.536	0.014	-0.22	0.65	9.11
	90	0.570	0.034	-0.33	1.82	7.65
	150	0.232	0.022	-0.42	2.31	3.78
	270	0.253	0.028	-0.50	3.02	3.62
S4-LB2	1	0.651	0.026	-0.13	0.73	14.87
	30	0.447	0.038	-0.23	1.17	12.99
	90	0.325	0.012	-0.37	1.17	4.29
	150	0.326	0.013	-0.46	1.46	3.72
	270	0.333	0.020	-0.50	2.78	2.94

to more than -0.350 mv showing the probability of corrosion reaction (Li et al. 2016). Thirdly, Table 4 also shows that corrosion current density is increased through time. Figure 3 illustrates the effect of time on Cl-B1, S2-LB3 and S4-LB2 corrosion current density.

Due to the direct relationship of corrosion potential with corrosion current density (Sherif et al. 2009, Li et al. 2016), it can be concluded that the increment in time, in addition to increasing the thermodynamic tendency to corrosion, also increases the kinetics of corrosion and therefore, the corrosion potential.

Corrosion current density can be classified in three categories by its degree of corrosion including low ($i_{corr} < 0.2$ μA/cm^2), medium (i_{corr}: 0.2 to 1.0 μA/cm^2) and high ($i_{corr} > 2.0$ μA/cm^2). The process of corrosion current density of C-LB1 is faster and it is drastically increased from 0.37 μA/cm^2 to 12.55 μA/cm^2 after 120 days, whereas, the two other sample had the same response and their corrosion current density is under 3.02 μA/cm^2 after 257 days. Consequently, it proves the corrosion potential of C-LB1 is higher than the other samples. In order to calculate the polarization resistance parameter, not only corrosion current density but also anode and cathode slope are needed. Figure 4 shows time-varying of the polarization resistance of specimen S2-LB3 and S4-LB2 presented a slight change after 257 days immersed. On the other hand, C-LB1 shows low polarization resistance with a downward trend that

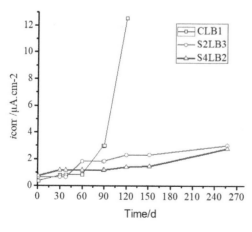

Figure 3. Time-varying curve of corrosion current density, i_{corr}.

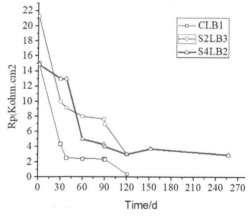

Figure 4. Time-varying curve of Polarization resistance, R_p.

continued for 120 days. Hence, specimen C-LB1 presents a high risk of corrosion compared to the other two specimens, which showed similar results.

7 CONCLUSION

In this paper, three steel cylinder embedded in cement-paste with different slag contents (0%, 20% and 40%) as Supplementary Cementitious Materials (SCMs) under coupled chloride and sulphate attack (*3.5%NaCl+5%Na₂SO₄*) were investigated. Besides, Tafel polarization method was used to study the electrochemical corrosion behavior of steel. This research shows how to improve the sustainability and durability of concrete structures, particularly steel reinforced concrete structures. Overall, it is concluded that after 270 days, the chemicals penetrated into the cement-paste and get through the protective layer which leads to corrosion of all three steels. However, S2-LB3 and S4-LB2 specimens shows more resistance than C-LB1 which is the specimen without any slag. The i_{corr} for C-LB1 is more than 12 $\mu A/cm^2$ after 120 days and for S2-LB3 and S4-LB2 is less than 2.31 $\mu A/cm^2$. Moreover, S2-LB3 (specimen with 20 % slag) showed the best results of corrosion resistance after 270 days compared to S4-LB2 and C-LB1.

ACKNOWLEDGMENT

This study was financially supported by National Science Foundation of China (51378262, 51778297) at Nanjing University of Science and Technology.

REFERENCES

Gartner, E.M. & Macphee, D. E. 2011. A physico-chemical basis for novel cementitious binders. *Cement and Concrete Research* 41: 736-749.

Hope, B.B. Page, J.A. & IP, A. K. 1986. Corrosion rates of steel in concrete. *Cement and Concrete Research* 16: 771-781.

Lee, C. 2006. Electrochemical behaviour of chromium nitride coatings with various preferred orientations deposited on steel by unbalanced magnetron sputtering. *Materials science and technology* 22: 653-660.

Li, J. Wan, K. Jiang, Q. Sun, H. Li, Y. Hou, B. Zhu, L. & Liu, M. 2016. Corrosion and discharge behaviors of Mg-Al-Zn and Mg-Al-Zn-In alloys as anode materials. *Metals* 6:65.

Okafor, P. A. Singh-Beemat, J. & Iroh, J.O. 2015. Thermomechanical and corrosion inhibition properties of graphene/epoxy ester–siloxane–urea hybrid polymer nanocomposites. *Progress in Organic Coatings* 88: 237-244.

Pradhan, B. & Bhattacharjee, B. 2007. Corrosion zones of rebar in chloride contaminated concrete through potentiostatic study in concrete powder solution extracts. *Corrosion Science* 49: 3935-3952.

Sanusi, M.S. Shamsudin, S.R. Rahmat, A. & Wardan, R. 2018. Electrochemical corrosion behaviours of AISI 304 austenitic stainless steel in NaCl solutions at different pH. *AIP Conference Proceedings AIP Publishing*, 020116.

Sherif, E.S.M. Potgieter, J. Comins, J. Cornish, L. Olubambi, P. & Machio, C. 2009. Effects of minor additions of ruthenium on the passivation of duplex stainless-steel corrosion in concentrated hydrochloric acid solutions. *Journal of Applied Electrochemistry* 39: 1385.

Sustainable Buildings and Structures: Building a Sustainable Tomorrow – Papadikis et al. (Eds)
© 2020 Taylor & Francis Group, London, ISBN 978-0-367-43019-1

Development of optimized waste-containing concrete

J. He
School of Civil Engineering and Transportation, South China University of Technology, Guangzhou, China

G. Gong
Department of Civil Engineering, Xi'an Jiaotong-Liverpool University, Suzhou, China

ABSTRACT: In order to minimize the environmental impact of carbon-dioxide emission in cement production, a study focusing on development of waste-containing concrete was carried out to create a formulation of cementitious material by introducing Ground Granulated Blast Furnace Slag (GGBS) and recycled aggregates into concrete to replace cement and aggregates respectively. Compressive test was performed for concrete specimens mixed with different proportions of recycled wastes. The results show that by introducing GGBS and recycled aggregates the compressive strength decreased considerably. Concrete containing 70% GGBS and 100% recycled coarse aggregate was considered as the optimum formulation, and the compressive strength reached 74.55% of that of normal concrete. Environmental and economic advantages were assessed for the optimum combination, with only a 6.17% increase in material cost, a 64.84% reduction in carbon-dioxide emission and a 51.80% reduction in energy consumption achieved.

1 INTRODUCTION

Enormous consumption of concrete in construction industry leads to a large amount of carbon-dioxide (CO_2) emission which greatly contributes to global warming. The direct (calcination) and indirect (fuel combustion) CO_2 emission from cement production together are considered responsible for around 8% of global emission (Olivier & Peters 2018) and this number is 15% in China (Wei & Cen 2019). Additionally, the rapid urbanization and growth of economy and population in China also result in a significant amount of construction and demolished waste (which leads to serious environmental pollution (Zheng et al. 2017). It is worth mentioning that CDW has high potential of recycling but nowadays only less than 5% of them has been reused (Huang et al. 2018). In order to save CO_2 emission and reduce environmental impact, developing an optimized use of construction and industrial wastes in concrete is no doubt a significant step toward a low carbon and sustainable future.

Replacement of raw materials in concrete by construction and industrial waste has been extensively studied. In previous research, ground granulated blast furnace slag (GGBS) and silica fume are commonly used as the substitutes of cement. GGBS, a by-product of iron smelting industry, can react with water to form hardened hydration product (Hawileh et al. 2017). Meanwhile, it can effectively enhance the durability and fire resistance of concrete (Hawileh et al. 2017). By introducing GGBS into concrete with low dosage, the durability of concrete can be improved and there is no much influence on compressive strength (Nath & Sarker 2014). However, when using a high dosage replacement (larger than 50%), the compressive strength will drop considerably (Cheah et al. 2019, Li et al. 2014). In addition, regardless of replacement ratio, early-age strength of concrete is usually compromised (Cheah et al. 2019, Higgins et al. 2011). Silica fume (SF) is a common mineral admixture used in concrete. On the one hand, it contains a large amount of silica (SiO_2) which can react with the hydration product of cement, calcium hydroxide ($Ca(OH)_2$), to gain a high strength (Lei et al. 2016). On the other hand, as a fine material, it can easily fill into the voids inside concrete which enhances the concrete strength and durability (Lei et al. 2016, Higgins et al. 2011). For the replacement of aggregates, recycled aggregates produced from crushed concrete were commonly used which results in porous surface of aggregates due to the existence of old mortar. As such, concrete containing recycled aggregates degrades obviously in its mechanical properties (Dong et al. 2019, Xu et al. 2018).

The replacement of both cement and aggregates are considered only in limited existing research. Andale et al. (2016) showed with 30% GGBS content, the compressive strength of concrete with 30% recycled aggregates and 100% recycled aggregates are 89% and 86% of that of normal concrete respectively, while the splitting tensile strength are only 85% and 74% respectively of that of normal concrete. Additionally, research on replacing both cement and aggregate with high dosage was rarely conducted.

This paper presents a parametric study of waste-containing concrete with raw materials replaced by GGBS and recycled aggregates with high dosage by assessing engineering, environmental and economic advantages. Uniaxial compressive test was conducted for a range of concrete mix proportions to evaluate the mechanical properties. Carbon-dioxide emission and energy consumptions throughout the production process of the waste-containing concrete were evaluated for the optimum concrete mix combination to reveal its sustainable features.

This study is a part of the bachelor's thesis of the first author (He 2017), more details can be found in the thesis.

2 EXPERIMENTAL PROGRAM

2.1 *Materials*

Ordinary Portland Cement (OPC) of type CEM I 42.5R was used. The fineness modulus and specific gravity were 370 m^2/kg and 2.89 t/m^3 respectively. GGBS used in this research was purchased from Hebei Province in China where the iron smelting industries is quite prosperous. The grade of GGBS was S95 which is classified in accordance with Chinese standard GB/T 18046-2017 (Standardization Administration of P.R.C 2017).

Four types of aggregates were used, i.e. natural coarse aggregates (NCAs), natural fine aggregates (NFAs), recycled coarse aggregates (RCAs) and recycled fine aggregates (RFAs). The pictures of them are shown in Figure 1.

The maximum particle size of NCAs were 10 mm, but due to the uncontrollable production process demolished concrete, the maximum particle of RCAs cannot be specified. Sieve analysis was conducted for NFAs and RFAs in accordance with BS EN 993-1:2012 (British Standard Institution 2012) and the particle size distribution is shown in Figure 2. The results of sieve analysis were used in concrete mix design.

For the purpose of improving concrete performance, three concrete additives were used, namely SF, Styrene Butadiene Copolymer (SBR) and Sodium Hydroxide (NaOH). As mentioned previously, SF can significantly improve the strength and durability of concrete. In addition to SF, SBR has proved to have positive effect on the strength of concrete. Bideci & Dogan (2016) tested concrete with different SBR content from 0% to 8% by weight of cement and found that 1% is the optimum dosage. Hence, 1% SBR was used in this research. Moreover, Yu et al. (2015) suggested that with 0.03% (by weight of GGBS) of NaOH as activator, the best effect on the improvement of concrete strength development can be achieved, which

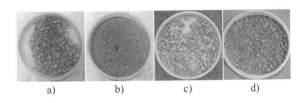

a)   b)   c)   d)

Figure 1.   Pictures of a) NCA b) NFA c) RCA d) RFA.

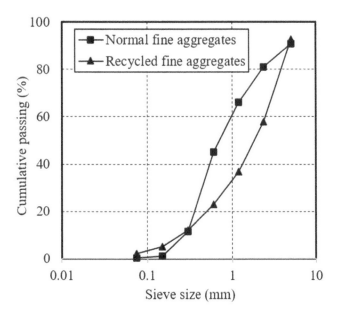

Figure 2.　Particle size distribution of NFA and RFA.

was employed in the research. Additionally, since the porous feature of recycled aggregates can result in poor workability of fresh concrete, superplasticizer with dosage of 1% (by weight of binder) was used in concrete mixing to improve workability. The pictures of SF, SBR, NaOH powder and superplasticizer are shown in Figure 3.

More information about material properties could be found in the bachelor's thesis of the first author (He 2017).

2.2　*Concrete mix design and specimen preparation*

The strength grade of normal concrete was designed to be C40/50. As high volume GGBS (more than 70%) can lead to more than 25% reduction in concrete strength, it is expected that the characteristic strength of high volume GGBS concrete in current study reaches 30MPa, which is commonly employed in construction industry. The concrete mix design was carried out in accordance with the content of a technical report published by Building Research Establishment (Teychenné et al. 1997).

The arrangement of the concrete mix proportion is shown in Table 1. A total of 11 unique batches of concrete mix were prepared, which included one batch of normal concrete (B1), five batches of concrete with 70% cement replaced by GGBS (B2-B5 & B11), four batches of concrete with 85% cement replaced by GGBS (B6-B9) and one batch of concrete with cement fully replaced by GGBS (B10). For the batches with 70% and 85% GGBS replacement, in addition to concrete mixes with natural aggregates only, recycled aggregates were gradually introduced in three steps: 50% RCAs, 50% RCAs + 50% RFAs and 100% RCAs + 100% RFAs.

a)　　　　　b)　　　　　c)　　　　　d)

Figure 3.　Pictures of a) SF b) SBR c) NaOH d) superplasticizer.

59

Table 1. Concrete mix proportion.

No.	Water kg	OPC kg	GGBS kg	NCA kg	NFA kg	RCA kg	RFA kg	NaOH, SBR g
B1	30.75 (100%)	59.90 (100%)	0 (0%)	127.28 (100%)	104.29 (100%)	0 (0%)	0 (0%)	0 (0%)
B2	30.75 (100%)	17.97 (30%)	41.93 (70%)	127.28 (100%)	104.29 (100%)	0 (0%)	0 (0%)	0 (0%)
B3	30.75 (100%)	17.97 (30%)	41.93 (70%)	63.64 (50%)	104.29 (100%)	63.64 (50%)	0 (0%)	0 (0%)
B4	30.75 (100%)	17.97 (30%)	41.93 (70%)	63.64 (50%)	52.15 (50%)	63.64 (50%)	52.15 (50%)	0 (0%)
B5	30.75 (100%)	17.97 (30%)	41.93 (70%)	0 (0%)	0 (0%)	127.28 (100%)	104.29 (100%)	0 (0%)
B6	30.75 (100%)	8.98 (15%)	50.91 (85%)	127.28 (100%)	104.29 (100%)	0 (0%)	0 (0%)	0 (0%)
B7	30.75 (100%)	8.98 (15%)	50.91 (85%)	63.64 (50%)	104.29 (100%)	63.64 (50%)	0 (0%)	0 (0%)
B8	30.75 (100%)	8.98 (15%)	50.91 (85%)	63.64 (50%)	52.15 (50%)	63.64 (50%)	52.15 (50%)	0 (0%)
B9	30.75 (100%)	8.98 (15%)	50.91 (85%)	0 (0%)	0 (0%)	127.28 (100%)	104.29 (100%)	0 (0%)
B10	30.75 (100%)	0 (0%)	59.90 (100%)	127.28 (100%)	104.29 (100%)	0 (0%)	0 (0%)	0 (0%)
B11	30.75 (100%)	17.97 (30%)	41.93 (70%)	0 (0%)	104.29 (100%)	127.28 (100%)	0 (0%)	12.6, 449 (3‰), (1%)

The last batch (B11) contains 70% GGBS and 100% RCAs. In addition, 10% SF was added in every batches containing GGBS, and concrete additives were added in the last batch.

For each batch, four cubic specimens with size of 150mm*150mm*150mm and four cylindrical specimens with size of φ150mm*300mm were prepared. After demolding, they were cured in water at room temperature for 28 days.

2.3 Testing

Compressive test was performed using a universal testing machine, in accordance with BS EN 12390-3 (British Standard Institution 2009). According to the standard, a stress rate of 0.6MPa/sec was used for both cubic and cylindrical specimens, and a strain gauge was installed at mid-height of specimen to measure axial strain. It should be noted that, because the height of strain gauge is 150mm, it was only installed on cylindrical specimens. Therefore, the axial strain was measured for cylindrical specimens, but for cubic specimens, only peak stress was obtained.

3 RESULTS AND DISCUSSION

The failure mode of specimens in compressive test is shown Figure 4. From the failure modes, it can be observed that the specimens are crushed and the failure mode follows the specification in the Eurocode. In addition, as mentioned previously, there was one batch containing 100% GGBS and it was extremely weak and decomposed during curing process.

The average compressive strength of each batch at 28 days is summarized in Table 2. The compressive strength of normal concrete (B1) was 44.5MPa, which was slightly lower than the target compressive strength (46.56MPa). With 70% cement replaced by GGBS, the compressive strength had a 17.7% decrease which is similar to the conclusion made by Li et al. (2014), and this number became 43.31% for concrete containing 85% GGBS. Under 70% dosage of GGBS, additional replacement of 50% RCA, 50% RCA + 50% RFA and 100% RCA + 100%

Figure 4.   Typical failure mode of specimens in compressive test.

RFA caused 23.03%, 35.45% and 38.20% decrease in compressive strength respectively. While under 85% dosage of GGBS, 50% RCA, 50% RCA + 50% RFA and 100% RCA + 100% RFA led to 51.52%, 51.91% and 53.63% decrease in compressive strength of concrete respectively.

Figure 5 shows the changing of compressive strength with dosage of recycled aggregates for 70% and 85% GGBS. Despite the variation of the test results, it is still clear that even the concrete with full replacement of aggregates for 70% GGBS had a higher compressive strength than the concrete with 50% RCA replacement for 85% GGBS, which indicates that for high-volume GGBS concrete, it is the amount of GGBS that governs the compressive strength rather than the amount of recycled aggregates. By comparing the strength reductions from no RA to full replacement of RA for 70% GGBS with that for 85% GGBS, the compressive strength of 70% GGBS concrete decreased more than that of 85% GGBS concrete, which means when the dosage of GGBS increases, the influence of recycled aggregates on the compressive strength of concrete became less significant. Moreover, it is noticed that, there is a sudden drop of compressive strength from 50% RCA to 50% RCA + 50% RFA for 70% GGBS, which indicates that the recycled fine aggregates content has more significant influence on the compressive strength than recycled coarse aggregates. As mentioned previously, the expected compressive strength of waste-containing concrete was 30MPa, therefor, B3 (70% GGBS + 50% RCA) was considered as the best mix proportion within B1 to B9.

Table 2.   28-days compressive strength.

Batch No.	Description	Compressive Strength (MPa)	Activity Index
B1	Normal Concrete	44.50	100%
B2	70% GGBS	36.63	82.30%
B3	70% GGBS + 50% RCA	34.25	76.97%
B4	70% GGBS + 50% RCA + 50% RFA	28.73	64.55%
B5	70% GGBS + 100% RCA + 100% RFA	27.50	61.80%
B6	85% GGBS	25.23	56.69%
B7	85% GGBS + 50% RCA	21.58	48.48%
B8	85% GGBS + 50% RCA + 50% RFA	21.40	48.09%
B9	85% GGBS + 100% RCA + 100% RFA	20.63	46.37%
B10	70% GGBS + 100% RCA	31.00	69.66%
B11	70% GGBS + 100% RCA + Additives	33.18	74.55%

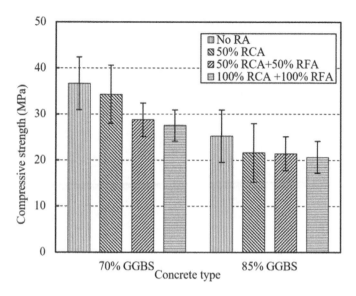

Figure 5. Compressive strength versus the dosage of recycled aggregates.

Since the compressive strength of B3 was 34.25MPa, which was still higher than the expected strength for the final formulation, additional RCA could be considered, which was included in B10. Figure 6 shows the optimization of compressive strengths of the considered mix proportion (70% GGBS + 100% RCA). It is noticed that, with additional 50% RCA replacement based on B3 the compressive strength decreased for 3.25MPa. After the addition of additives in B11, the strength of concrete became 33.18MPa, which suggests that the employed additives (1% SBR and 3% NaON) are able to enhance the compressive strength of concrete containing high-volume GGBS and RCA.

In environmental aspect, based on a study of environmental assessment of GGBS, which was done by Higgins (2007), the CO_2 emission and energy use of the optimum mix proportion were calculated to be only 35.16% and 48.20% of those of normal concrete, and more

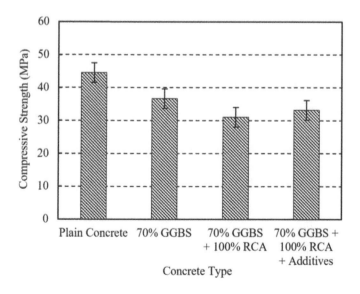

Figure 6. Optimization of compressive strength.

detail could be found in the bachelor's thesis of the first author (He 2017). Despite 6.17% higher material cost of the optimum mix proportion, which was calculated based on the price and quantity of each material, it can be considered as a significant achievement in sustainable construction.

4 CONCLUSIONS

In conclusion, a new formulation of cementitious materials was developed with raw materials in concrete replaced by GGBS and recycled coarse aggregate. Uniaxial compressive test was conducted to assess the mechanical properties of concrete and environmental evaluation was performed to reveal its sustainable features. The results show that the compressive strength decreased with increasing GGBS and recycled aggregate content. With the optimum mix proportion (70% GGBS + 100% RCA + additives), the compressive strength of concrete reached 74.55% of that of the normal concrete. In environmental aspect, with only 6.17% higher material cost, the optimum mix proportion can save 64.84% CO_2 emission and 51.80% energy consumption, which has the potential in making significant contribution to sustainable construction.

ACKNOWLEDGEMENTS

The first author would like to thank Prof. Chin Chee at Xi'an Jiaotong-Liverpool University for his supervision of the relevant work and the second author would like to thank the funding supported by Xi'an Jiaotong-Liverpool University (RDF-15-01-38, RDF-14-02-44 and RDF-18-01-23). Also, the funding supported by Key Program Special Fund in XJTLU (Grant No: KSF-E-19) is greatly appreciated.

REFERENCES

Andal, J. Shehata, M. & Zacarias, P. 2016. Properties of concrete containing recycled concrete aggregate of preserved quality. *Construction and Building Materials* 125:842-855.

Bideci, A. & Dogan, M. 2016. Effect of styrene butadiene copolymer (SBR) admixture on high strength concrete. *Construction and Building Materials* 112: 378-385.

British Standard Institution. 2009. *Testing hardened concrete*. Part 3: Compressive strength of test specimens BS EN 12390-3.

British Standard Institution. 2012. *Test for geometrical properties of aggregates*. Part 1: Determination of particle size distribution-sieving method BS EN 933-1.

Cheah, C. Tiong, L. Ng, E. & Oo, C. 2019. The engineering performance of concrete containing high volume of ground granulated blast furnace slag and pulverized fly ash with polycarboxylate-based superplasticizer. *Construction and Building Materials* 202:909-921.

Dong, H. Song, Y. Cao, W. Sun, W. & Zhang, J. 2019. Flexural bond behavior of reinforced recycled aggregate concrete. *Construction and Building Materials* 213: 514-527.

Hawileh, R. Abdalla, J. Fradmanesh, F. Shahsana, P. & Khalili, A. 2017. Performance of reinforced concrete beams cast with different percentages of GGBS replacement to cement. *Archives of Civil and Mechanical Engineering* 17(3): 511-519.

He, J. 2017. *Development of optimized waste-containing concrete*. Thesis (BEng), Xi'an Jiaotong-Liverpool University.

Higgins, D. 2007. Briefing: GGBS and sustainability. *Proceedings of the Institution of Civil Engineers – Construction Materials* 160(3): 99-101.

Higgins, D. Sear, L. King, D. Price, B. Barnes, R. & Clear, C. 2011. *Cementitious materials: the effect of ggbs, fly ash, silica fume and limestone fines on the properties of concrete*. Technical Report No. 74, The Concrete Society.

Huang, B. Wang, X. Kua, H. Geng, Y. Bleischwitz, R. & Ren, J. 2018. Construction and demolition waste management in China through the 3R principle. *Resources, Conservation and Recycling* 129: 36-44.

Lei, D. Guo, L. Sun, W. Shu, X. & Guo X. 2016. A new dispersing method on silica fume and its influence on the performance of cement-based materials. *Construction and Building Materials* 115:716-726.

Li, Q. Yuan, G. Xu, Z. & Dou, T. 2014. Effect of elevated temperature on the mechanical properties of high-volume GGBS concrete. *Magazine of Concrete Research* 66(24):1277-1285.

Nath, P. & Sarker, P. 2014. Effect of GGBFS on setting, workability and early strength properties of fly ash geopolymer concrete cured in ambient condition. *Construction and Building Materials* 66:163-171.

Olivier, J. & Peters, J. 2018. *Trends in global CO_2 and total greenhouse gas emissions: 2018 report.* PBL publication number: 3125, PBL Netherlands Environmental Assessment Agency.

Standardization Administration of the P.R.C. 2017. *Ground granulated blast furnace slag used for cement, mortar and concrete.* GB/T 18046-2017.

Teychenné, D. Franklin, R. & Erntroy, H. 1997. *Design of normal concrete mixes.* Building Research Establishment L td, Garston, Watford, Second edition.

Wei, J. & Cen, K. 2019. Empirical assessing cement CO_2 emissions based on China's economic and social development during 2001-2030. *Science of the Total Environment* 653:200-211.

Xu, G. Shen, W. Zhang, B. Li, Y. Ji, X. & Ye, Y. 2018. Properties of recycled aggregate concrete prepared with scattering-filling coarse aggregate process. *Cement and Concrete Composites* 93:19-29.

Yu, L. Lin, M. & Wang, Q. 2015. Research on the effect to different activators of steel slag and blast furnace slag powder. *Fly Ash Comprehensive Utilization* 5:9-11.

Sustainable Buildings and Structures: Building a Sustainable Tomorrow – Papadikis et al. (Eds)
© 2020 Taylor & Francis Group, London, ISBN 978-0-367-43019-1

A seismic integrity evaluation method of structures under multi-stage earthquake

H. Hou
School of Civil Engineering, Qingdao University of Technology, Qingdao, Shandong, China

W. Liu
School of Civil Engineering, Qingdao University of Technology, Qingdao, Shandong, China
Collaborative Innovation Center of Engineering Construction and Safety in Shandong Blue Economic Zone, Qingdao, China

ABSTRACT: As there is no mature method of evaluation of seismic behavior under multi-stage seismic actions that can be referenced internationally, the seismic integrity analysis method and evaluation parameters are put forward. The nonlinear static analysis (NSA) and nonlinear time-history dynamic analysis (NTDA) are used in seismic integrity analysis. The parameterized hinge rotations (HR) and story drifts (SD) are adopted as evaluation parameters. Based on a case study of a reinforced concrete frame structure, the evaluation of structural seismic integrity under multi-stage earthquake is conducted.

1   INTRODUCTION

The seismic performance of structures has multiple objectives. How to predict, analyze and evaluate the performance of structures under earthquake, that is, the evaluation of the seismic performance is an important means to check the rationality of the seismic design of structures (Cimellaro et al. 2016, Riahi et al. 2015). Time-history analysis and response spectrum analysis are the two most common basic methods, but they can only consider a single performance target of structures (Alembagheri & Ghaemian 2013, Kosic et al. 2014). In recent years, performance-based seismic has put forward higher requirements for multi-objective seismic performance evaluation of structures.

The extremely rare earthquake refers to earthquake with higher seismic fortification intensity than rare earthquake, which are also known as the extra-large earthquake. Nearly five decades, one of salient features of seismic hazard around the world is the occurrence of earthquakes with super fortification intensity (Dong & Frangopol 2016). To name a few, Tangshan (China, in 1976), Northridge (America, in 1994), Kobe (Japan, in 1995), Chi-Chi (Taiwan, in 1999), Sumatra-Andaman (Indonesia, in 2004), Peru (Peru, in 2007), Wenchuan (China, in 2008), Port-au-Prince (Haiti, in 2010), Concepcion (Chile, in 2010), Yushu (China, in 2010), Japan's northeast Pacific waters (Japan, in 2011), Lushan (China, in 2013), Ludian (China, in 2014). The peak acceleration of these earthquakes is as much as ten times as those basic ones, in particular Tangshan Earthquake is nearly 20 times. Chief among the causes of casualties and economic losses is earthquake-induced collapse and damage of structures which is characterized by its wide-ranging power, far-reaching impact on politics, economy and public safety in our society with ensuing social concerns (Khorami et al. 2017).

For the first time, ASCE 7-10 takes the risk of seismic inversion of structures within 50 years of the design benchmark period no higher than 1% as the seismic fortification target (American Society of Civil Engineers 2010). The extremely rare earthquake is added into *Chinese Seismic Ground Motion Parameter Zonation Map* GB18306-2015 has been put into effect since June 1, 2016, thereby a seismic action system is formed covering frequent earthquake, design earthquake, rare earthquake and extremely rare earthquake (State Administration for Quality Supervision Inspection Quarantine and Standardization of China 2016). Structures go through the whole process of seismicity evolution including elasticity, yielding, plasticity, ultimate and even collapse under seismic actions of frequent, design, rare and extremely rare earthquakes. Though there is no mature method of evaluation of seismic behavior under multi-stage seismic action can be referenced internationally, this paper puts forward the seismic integrity analysis and evaluation parameters, and carries out seismic integrity analysis by nonlinear static analysis (NSA) and nonlinear time-history dynamic analysis (NTDA), and adopts parameterized hinge rotations (HR) and story drifts (SD) as the parameters. In addition, evaluation of structural seismic integrity under multi-stage seismic actions is conducted based on a case study of reinforced concrete frame structure.

2 EVALUATION METHOD OF STRUCTURAL SEISMIC INTEGRITY UNDER MULTI-STAGE SEISMIC ACTIONS

2.1 *Multi-stage earthquake level*

The multi-stage earthquake level refers to earthquake intensity defined by seismic environment and presupposed exceeding probability of the emergency of earthquakes, which is usually expressed in seismic intensity or ground motion parameters (such as, ground motion peak acceleration and seismic influence coefficient). The extremely rare earthquake added into *China Seismic Ground Motion Parameter Zonation Map* GB18306-2015 defines annual exceeding probability as 10^{-4}, and its ground motion peak acceleration is as much 2.7 times to 3.2 times as basic one. With regard to this, China adopts four-grade earthquake level system, including frequent earthquake, design earthquake, rare earthquake and extremely rare earthquake, the details can be seen in Table 1.

2.2 *Evaluation standard and quantitative parameters of seismic integrity*

The evaluation standard of seismic integrity refers to the maximum damage expected of buildings, which can be described by damage degree of structures, structure functions and personnel security (Cimellaro et al. 2016). As a macroscopic concept, evaluation standard of seismic integrity cannot be used for structural seismic design and evaluation directly, thus making quantitative parameters necessary. Massive earthquake damage investigations and theoretical researches have proven that, oversize deformation after structural yielding is the main cause for the damage of structural and non-structural elements (Murao 2017, Soleimani et al. 2017). Story drifts angle reflects the synthesis results of elements deformation among different layers of RC frame structures and has a good dependency with damage degree of structures. Generally, countries now adopt the story drifts angle as a quantitative parameter of overall seismic behavior and the beam-column plastic rotation angle as a quantitative index of seismic performance of elements when it comes to seismic design

Table 1. Multi-stage earthquake level.

Seismic fortification level	Multi-stage seismic action	Exceedance probability
1- level	Frequent earthquake	63.2% (50y)
2- level	Basic earthquake	10% (50y)
3- level	Rare earthquake	2% (50y)
4- level	Extremely rare earthquake	10-4 (1y)

Table 2. Seismic integrity evaluation standard and quantitative parameters of Chinese RC frame structures.

Evaluation standard	Damage description	Possibility of continued use	Limits of story drift angle
Basically intact	The bearing members are intact; the individual nonbearing members are slightly damaged; the attached members have different degrees of damage.	Generally do not need to repair and continue to use.	1/550
Slight damage	Individual bearing components are slightly cracked; individual nonbearing members are obviously damaged; ancillary components have varying degrees of damage.	No need to repair or need a little repair, still can continue to use.	1/250
Medium damage	Most of the bearing members are slightly cracked, and some cracks are obvious.	General repair, take appropriate safety measures can be used.	1/120
Serious damage	Most of the load-bearing members are severely damaged or partially collapsed.	Should for overhaul, partial demolition.	1/60
Collapse	Collapse of most load-bearing members.	Need removal.	1/50

Table 3. Seismic integrity evaluation standard and quantitative index of American RC frame structures.

Performance level	IO	LS	CP	
Beam plastic rotation angle	0.01	0.02	0.025	
Column plastic rotation angle	0.005	0.015	0.02	

(TBI Version 2010, American Society of Civil Engineers 2013, Federal Emergency Management Agency 2012). This paper will evaluate seismic integrity from perspectives of overall existing structure damage state and the damage degree of elements. The multi-stage earthquake level and permissible limits of its quantitative parameters are respectively consistent with *Chinese Seismic design code for Buildings* (GB50011-2010) (2012) and FEMA400 (2016). Tables 2 and 3 show more details.

2.3 *Evaluation method of structural seismic integrity under multi-stage seismic actions*

The steps of evaluation method of structural seismic integrity under multi-stage seismic actions are as follows: (1) defining four-stage seismic ground motion parameters according to seismic environment and multi-stage earthquake level; (2) establishing finite element analytical models; (3) obtaining seismic behavior of story drifts and hinge rotations of elements through seismic calculation under four-stage design earthquake based on NSA and NTDA respectively; (4) under four-stage design earthquake, making a comparison between seismic performance behavior parameters and presupposed evaluation standard parameters, and if the result is satisfied, evaluation of structural seismic integrity under the earthquake can be made, if not, structural design should be adjusted or strengthened; (5) evaluating structural seismic integrity step by step under stage one to stage four earthquake, and finally, reporting the evaluation of structural seismic integrity. The details can be seen in Figure 1.

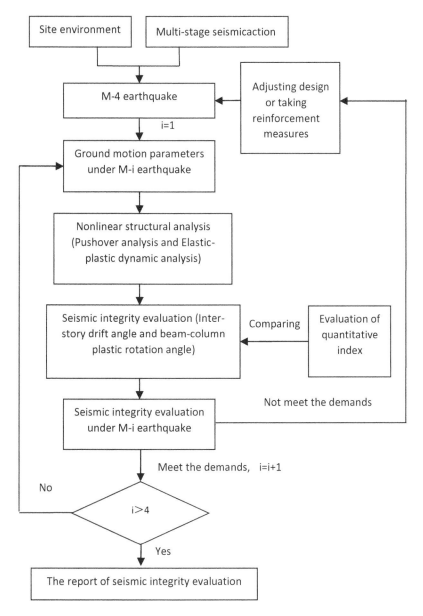

Figure 1. Evaluation steps of structural seismic integrity under multi-stage seismic actions.

3 ENGINEERING EXAMPLE

3.1 *Engineering overview*

A four-story RC frame structure with each story in 3.6m has vertical span with length of 5.1m that cuts in six, and transverse span cuts in three, side-span with length of 6m and a mid-span with length of 3m. Column cross-section is 500mm × 500mm and beam cross-section is 250mm×500mm. Seismic fortification intensity is eight (0.2g), design earthquake group belongs to the third group, site classification is II. Concrete strength is C40, longitudinal reinforcement strength is HRB400 and stirrup strength is HRB335. Reference wind pressure is 0.4kN/m², ground roughness is C. Pkpm is used to design structure and it needs to ensure that all design controlling parameters are satisfied and verify the accuracy of SAP2000 model. SAP200 is used to calculate seismic behavior of RC frame structures.

3.2 *Pushover nonlinear static analysis (NSA)*

Pushover analysis is carried out in Y-direction of two structures through the first load mode, steps are as follows:

(1) Establishing structure models and defining the property of material, member section and plastic hinge.
(2) Applying vertical load on structures, and calculating internal force in members and the natural vibration period and mode of vibration of structures.
(3) Selecting reasonable lateral load pattern and increasingly applying load on structures.
(4) Changing the stiffness of yield components and continuing to load. The third step should be repeated until structures reach its evaluation standard displacement, damage or collapse. Internal force in member, displacement and plastic hinge in each load step shall be recorded.
(5) Plotting a curve relationship between base shear and top displacement and obtaining the Pushover curve.
(6) FEMA440 (2005) capacity spectrum method is adopted to find out the evaluation points of structure under frequent earthquake, design earthquake, rare earthquake and extremely rare earthquake, steps are as follows.
(7) Establishing corresponding response spectrum based on multi-stage seismic actions and building demand curve under multi-stage seismic actions.
(8) Obtaining capacity curve through Pushover.
(9) By the conversion between multi-degree of freedom (MDOF) system and single-degree of freedom (SDOF) system, plotting demand curve and capacity curve within displacement-acceleration coordinate system of SDOF.
(10) Choosing an initial point of displacement spectrum on capacity spectrum and calculating equivalent period and equivalent damping ratio based on the point's ductility factor.
(11) Inducing equivalent period and equivalent damping ratio into the response spectrum of multi-stage earthquakes and then spectral acceleration and spectral displacement can be solved.
(12) If the difference of two spectral displacements is less than a admissible value, it's the performance point of evaluation standard, or otherwise, above steps should be repeated.
(13) Turning the spectral displacement of SDOF into the top displacement MDOF and calculating story drifts based on evaluation standard under multi-stage earthquakes and observing the state of elements' plastic hinge.

According to NSA, Figure 2 shows the performance point of evaluation standard under multi-stage earthquakes and Figure 3 shows the state of plastic hinge notations, Figure 4 is the evaluation law of story drifts.

3.3 *Nonlinear dynamic analysis (NDA)*

The Wilson-θ nonlinear dynamic analysis is adopted to calculate structures' acceleration, speed and displacement under earthquake. It's also an effective step-by-step integration method based on linear acceleration method. Specific steps of nonlinear dynamic analysis (NDA) are as follows:

(1) Selecting ground motion records that suit the site environment. Statistically, response spectrum curve of mean acceleration of ground motion records is consistent with standard response spectrum. Figure 5 illustrates the result consistency of seven ground motion records selected with design response spectrum.

69

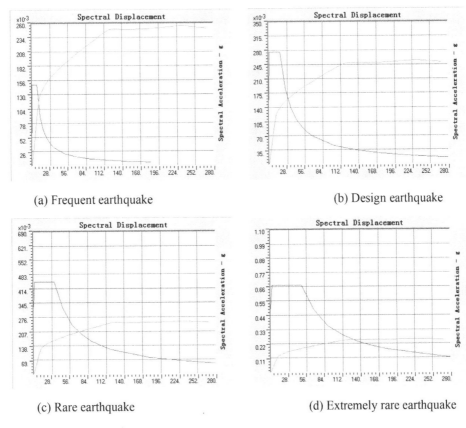

(a) Frequent earthquake

(b) Design earthquake

(c) Rare earthquake

(d) Extremely rare earthquake

Figure 2. The performance point of evaluation standard under multi-stage earthquakes (NSA).

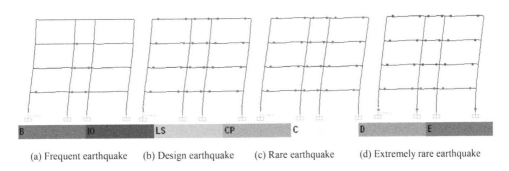

(a) Frequent earthquake   (b) Design earthquake   (c) Rare earthquake   (d) Extremely rare earthquake

Figure 3. The state of plastic hinge notations (NSA).

(2) Establishing incremental differential equation based on structural elastic-plastic dynamic analysis.
(3) Solving structural elastic-plastic dynamic analysis by Wilson-θ method in the form of step-wise iterated and calculating structures' acceleration, speed and displacement under seismic motions.
(4) Getting the story drift based on evaluation standard under multi-stage seismic motions and deciding the state of element's plastic hinge.

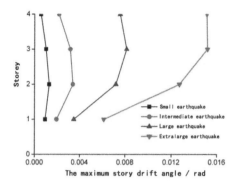

Figure 4. Evaluation law of story drifts (NSA).

Figure 5. Comparison of mean response spectrum and design response spectrum.

According to the NDA, the peak acceleration of seven multi-stage seismic motions is 70cm/s², 200cm/s², 400cm/s² and 600cm/s² respectively. Figure 6 shows the evaluation law of story drift under multi-stage seismic motions. Figure 7 depicts the state of plastic hinge notions.

3.4 Evaluation of seismic integrity

Figures 2-7 show that NSA and NDA demonstrate the evolutionary process of seismic behavior of elastic and plastic structures. Parameterized plastic hinge and story drift can be

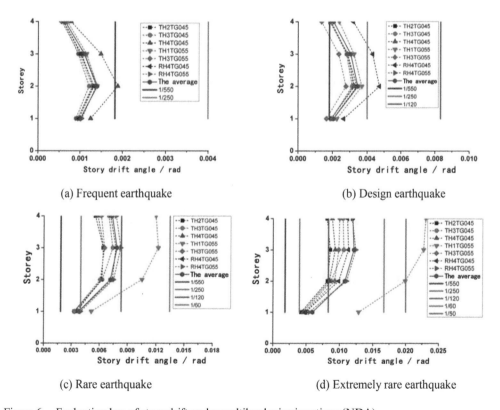

(a) Frequent earthquake

(b) Design earthquake

(c) Rare earthquake

(d) Extremely rare earthquake

Figure 6. Evaluation law of story drift under multilevel seismic actions (NDA).

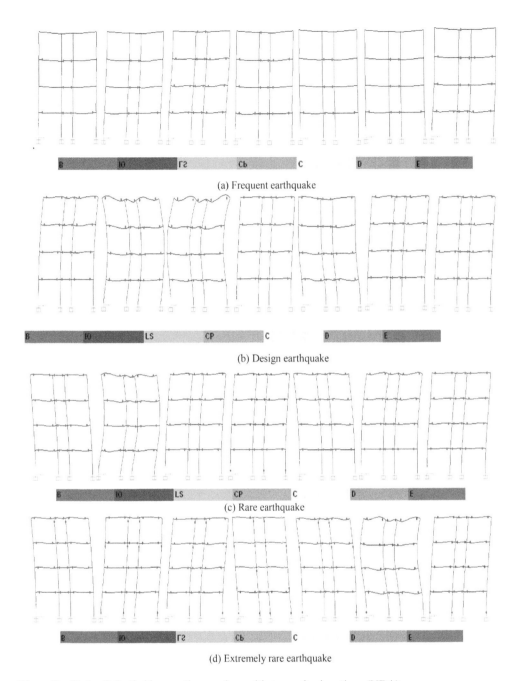

(a) Frequent earthquake

(b) Design earthquake

(c) Rare earthquake

(d) Extremely rare earthquake

Figure 7.   State of plastic hinge notions under multi-stage seismic actions (NDA).

achieved, which represent structural seismic behavior under multi-stage seismic motions and evaluation parameters of structural seismic integrity.

Tables 4 and 5 illustrate the evaluation of overall structure and elements seismic integrity by NSA and NDA. With the largest story drift achieved by NTDA method as a standard, the error rates of the largest story drift angle by pushover NSA method under four-grade earthquakes are 8.68%, 2.22%, 3.55% and 20.20% respectively. Under frequent earthquakes, the number of beam hinges achieved by pushover NSA method is less than that of NTDA method. Under extremely rare earthquakes, plastic hinges only appear at the bottom column

Table 4. The evaluation of overall structure seismic integrity.

Analysis method	Seismic action	Overall performance		
		Storey of maximum deformation	Maximum storey drift angle	Performance level
NSA	Frequent earthquake	2	1/778	Basically intact
	Design earthquake	2	1/296	Slight damage
	Rare earthquake	3	1/123	Medium damage
	Extremely rare earthquake	3	1/66	Serious damage
NDA	Frequent earthquake	2	1/714	Basically intact
	Design earthquake	2	1/289	Slight damage
	Rare earthquake	3	1/127	Medium damage
	Extremely rare earthquake	3	1/83	Serious damage

Table 5. The evaluation of structural elements seismic integrity.

Analysis method	Seismic action	Performance of beam				Performance of column			
		A-B	B-IO	IO-LS	LS-CP	A-B	B-IO	IO-LS	LS-CP
NSA	Frequent earthquake	110	34	0	0	192	0	0	0
	Design earthquake	0	144	0	0	192	0	0	0
	Rare	0	144	0	0	192	0	0	0
	Extremely rare earthquake	0	36	108	0	168	24	0	0
NDA	Frequent earthquake	79	65	0	0	192	0	0	0
	Design earthquake	1	143	0	0	192	0	0	0
	Rare earthquake	0	129	15	0	190	2	0	0
	Extremely rare earthquake	0	99	30	15	158	31	3	0

base through pushover NSA method. Five of seven ground motion records have column hinge at upper layers through NTDA.

Though the number of plastic hinges achieved through pushover analysis is slightly less than that of NTDA, the sequence, distribution and plastic degree of plastic hinge are essentially the same. The sequence is that frame beams yield first and then a few frame columns yield. Generally, columns yielded will occur after rare earthquake happens, the largest story drift will occur at the bottom of structures. Yielded structures with large plastic deformation have better ductility and security reserve capacity. Although there are some errors between structural seismic response obtained by Pushover method and the real one, the accuracy meets the engineering requirements from the perspective of the evaluation result and the one achieved through nonlinear analysis method that show a good consistency. Meanwhile, the workload of evaluating structural seismic behavior through pushover method is far less than that of NTDA. Meanwhile, the seismic integrity evaluation method of structures under Multi-stage Earthquake can increase the sustainability of RC structures by optimizing the amount of materials under different levels of demand.

4 CONCLUSIONS

The results show that, (1) The NSA and the NTDA, which have basically consistent conclusions, describe the seismic evolution process of the structure from elasticity to plasticity. Also the parametric plastic hinge and the story drift are obtained from the process. The NTDA is a better way to respond to the damage accumulation effect of earthquake, but the work of the NSA is much less than that of NTDA. The accuracy of the two methods meet the engineering requirements. (2) The seismic behaviors of the structure under multi-stage seismic action are represented by the parametric plastic hinge and the story drift, which are two parameters expression of seismic integrity evaluation index. (3) For reinforced concrete frame structures, the results of NSA and NTDA are the same in the case of frequent earthquake or extremely rare earthquake. The conclusions of seismic integrity analysis are basically intact in frequent earthquake, slight damage in design earthquake, medium damage in rare earthquake and serious damage in extremely rare earthquake. Under multi-stage seismic actions, the frame beams yield first, only a small number of frame columns yield after all the beams yield. Columns yield usually after rare earthquake. The maximum story drift occurs at lower floor of the structures. The structures have large plastic deformation after yielding, and have good ductility and safety reserve capacity. (4) The seismic integrity evaluation method of structures under multi-stage earthquake can optimize the amount of materials under different levels of demand, to increase the sustainability of RC structures.

REFERENCES

Alembagheri, M. & Ghaemian, M. 2013. Damage assessment of a concrete arch dam through nonlinear incremental dynamic analysis. *Soil Dynamics and Earthquake Engineering* 44(1):127-137.

American Society of Civil Engineers. 2010. *Minimum Design Loads for Buildings and Other Structures: ASCE7-10*. Reston, VA: American Society of Civil Engineers.

American Society of Civil Engineers. 2013. *ASCE/SEI 41-13. Seismic evaluation and retrofit of existing buildings*. Reston: Virginia.

Cimellaro, G.P. Malavisi, M. & Mahin, S. 2016. Using discrete event simulation models to evaluate resilience of an emergency department. *Journal of Earthquake Engineering* 21(2):203-226.

Cimellaro, G.P. Renschler, C. Reinhorn, A.M. & Arendt, L. 2016. A framework for evaluating resilience. *Journal of Structural Engineering* 142 (10): 1-13.

Dong, Y. & Frangopol, D.M. 2016. Performance-based seismic assessment of conventional and base isolated steel building including environmental impact and resilience. *Earthquake Engineering and Structural Dynamics* 45(5):739-756.

Federal Emergency Management Agency. 2005. *FEMA 440. Improvement of nonlinear static seismic analysis procedures*. Washington D.C.

Federal Emergency Management Agency. 2012. *FEMA P58. Seismic performance assessment of buildings*. Washington D.C.

Khorami, M. Alvansazyazdi, M. Shariati, M. Zandi, Y. Jalali, A. & Tahir, M.M. 2017. Seismic performance evaluation of buckling restrained braced frames (BRBF) using incremental nonlinear dynamic analysis method (IDA). *Earthquakes and Structures* 13(6): 531-538.

Kosic, M. Dolek, M. & Fajfar, P. 2017. Pushover-based risk assessment method: a practical tool for risk assessment of building structures. *16th World Conference on Earthquake Engineering. Tokyo, Japan*: International Association for Earthquake Engineering No. 1523.

Ministry of Construction of the People's Republic of China. 2016. *GB 50011-2010 Code for seismic design of buildings (2016 Edition)*. Beijing: China Architecture and Building Press.

Murao, O. 2017. Recovery curves for permanent houses after the 2011 great east Japan earthquake. 16th World Conference on Earthquake Engineering. Tokyo, Japan: International Association for Earthquake Engineering No. 1865.

Riahi, H.T. Amouzegar, H. & Falsafioun, M. 2015. Seismic collapse assessment of reinforced concrete moment frames using endurance time analysis. *The Structural Design of Tall and Special Buildings* 24(4): 300-315.

Soleimani, S. Aziminejad, A. & Moghadam, A.S. 2017. Approximate two-component incremental dynamic analysis using a bidirectional energy-based pushover procedure. *Engineering Structures* 157: 86-95.

State Administration for Quality Supervision Inspection Quarantine and Standardization of China. 2016. *Seismic Ground Motion Parameter Zonation Map of China (GB18306-2015)*. Beijing: China Standards Press.

TBI Version 1.0. 2010. *Guidelines for performance-based seismic design of tall buildings*. Pacific Earthquake Engineering Research Center.

Sustainable Buildings and Structures: Building a Sustainable Tomorrow – Papadikis et al. (Eds)
© *2020 Taylor & Francis Group, London, ISBN 978-0-367-43019-1*

Improving properties of recycled coarse aggregate (RCA) by biomineralization method

Z.W. Liu, C.S. Chin & J. Xia
Department of Civil Engineering, Xi'an Jiaotong-Liverpool University, Suzhou, Jiangsu, China
Institute for Sustainable Materials and Environment (ISME), Xi'an Jiaotong-Liverpool University, Suzhou, Jiangsu, China

V. Achal
School of Ecological and Environmental Sciences, East China Normal University, Shanghai, China

ABSTRACT: Due to the growing concern about the environmental degradation and increasing awareness of green infrastruture, sustainable development became an indispensable requirement particularly in construction area. Reusing recycled wastes coarse aggregate (RCA) into concrete production, as a sustainable way, has aroused increasingly interest by researchers in recent years. However, the low performance of RCA has limited its largescaled application. The research presented in this paper was aimed to apply biomineralization method to improve the properties including density, water absorption and crushing value of RCA. Three ureolytic bacteria strains including Lysinibacillus fusiformis, Bacillus megaterium and Sporosarcina pasteurii were used for the purpose of biomineralization in RCA/concrete. It was found that after soaking the RCA into bacterial solution, the water absorption was decreased and the apparent density was increased, while this method did not show effect on crushing value. Moreover, replacing 30% of natural aggregates by mass with the bio-treated RCA, the compressive strength of concrete would not decrease, Bacillus megaterium group even showed higher compressive strength compared to concrete only with nature aggregates.

1 INTRODUCTION

Ultilizing recycled aggregate (RA) in concrete, as a sustainable way, has aroused increasing interest by researchers in recent years (Hendriks & Pietersen 2000). It is estimated that there are 300 million tons of construction and demolishment wastes created every year in China (Chen et al. 2011). At the same time, natural aggregates with a limited amount on the earth have been massively exploited due to the increasing demand for concrete (Kim et al. 2018). In this sense, reusing these wastes into concrete production can not only save natural material but also relieve the severe stress of wastes disposal. Hence, many studies paid attention to replacing natural aggregates with the RA in concrete production.

However, high water absorption and weak and loose bonding ability with new mortar are two main drawbacks of RA that need to be overcomed (Ryu 2002). The water absorption of RA ranges from 3% to 13% (Katz 2003), which is much higher than that of natural aggregate. For concrete with the same mixing design, when RA replaces its natural aggregates, the water absorption usually becomes much higher. Moreover, the poor bonding ability with new mortar is caused by the increase of the interfacial transition zone (ITZ) (Ryu 2002). ITZ is usually considered as the weakest part in concrete, on account of its local high water cement ratio in this zone (Xuan et al. 2009). Furthermore, due to the former aggregate coated by the old mortar, there will be two layers of ITZ in new concrete, which will cause a further decrease in bonding ability between new matrix and aggregates particles (Poon et al. 2004, Li et al. 2012).

Different techniques, such as ultrasonic treatment (Katz 2004), chemical treatment (Ismail & Ramli 2014) and microwave treatment (Bru et al. 2014), were used to enhance the RA. However, these methods aimed to knock off or resolve the old mortar layer left on the origin aggregate particles. This may consume a large amount of energy or risk chemical agent (García-González et al. 2017), which is opposite to the primary objective of using RA in concrete production. Treating the RC by some bio-method might be an environment-friendly approach to improve the properties of RC. With the discovery of biomineralization technique, many ureolytic bacteria were used for improving the performance of concrete because of their ability of secrete urease enzyme which can hydrolyze urea into carbonate ion and induced the production of calcium carbonate (Chahal et al. 2012). Moreover, the diameter of bacteria is only around 1μm, which means it is possible that bacteria permeate into the old mortar layer with urea and calcium ion and create calcium carbonate prescription to fill and bond the loose layer (Wong 2015). If the calcium carbonate prescription fills the old mortar, the water absorption will decrease, and the density will increase (Singh et al. 2018). Hence, it will be a feasible way to utilize the biomineralization method in RCA treatment.

This research aimed to study the biomineralization process to improve the RCA with low energy consumption. RCA was soaked in the bacterial solution and its properties were measured. Compressive strength of concrete cast by the treated RCA was also investigated. The results of the bio-treated RCA were compared with those from untreated RCA, and the effectvieness was reported in this paper.

2 EXPERIMENTAL PROGRAM

2.1 Material preparation

2.1.1 Bacterial culturing and preparation

Three bacterial strains were chosen in this research. *Lysinibacillus fusiformis* (LF) was isolated from sludge and identified based on 16S rRNA gene sequencing and sequences were deposited in GenBank under accession number MN097919. *Bacillus megaterium de Bary* (BM) (ATCC 25300) and *Sporosarcina pasteurii* (SP) (ATCC 11859) were procured from American Type Culture Collection (ATCC), United States. The bacteria were subcultured in nutrient agar medium (Figure 1(a)) and nutrient broth media (Figure 1(b)). The media contained 1 g of "Lab-Lemco" powder, 2 g of yeast extract, 5 g of peptone and 10 g NaCl dissolved into 1 L distilled water. The spectrophotometer was used to measure the optical density (OD) value at 600 mm, which can be used to judge the concentration of the bacteria solution. In the research, bacteria were cultured in 30 °C until the OD600 arrived 0.6 trans^{-1} and then centrifuged bacteria solution in 20 °C and 6000 rpm to divide the bacteria cells and medium.

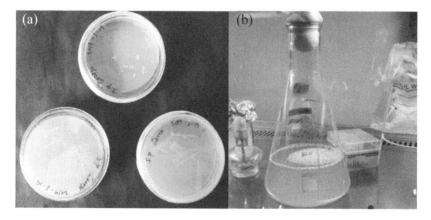

Figure 1.   Bacteria subcultured in (a) agar medium and (b) liquid medium.

Figure 2. The photo (a) and particle size distribution (b) of recycled coarse aggregate.

2.1.2 Recycled coarse aggregates

After acquiring the RCA, the sieving test was carried out first. The photo and particle size distribution are shown in Figure 2.

2.2 Experimental method

2.2.1 RCA treatment method

Before the RCA was soaked, aggregates were washed by spraying water until the dust and mud on their surface were cleaned up. After that, RCAs were put into four tanks, and three kinds of bacterial solution and water were poured into these tanks respectively, until the liquid submerged all the RCA particles. The soaking continued at 30 °C for 72 h. The compositions of bacterial treatment solution are shown in Table 1.

2.2.2 Water absorption, apparent density test, and crushing value test

The water absorption and apparent aggregate density test were referenced to BS EN 1097-6:2013 (BSI 2013) while crushing value test was referenced to ISO/DIS 20290-3 (BSI 2018).

2.2.3 Concrete strength test

The reminded RCA was also tried to replace natural coarse aggregate. In this research, the natural aggregates were replaced with 30% of RCA by mass. The mixing designs are shown in Table 2.

Table 1. Composition of four kinds of soaking solution.

Name	Nutrition (g/L)	Urea (g/L)	CaCl$_2$ (g/L)	Bacteria strain	Concentration of bacteria solution (Cells/L solution)
NC	0	0	0	No bacteria	0
LF	0.26	20	11.1	LF	10^7
BM	0.26	20	11.1	BM	10^7
SP	0.26	20	11.1	SP	10^7

Table 2. Mix designs of concrete with recycled coarse aggregate.

Mixing design	Cement (kg/m³)	RCA (kg/m³)	Fine aggregate (kg/m³)	Natural coarse aggregate (kg/m³)	Water (kg/m³)
RC*	415	0	620.6	1153.4	195.05
NC	415	345.7	620.6	806.7	195.05
LF	415	345.7	620.6	806.7	195.05
BM	415	345.7	620.6	806.7	195.05
SP	415	345.7	620.6	806.7	195.05

* RC means reference concrete without recycled coarse aggregate and bio-agent.

3 TESTING RESULTS

3.1 Influence on water absorption, apparent density and crushing value of RCA

The water absorption and apparent density of four groups of aggregates are shown in Figure 3(a) and (b). It can be found that soaking recycled coarse aggregate in bacterial solution could increase the density and decrease the water absorption of RCA. Reductions of 15.8%,18.4% and 15.7% in water absorption were observed in LF, BM, and SP groups, respectively. And for apparent density, there were 1.72%, 2.02% and 2.35% improvements in LF, BM and SP groups, respectively. Ureolytic bacteria can hydrolyze urea into ammonium and carbonate ions, and at the same time, the cytomembrane of bacteria with negative charge can attract the calcium ion which can make a bacteria cell a micro calcium carbonate filler to fill the void and hole in cementitious materials (Hammes et al. 2002). Hence, the biomineralization ability of bacteria precipitated calcium

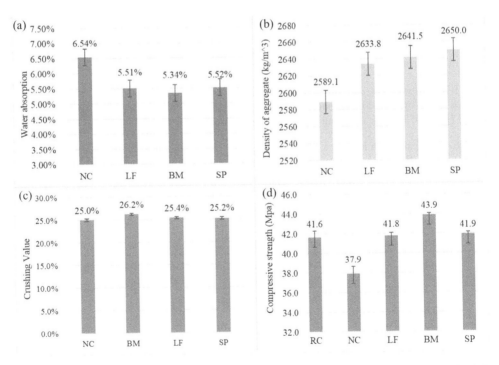

Figure 3. Tests results of (a) water absorption; (b) apparent density; (c) crushing value and (d) 28-day compressive strength.

carbonate in a mortar with RCA to improve the engineering properties of resulting materials. However, little influence was found in crushing value, as Figure 3(c) shows. This result may be caused by insufficient soaking time. Calcium carbonate forming on the surface of RCA mainly influences the absorption, while calcium carbonate forming inside of RCA mainly influences the crushing value (Wu et al. 2018). In this sense, enhancing crushing value needs a longer soaking term. In a study by Wu et al. (2018), a 20-day bio-soaking successfully decreased the crush valve by 15% compared to their control group.

3.2 Influence on compressive strength of concrete

The 28-day compressive strength test results of each group are presented in Figure 3(d). It was found that replacing 30% of natural aggregate by RCA can decrease the compressive strength by 8.8%. Compared with NC group, the compressive strength increased by 9.3%, 13.6% and 9.5% in LF, BM and SP group, respectively, and the compressive strength was even slightly higher than concrete without RCA. In this sense, replacing 30% of natural coarse aggregate with the bacteria soaked RCA do not have a negative influence on the compressive strength of concrete.

4 CONCLUSION

Based on the test results, the following conclusion can be drawn.

- Applying the biomineralization method can improve the properties of RCAs. After a 72 h soaking, all the three bacterial strains decrease the water absorption of RCA and increase the apparent density of RCA.
- No obvious influence was observed on crushing value up to a 72 h soaking.
- For concrete in this research, replacing 30% natural aggregate with RCA will cause an 8.8% loss of compressive strength. However, if the RCA is soaked by a bacterial solution, the compressive strength will not lower than the natural aggregate concrete.

All three strains of ureolytic bacteria in this research can improve the properties of RCA. Using the biomineralization method to improve the RCA does not need a hazardous chemical agent or a large amount of energy. Moreover, bacteria can reproduce themselves. Therefore, this method is deemed to be sustainable and has enormous potential to be applied in the construction area.

REFERENCES

British Standards Institution BSI. 2013. BSI Standards Publication Tests for mechanical and physical properties of aggregates Part. 6. *Determination of particle density and water absorption.*
British Standards Institution BSI. 2018. Aggregates for concrete — Test methods for mechanical and physical properties — Part 3. *Determination of aggregate crushing value (ACV).*
Bru, K. Touzé, S. Bourgeois, F. Lippiatt, N. & Ménard, Y. 2014. Assessment of a microwave-assisted recycling process for the recovery of high-quality aggregates from concrete waste. *International Journal of Mineral Processing* 126: 90–98.
Chahal, N. Siddique, R. & Rajor, A. 2012. Influence of bacteria on the compressive strength, water absorption and rapid chloride permeability of concrete incorporating silica fume. *Construction and Building Materials* 37 (1): 645–651.
Chen, M.Z. Lin, J.T. Wu, S.P. & Liu, C.H. 2011. Utilization of recycled brick powder as alternative filler in asphalt mixture. *Construction and Building Materials* 25 (4): 1532–1536.
García-González, J. Rodríguez-Robles, D. Wang, J. De Belie, N. Morán-del Pozo, J.M. Guerra-Romero, M.I. & Juan-Valdés, A. 2017. Quality improvement of mixed and ceramic recycled aggregates by biodeposition of calcium carbonate. *Construction and Building Materials* 154: 1015–1023.
Hammes, F. Verstraete, W. & Verstraete, W. 2002. Key roles of pH and calcium metabolism in microbial carbonate precipitation. *Reviews in Environmental Science & Bio/Technology* 1 (Morita 1980): 3–7.

Hendriks, C.F. & Pietersen, H.S. 2000. *Report 22: SUSTAINABLE raw materials: construction and demolition waste–state-of-the-art report of RILEM technical committee 165-SRM*. RILEM publications.

Ismail, S. & Ramli, M. 2014. Mechanical strength and drying shrinkage properties of concrete containing treated coarse recycled concrete aggregates. *Construction and Building Materials* 68: 726–739.

Katz, A. 2003. Properties of concrete made with recycled aggregate from partially hydrated old concrete. *Cement and Concrete Research* 33(5): 703–711.

Katz, A. 2004. Treatments for the Improvement of Recycled Aggregate. *Journal of Materials in Civil Engineering* 16(6): 597–603.

Kim, I.S. Choi, S.Y. & Yang, E.I. 2018. Evaluation of durability of concrete substituted heavyweight waste glass as fine aggregate. *Construction and Building Materials* 184: 269–277.

Li, W. Xiao, J. Sun, Z. Kawashima, S. & Shah, S.P. 2012. Interfacial transition zones in recycled aggregate concrete with different mixing approaches. *Construction and Building Materials* 35:1045–1055.

Poon, C.S. Shui, Z.H. & Lam, L. 2004. Effect of microstructure of ITZ on compressive strength of concrete prepared with recycled aggregates. *Construction and Building Materials* 18(6): 461–468.

Ryu, J.S. 2002. An experimental study on the effect of recycled aggregate on concrete properties. *Magazine of Concrete Research* 54(1): 7–12.

Singh, L.P. Bisht, V. Aswathy, M.S. Chaurasia, L. & Gupta, S. 2018. Studies on performance enhancement of recycled aggregate by incorporating bio and nano materials. *Construction and Building Materials* 181: 217–226.

Wong, L.S. 2015. Microbial cementation of ureolytic bacteria from the genus Bacillus: A review of the bacterial application on cement-based materials for cleaner production. *Journal of Cleaner Production* 93: 5–17.

Wu, C.R. Zhu, Y.G. Zhang, X.T. & Kou, S.C. 2018. Improving the properties of recycled concrete aggregate with bio-deposition approach. *Cement and Concrete Composites* 94: 248–254.

Xuan, D.X. Shui, Z.H. & Wu, S.P. 2009. Influence of silica fume on the interfacial bond between aggregate and matrix in near-surface layer of concrete. *Construction and Building Materials* 23(7): 2631–2635.

Using travertine as pervious pavements to control urban-flooding and storm water quality

H.R. Rahimi, X. Tang & P.K. Singh
Department of Civil Engineering, Xi'an Jiaotong-Liverpool University, Suzhou, China

S. Rahimi
Department of Health, Safety and Environment, Shahid Beheshti University, Tehran, Iran

ABSTRACT: Mineral rock of Travertine can be used as pervious pavement systems. Travertine is suitable for low traffic loads and has the capability of acting as the drainage system. Travertine, as a 'spongy' material, allows water to pass through surface and get into the underlying course and permeable sub-base where the water is stored and released slowly into the sub-grade. This paper studies the quantity and quality of infiltrated water indicating that Travertine can reduce the amount of runoff as well as pollution from runoff when water passes through the underlying course and sublayer material. The results are useful for designers to control surface runoff and provide a reference criterion for sustainable design in Travertine pavement via the comparison between total volumes of rainfall and real storage capacity in Travertine pavement.

1 INTRODUCTION

In the last decades, the increasing construction of impervious surfaces has been becoming one of the main problems in controlling urban flooding due to the rapid urbanization, which negatively alter the urban water cycle (Figure 1). Impervious surfaces impede rainwater infiltration into the ground, thus resulting the significant increase of runoff. The increased runoff can cause many urban water problems associated with quantity, quality and urban services (Acioli 2005, Gilbert 2006). The increased runoff resulting from the higher imperviousness may also lead to the overloading of conventional drainage systems, thus forming superficial water courses or causing flooding. This becomes part of the quantity-related urban water problems. Moreover, the increased runoff can also cause water quality problems because the runoff accumulates the pollution found along the urban pavement surfaces (Brattebo 2003, Gomez-Ullate 2010).

Urban water management needs to be improved to adapt rapid growing urbanization, which involves both the management of wastewater and the groundwater and rainwater. Recovering rainwater by collecting runoff is now becoming a water management strategy, which is increasingly used in many countries. For example, in Australia there are already storm-water treatment and recycling practices. Among those practices, there are some techniques that recycle runoff from all urban surfaces for non-potable demands (Hatt et al. 2005). This also occurs in Germany, where rainwater harvest has been widespread since the 1980s. Ever since then, the techniques of recycling runoff from every urban surface have been developed (Nodle 2007). Within this framework many new techniques for rainwater harvest emerge to make urban drainage more sustainable. Pervious pavement is a type of these techniques to treat water from the infiltration of rainfall to the passive stored water (Castro Fresno et al. 2005).

A pervious pavement is a load bearing pavement structure that is permeable to water overlying a reservoir storage layer. Compared with a conventional pavement, a pervious pavement structure has a higher permeability of the surface short period (Jayasuriya et al. 2007).

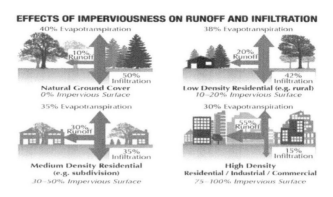

Figure 1.   Effect of urbanization on runoff and urban flooding (Arnolad & Gibbons 1996).

Pervious pavements can be classified as either porous or permeable pavements, based on the surface layout and the surface layer materials, and there is a significant difference between porous and permeable pavements. For better understanding; Argue & Pezzaniti (2005) defined porous pavements and permeable pavements as follows: Porous pavements are a thick porous layer with a high infiltration capacity; porous asphalt is a typical example of porous pavement. On the other hand, permeable pavement surfaces are normally constructed using an impervious paver with infiltration voids between the blocks; pervious concrete blocks are the most famous type of these pavements.

It should be noticed that both porous asphalt and pervious concrete blocks are not economically affordable in many developing countries. Therefore, this proposal, for the first time, introduces a new pavement called Travertine, a mineral and natural pavement, which is classified as a porous pavement and can naturally infiltrate runoff thorough its voids. Travertine is a form of limestone deposited by mineral springs, especially hot springs. Travertine often has a fibrous or concentric appearance and exists in white, tan, cream-colored, and even rusty varieties. It is formed by a process of rapid precipitation of calcium carbonate, often at the mouth of a hot spring or in a limestone cave. In this study, Travertine has been under experimental studies for evaluating its ability to control the quality and quantity of storm water runoff. The collected data has been compared to porous asphalt and concrete blocks as most famous and useful types of pervious pavements.

2   METHODOLOGY

The results of this study are based on experimental data obtained from the FUM and XJTLU hydraulic engineering laboratory. The study on quality and quantity of infiltrated water requires the experimental setup which has been built as below (Figures 2(a)-(g)).

The experiments on the different types of pavement were conducted in a 2m×1m×0.6m steel box in which the bottom plate with holes was constructed for water to pass through (Figures 2(a) & 2(b)). A bottom plate was made of galvanized steel, which has a good resistance against corrosion (Figure 2(b)). A rainfall simulator with 8 spaced sprays (Figure 2(c)) was set up above the pavement surface to cover an area of 2m × 1m. The rainfall intensities were controlled by a flow meter (Figure 2(d)). The water flowing through the pavement was collected from an inclined steel plate (Figure 2(e)) that is underneath the pavement and connected only to the central 2m × 1m area. Plastic cover has been used to keep all the water inside the setup (Figure 2(f)).

For better simulation it was necessary to use sub-structure materials under the travertine pavement. Sub-structure material selection is an important factor for pavement engineers who construct pervious pavements. The load bearing capacity and hydraulic conductivity of the selected material are important factors that determine the success of the system (Zhang 2006).

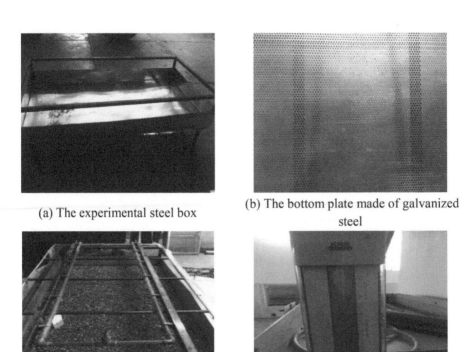

(a) The experimental steel box

(b) The bottom plate made of galvanized steel

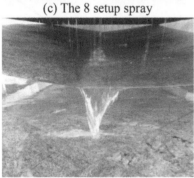

(c) The 8 setup spray

(d) The flow meter apparatus

(e) Infiltrated water collecting plate

(f) The plastic cover

Figure 2.   Experimental setup.

The sub-structure materials that were used in this research were divided to two parts: a base layer and a subbase layer, which are exactly the same as a pavement construction. In our study, the aggregate sizes of base materials and sub base materials are 2.36-4.75 mm and 4.75-10 mm, respectively. While for quality aspect, water has been sprayed to the pavement surface for an hour and with 5 m^3/min inlet intensity, quantity of infiltrated water has been measured on three different intensity. Measurements have been done every 15 minutes in order to evaluate the quantity and quality of infiltrated water.

3   RESULTS

Pervious pavement has been suggested to allow infiltration of storm water through the surface into the soil below where the water is naturally filtered and pollutants are removed.

Porous asphalt is produced and placed using the same methods as conventional asphalt concrete; it differs in that fine (small) aggregates are omitted from the asphalt mixture. The remaining large, single-sized aggregate particles leave open voids that give the material its porosity and permeability. To ensure pavement strength, fiber may be added to the mix or a polymer-modified asphalt binder may be used. Generally, porous asphalt pavements are designed with a subsurface reservoir that holds water that passes through the pavement, allowing it to evaporate and/or percolate slowly into the surround soils. Permeable Concrete Blocks are blocks which has been made by concrete and there are some seams and gaps between them, runoff is able to pass through these gaps and reduce the amount of runoff.

3.1 *Water quantity*

The methodology used to analyze the water quantity is based on comparing the volume of input water with infiltrated water through the Concrete Blocks (CB), Travertine (TR) and Porous Asphalt (AS) under three different inlet intensity, 5, 10 and 15 L/min. (Figure 3). As it is clear in Figure 3, Concrete Blocks has the best efficiency in passing the water, it might be because of cracks between each blocks. While porous asphalt experienced the lowest efficiency

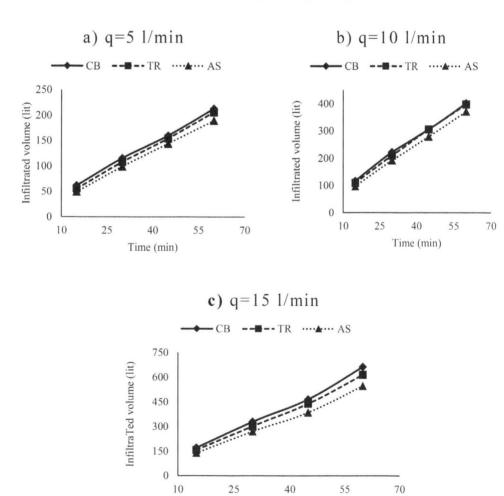

Figure 3.   Input and output volumes of three different types of pervious pavements under different discharges: a) 5 liters/minutes, b) 10 liters/minutes & c) 15 liters/minutes.

in compare to two other pavements, Travertine reduce the runoff to the acceptable amount. Figure 3 indicates how each pavement behave based on different inlet intensities; as much as the inlet intensity increases the efficiency of pavement would decrease.

Considering the inlet volume of water, Figure 3 shows that the travertine pavement is able to reduce the volume of storm water up to 68.66% during 60 minutes experiment. It should also be mentioned that as same as other pervious pavements, the efficiency of travertine experienced a reduction as the inlet intensity increases from 5 L/min to 15 L/min. Generally, Figure 3 reveals an acceptable trend of Travertine based on two other famous types of pervious pavements.

3.2 *Water quality*

The main aim of water quality analyses is to verify if the quality of the water passed through Travertine was good enough to be used for non-potable demands.

Water quality parameters could be selected according to different aspects. In our study which has been held in Iran, the parameters selected based on Iran non-potable Standard Water which could be considered as high quality water. Figure 4 indicates the concentration of important factors of water quality, Copper, Zinc, Lead and Cadmium, over an hour experiment.

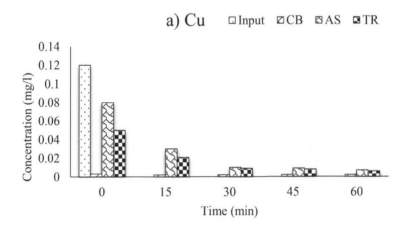

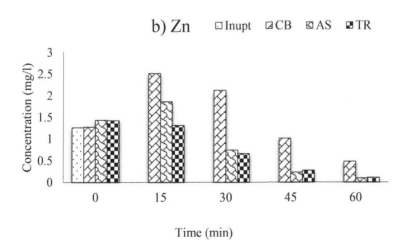

Figure 4. (a) Copper (b) Zinc (c) Lead (d) Cadmium.

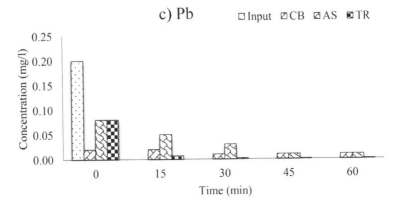

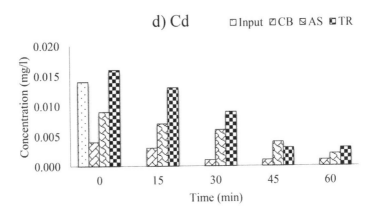

Figure 4. (Cont.)

Figure 4 shows the input and output concentrations of the pollutants over an hour experiment: 95% of the Copper and up to 99% of Lead were retained on the travertine pavement, although the concentration of Zinc was more than the input concentration levels in the two first trials., which may be due to the Zinc in the galvanized metal pavement structure (Jayasuriya 2007). The concentration of Cadmium experienced a similar fluctuation in the first trial. Such fluctuation again may be due to the travertine material and its component, which needs further investigation to examine this phenomenon.

4 CONCLUSION

The filtration characteristics of travertine pavement has been studied through quantitative and qualitative experiments. The experimental results showed that regarding quantity, three pavements were differentiated with distinct behavior. Although concrete blocks had almost the best performances during an hour experiment, the Travertine pavement infiltrated more runoff than porous asphalt in all three inlet intensities.

Regarding quality, it is really hard to generalize the data to choose the best pavement but it was almost clear that in initial period of experiments rainfall washed the pavement's materials after the simulation process and it caused that the passed water quality was not enough suitable to be used in beginning of experiments.

Travertine pavement which used for the first time, could be useful material in water sensitive urban design as it helps to improve storm water quality, reduce peak flow and if properly designed and constructed, allows the infiltrated water to be reused.

Therefore, the travertine material had a great potential in many practices, particularly in the reuse of storm water.

REFERENCES

Acioli, L.A. Da Silveira, A.L.L. & Goldenfun, J.A. 2005. August. Experimental study of permeable reservoir pavements for surface runoff control at source. In *Proceedings of 10th International Conference on Urban Drainage*.

Arnold Jr, C.L. & Gibbons, C.J. 1996. Impervious surface coverage: the emergence of a key environmental indicator. *Journal of the American planning Association* 62(2):243-258.

Brattebo, B.O. & Booth, D.B. 2003. Long-term stormwater quantity and quality performance of permeable pavement systems. *Water research* 37(18):4369-4376.

Castro Fresno, D. Rodríguez Bayón, J. Rodríguez Hernández, J. & Ballester Muñoz, F. 2005. Sistemas urbanos de drenaje sostenible (SUDS). *Interciencia* 30(5):255-260.

Gomez-Ullate, E. Novo, A.V. Bayon, J.R. Hernandez, J.R. & Castro-Fresno, D. 2011. Design and construction of an experimental pervious paved parking area to harvest reusable rainwater. *Water Science and Technology*, 64(9):1942-1950.

Gilbert, J.K. & Clausen, J.C. 2006. Stormwater runoff quality and quantity from asphalt, paver, and crushed stone driveways in Connecticut. *Water research* 40(4):826-832.

Hatt, B.E. Deletic, A. & Fletcher, T.D. 2006. Integrated treatment and recycling of stormwater: a review of Australian practice. *Journal of environmental management* 79(1):102-113.

Jayasuriya, N. Kadurupokune, N. Othman, M. & Jesse, K. 2007. Managing Stormwater Productively Using Pervious Pavements. *NOVATECH*.

Zhang, J. 2006. *Laboratory scale study of infiltration from Pervious pavements; School of Civil, Environmental and Chemical Engineering, RMIT University* (Doctoral dissertation, Thesis).

Sustainable Buildings and Structures: Building a Sustainable Tomorrow – Papadikis et al. (Eds)
© 2020 Taylor & Francis Group, London, ISBN 978-0-367-43019-1

Compressive strength of recycled coarse aggregate pervious concrete containing cement supplementary material

L.N. Jia, J. Xia & M.D. Liu
Department of Civil Engineering, Xi'an Jiaotong Liverpool University, Suzhou, Jiangsu, China
Institute for Sustainable Materials and Environment (ISME), Xi'an Jiaotong-Liverpool University, Suzhou, Jiangsu, China

ABSTRACT: The pervious concrete utilising recycled coarse aggregate (RCA) is known as sustainable pavement material which has 15-35% void content of the total volume. This study aims to improve the compressive strength of pervious concrete containing 100% RCA as coarse aggregate. Three pre-treatment methods on RCA were applied in the experiment. Ground granulated blast furnace slag (GGBS) and silica fume were used in the pervious concrete to replace ordinary Portland cement at 30% and 5% weight replacement level, respectively. Two-stage mixing approach (TSMA) was used as a pre-treatment method to improve the quality of pervious concrete in concrete specimens. It was found that all three pre-treatment methods can improve the qualities of RCA while the effectiveness of TSMA on improving the compressive strength of pervious concrete needs further investigation.

1   INTRODUCTION

The landfill of construction and demolition wastes (CDW) is costly and takes a lot of land space. Recycled coarse aggregate (RCA) is produced from DCW with the process of storage, management and transformation (Shi et al. 2016). The recycled aggregate is mainly used to landfills and roadbed in the past (Dilbas 2014). As an essential approach for sustainable development, using recycled aggregates in concrete is more and more popular in the world. However, concrete using recycled aggregates is known for leading to lower durability and mechanical performance comparing to the natural aggregates. By removing or strengthening the attached mortar on recycled aggregate particles, the performance of concrete containing RCA can be improved.

The pervious concrete is known as a sustainable pavement material which has 15-35% void content of the total volume. The high volume of connected pores gives permeability to pervious concrete, which can help mitigate the stormwater rainfall issue and improve its quality in an urban area (Barnhouse et al. 2016). Surface treatments of RCA and mix design adjustments by incorporating cement supplementary material, such as silica fume and GGBS, have the potential to enhance the mechanical performance of the pervious blocks. Microwave treatment can also be adopted as a method to remove old mortar. For manufacture methods improvement, the two-stage mixing approach (TSMA) was reported in the literature, which can help to improve the compressive strength when mineral admixture was added. The combination of pervious concrete and recycled aggregate can help to achieve the sustainable development of new urbanisation, which will have a bright future in the future (Chandrappa & Biligiri 2016).

This research aims to find an appropriate approach to manufacture a high-quality pervious concrete material which has suitable permeability and mechanic properties when using one hundred percentage RCA. The effectiveness of utilization of cement supplementary material and different manufacture methods will be investigated.

2 LITERATURE REVIEW

There have been many research efforts on improving the performance of RCA in the past years. For example, Dimitriou et al. noted that the mechanical grinding has the most uncomplicated procedure, including grinding by rolling and vibration effects from a high-speed rotating machine. The 62% of adhered mortar can be removed through mechanical grinding, and sounder and slightly rounder aggregates were produced (Dimitriou et al. 2018). Wang et al. (2017) found that pre-soaking in water can improve the compressive strength about 7% at 28 days but only suit for the weak adhered mortar because it cannot remove the strong mortar. In addition, the microwave is used to heat and weak the connection of old adhered mortar where the high temperatures can help to achieve the embrittlement of adhered mortar (Bru et al. 2014). The relationship between the microstructure of concrete, especially the influence of interfacial active agent and microwave selective brittleness and release, is a critical point in the development of microwave treatment process (Lippiatt & Bourgeois 2012). Therefore, microwave-assisted treatment can significantly improve the mechanical properties of RCA concrete (Akbarnezhad et al. 2011). Spaeth & Tegguer (2014) evaluated the effect of recycled concrete aggregate properties by polymer treatments. The test result showed that the polymer treatment could improve the quality of RCA and improve the workability and durability of concrete. However, the compressive strength of concrete was reduced. Recent research has shown that calcium carbonate bio deposition using the bacteria to precipitate calcium carbonate on the RCA surface can fill the cell wall and improve the quality of RCA. Experimental results showed that bio deposition could decrease the water absorption of RCA because of the lower porosity of the surface (Bui et al. 2018).

Except for the above pre-treatment RCA methods, some researchers have proposed other approach to improve the properties of RCA in concrete. For example, silica fume (SF) can be added to the mix design of pervious block concrete to improve the performance and to remove the bonded mortar (Dimitriou et al. 2018). Besides adding silica fume, fly ash also improves the compressive strength of cement and improve the relatively weak durability of recycled aggregate concrete (Kou & Poon 2012). The optimum mix ratio of RAC mixture was proposed by replacing natural aggregate with 50% RA and replacing 25% ordinary Portland cement (OPC) by fly ash (Kou & Poon 2013).

Generally, the typical compressive strength of standard PC with porosities between 15% and 30% ranges from 7 to 25 MPa. (Zhong et al. 2018). Research has shown that increasing the particle size of aggregate can improve the total porosity and permeability of pervious concrete, but reduce the compressive strength and density of concrete (Sun et al. 2018). Aliabdo et al. (2018) used polypropylene fibre to enhances the tensile strength and degradation performance of permeable concrete. However, the compressive strength was slightly reduced. Moreover, the addition of mineral admixture to permeable concrete does not increase the strength of concrete but significantly improves durability with sufficient permeability (Ramkrishnan et al. 2018).

3 PROPERTIES OF MATERIALS

The No.8 RCA has a particle size range between 2.36 and 9.5 mm. The sieving analysis result is shown in Figure 1. The No.8 is selected based on the aggregates used in ACI 522R-10, for which size, the mechanical properties of pervious concrete are available (ACI Committee 522 2013). This project will examine whether the standards for natural aggregate in ACI suit for the RCA. Both natural aggregate and recycled aggregate were satisfied with the requirement of No.8 size grading in this study. The loose bulk density of RCA is 1165.63 kg/m^3, and dense bulk density is 1376.57kg/m^3. The moisture content of RCA is three %in air-dry conditions, and the water absorption is 6.7%. The loose bulk density of natural aggregate is 1376.57 kg/m^3, and dense bulk density is 1643.33kg/m^3. The moisture content of natural aggregate is 0% in air-dry condition, and the water absorption is 1.2%.

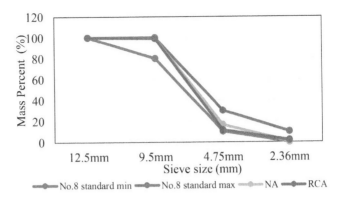

Figure 1. Sieving analysis.

Table 1. RCA surface treatment methods.

Treatment method	Crushing Index	Water Absorption
RCA	15.70%	6.61%
wrapping mortar: RCA=1:7	13%	5%
wrapping mortar: RCA=1:7.5	14%	8%
wrapping mortar: RCA=1:9.7	14%	8%
Dry microwave	12%	5%
SSD microwave	14%	5%

Since the water absorption of RCA is much higher than that of the natural aggregates, several surface treatment methods were used. The results of crushing index and water absorption are shown in Table 1. Improvement on both indices was identified for microwave treatment, while the mortar wrapping method only has improvement on crushing index. The weight ratio of wrapping mortar and RCA has some impact on water absorption improvement. Since the improvement of the surface treatment methods tried in this study are not significant, the RCA used in the following stage is therefore not pre-treated.

4 MIX DESIGN STEP BY STEP FOR PERVIOUS CONCRETE

The mix design of pervious concrete followed the ACI 522R-10, which is the standard for pervious concrete using natural aggregate (ACI Committee 522 2013). Detailed adjustments were enforced in the design procedures to ensure the design was more suitable for RCA to consider its high water absorption. The mixture had water versus cement ratio of 0.28. Used No. 8 recycled coarse aggregate at the dry condition. In addition, vibration is not required in the casting since the vibration usually causes the sediment of cement paste on the bottom of the specimen.

Table 2. Group 1 22.5% paste content (unit: kg/m^3).

Mix No.	Mix design	Cement	Free water	Adjust Water	RCA	SF
1	Control RCA	376.59	105.46	48.91	1376	/
2	Control NA	376.59	105.46	19.70	1643	/
3	RCA+Washing Ovendry	376.59	105.46	48.91	1376	/
4	RCA+5%SF	357.76	105.46	48.91	1376	18.83
5	RCA+5%SF TSMA	357.76	105.46	48.91	1376	18.83

Table 3. Group 2 16.5% paste content (unit: kg/m³).

Mix No.	Mix designation	Cement	Free water	Adjust Water	RCA	GGBS	Sand	SF	Super plasticiser
1	30% GGBS	193.32	77.34	48.91	1376	82.85	/	/	/
2	30% GGBS TSMA	193.32	77.34	48.91	1376	82.85	/	/	/
3	7% sand	276.17	77.34	48.91	1280	/	89.58	/	/
4	RCA+5%SF +1%S.P.	262.36	77.34	48.91	1376	/	/	13.81	2.76
5	RCA+5%SF +1%S.P. TSMA	262.36	77.34	48.91	1376	/	/	13.81	2.76
6	Control 16.5% paste	276.17	77.34	48.91	1376	/	/	/	/

Table 4. Group 3 16.5% paste content (unit: kg/m³).

Mix No.	Mix designation	Cement	Free water	Adjust Water	RCA	GGBS	Sand	SF	Super plasticiser
1	30% GGBS, TSMA	193.319	77.33	48.91	1376	82.851	/	/	/
2	30% GGBS	193.319	77.33	48.91	1376	82.851	/	/	/
3	5% SF, TSMA	262.3615	77.33	48.91	1376	/	/	13.80	1.380
4	5% SF,	262.3615	77.33	48.91	1376	/	/	13.80	1.380
5	Control, TSMA	276.17	77.33	48.91	1376	/	/	/	/
6	Control	276.17	77.33	48.91	1376	/	/	/	/

Specimens in batch one were designed as light-compacted concrete with 22.5% paste content since vibration was not applied. However, batch 2 and batch 3 were designed as well-compacted concrete with 16.5% paste content. All mixing design for three batches is shown in Table 2, Table 3, and Table 4, respectively.

5 SAMPLE PREPARATION AND EXPERIMENT METHODS

The concrete mixing was operated in the mixer, as shown in Figure 2(a). The solid materials, including cement, RCA and other solid materials, were mixed for 1 min. Then all the water and superplasticiser (if needed) were added into the mixer and mixed for another 2 min. For TSMA, half of the total water was added initially into the mixer and mixed with the solid materials for 1 min. Then, the other half of the total water was added in the mixer and mixed for one more minute. The total concrete mixing time was around 3-4 min. The actual states

a) Mixer   b). 1 min after Stage 1   c). 1 min after Stage 2

Figure 2. Mixing procedure and condition of pervious concrete.

for two stages were shown in Figure 2b) and Figure Figure 2c), respectively. After mixing for 3 minutes, the concrete can be used for specimen casting. The workability can be identified by whether it can form a concrete ball. The concrete was put in moulds and hand compacted with iron dumbbells. The 100 mm cube mould was used for the compressive strength tests. The concrete mixing in TSMA had low water/cement ratio for first stage mixing, which would give a dense cement wrapping on the surface of RCA.

6 RESULTS

6.1 Density and void content

The results of the density and void content tests on specimens of batch three at 28days were shown in Table 5. The densities were the average value from five 100 mm cubes. The void contents of each group were the average value of five specimens, which were measured by water drain method. The measured density is close to the design value.

6.2 Compressive strength at 7 and 28-day

The compressive strength of pervious concrete specimens from batch 1 and 2 at 7 days are presented in Figure 3. The compressive strength at an early age mostly depended on the void content, cement mortar strength and strength of interfacial transition zone. Therefore, it is reasonable that the compressive strength of specimens from batch 1 is higher than those from batch 2 because its lower paste content. There are no significant differences regarding compressive strength for the specimen with 100% nature aggregates or RCA. This insensitivity on the types of coarse aggregate can be attributed to the actual failure pattern of the specimens. It was observed that the failure was more attribute to the bond failure instead of aggregate crushing. Specimen with washed and dried RCA only lead to slightly higher

Table 5. Density and void content.

Age: 28 day	Density (kg/m³)	Design Density	Void
G1,30% GGBS, TSMA	1768.3	1778	/
G2,30% GGBS	1763.3	1778	/
G3,5% SF, TSMA-28	1670.0	1778	22.67%
G4, 5% SF-28	1646.7	1778	23.40%
G5,Control, TSMA-28	1696.7	1778	20.40%
G6, Control-28	1755.0	1778	17.67%

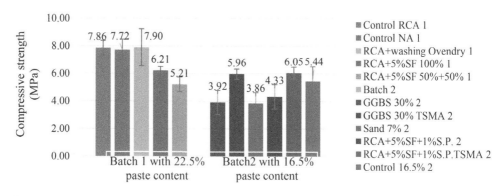

Figure 3.   Compressive strength of batch 1 and batch 2 in 7 days.

average compressive strength (less than 5%) and the results have a higher dispersion compared to the specimens with 100% recycled aggregates without washing and drying. Considering that washing and oven-dry treatment demand both time and energy, this pre-treatment method may not be valuable in practical production. Specimens with additional SF resulted in lower compressive strength at 6.21 MPa and 5.21 MPa, no matter if the water was added in one batch or two batches.

Two-Stage Mixing Approach with cement supplementary materials, such as GGBS or SF, can lead to higher compressive strength than the control group for specimens in batch 2. The best group is the group with 5% SF replacement, followed by specimens with 30% GGBS replacement and using TSMA. Specimens with sand addition post the lowest compressive strength at 7 days. The specimen with one batch mixing method has the compressive strength of 6.21 MPa, and the specimens with two-stage mixing lead to an even lower compressive strength of 5.21 MPa. The lower than expected compressive strength may be attributed to the lower flowability due to the addition of SF.

The specimens in batch 3 were tested at both 7 and 28 days. The results from different ages were presented side by side for comparison purpose for each specimen group. Either GGBS (30% weight replacement of cement) or SF (10% weight replacement of cement) was used in the specimen design. Impact of one stage batching and TSMA were investigated. It is obvious that compressive strength at 28 days is higher than 7 day's strength for all groups, except group 3, for which no significant strength increase was observed. It might be due to the fact that the flowability of the cement paste was reduced by addition of SF. Specimens contain 5% SF and mixing using TSMA has the highest strength at 7.76 MPa among all groups at 7 days while the group with 30% GGBS replacement and using normal mixing method achieved the highest compressive strength of 8.61 MPa at the age of 28 days. This group also achieved the highest compressive strength increase of 22%. While the single batch mixing protocol leads to high strength for the group with GGBS at 28 days's age, it has an adverse effect on the 7 day's

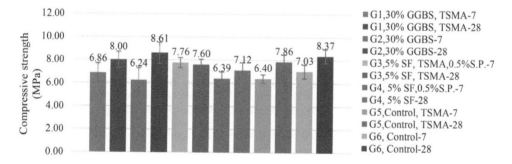

Figure 4.   Compressive strength of batch 3 at 7 and 28d.

Figure 5.   Failure model of the specimen.

strength. For groups with SF, TSMA method lead to higher compressive strength at both 7 and 28 days' age, however, TSMA is not that effective for the control groups. Typical failure pattern of the specimen can be found in Figure 5. The specimen failed typically with local crushing and spalling.

7 CONCLUSION & DISCUSSION

This study tested three batches of specimens for the purpose to find a more effective method to improve the compressive strength of pervious concrete. Several general conclusions can be developed as follows:

1. The paste content in mixing design influence the compressive strength of pervious concrete containing RCA. The strength of pervious concrete mostly depends on the cement mortar strength and interfacial transition zone in the early stage. Pervious concrete using 100% RCA has an equivalent compressive strength to those using nature aggregates. Washed and oven-dried RCA does not enhance the compressive strength of pervious concrete significantly, therefore, is not effective enough to compensate the time and energy demand.
2. The effectiveness of TSMA on improving the compressive strength of pervious concrete needs further investigation. Based on the current study, both positive and negative impacts were observed. Replacement cement up to 30% weight of GGBS can lead to similar compressive strength while replacing 5% weight of cement by SF lead to lower compressive strength due to the reduction of workability of cement paste.
3. The cement paste/mortar bonding the aggregates play the most critical role in low-strength pervious concrete. Increasing of mortar among and strength lead to the higher compressive strength of pervious concrete.

REFERENCES

ACI Committee 522. 2013. *Report on pervious concrete ACI 522R-10*. Farmington Hills, MI, USA: American Concrete Institute.

Akbarnezhad, A. Ong, K.C.G. Zhang, M.H. Tam, C.T. & Foo, T.W.J. 2011. Microwave-assisted beneficiation of recycled concrete aggregates. *Construction and Building Materials* 25(8): 3469–3479.

Aliabdo, A. A. M. Abd Elmoaty, A.E. & M.Fawzy, A. 2018. Experimental investigation on permeability indices and strength of modified pervious concrete with recycled concrete aggregate. *Construction and Building Materials* 193:105–127.

Barnhouse, P. W. Srubar, W.V. & Iii, W.V.S. 2016. Material characterization and hydraulic conductivity modelling of macroporous recycled-aggregate pervious concrete. *Construction and Building Materials* 110: 89–97.

Bru, K. Touze, S. Boutgeois, F. Lippiatt, N. & Menard, Y. 2014. Assessment of a microwave-assisted recycling process for the recovery of high-quality aggregates from concrete waste. *International Journal of Mineral Processing* 126: 90–98.

Bui, N. K. Satomi, T. & Takahashi, H. 2018. Mechanical properties of concrete containing 100% treated coarse recycled concrete aggregate. *Construction and Building Materials* 163:496–507.

Chandrappa, A. K. & Biligiri, K.P. 2016. Pervious concrete as a sustainable pavement material – Research findings and future prospects : A state-of-the-art review. *Construction and Building Materials* 111: 262–274.

Dilbas, H. Simsek, M. & Cakir, O. 2014. An investigation on mechanical and physical properties of recycled aggregate concrete (RAC) with and without silica fume. *Construction and Building Materials* 61: 50–59.

Dimitriou, G. Savva, P. & Petrou, M.F. 2018. Enhancing mechanical and durability properties of recycled aggregate concrete. *Construction and Building Materials* 158:228–235.

Kou, S. C. & Poon, C. S. 2012. Enhancing the durability properties of concrete prepared with coarse recycled aggregate. *Construction and Building Materials* 35: 69–76.

Kou, S. & Poon, C. 2013. Long-term mechanical and durability properties of recycled aggregate concrete prepared with the incorporation of fly ash. *Cement and Concrete Composites* 37: 12–19.

Lippiatt, N. & Bourgeois, F. 2012. Investigation of microwave-assisted concrete recycling using single-particle testing. *Minerals Engineering* 31:71–81.

Ramkrishnan, R. Abhilash, B. Trivedi, M. Varsha, P. Varun, P. & Vishanth, S. 2018. Effect of mineral admixtures on pervious concrete. *Materials Today: Proceedings* 5(11): 24014–24023.

Shi, C. Li, Y. Zhang, J. Li, W. Chong, L. & Xie, Z. 2016. Performance enhancement of recycled concrete aggregate - A review. *Journal of Cleaner Production* 112:466–472.

Spaeth, V. & Tegguer, A.D. 2013. Improvement of recycled concrete aggregate properties by polymer treatments. *International Journal of Sustainable Built Environment* 2(2): 143–152.

Sun, Z. Lin, X. & Vollpracht, A. 2018. Pervious concrete made of alkali activated slag and geopolymers. *Construction and Building Materials* 189:797–803.

Wang, L. Wang, J. Qian, X. Chen, P. Xu, Y. & Guo, J. 2017. An environmentally friendly method to improve the quality of recycled concrete aggregates. *Construction and Building Materials* 144:432–441.

Zhong, R. Leng, Z. & Poon, C. 2018. Research and application of pervious concrete as a sustainable pavement material : A state-of-the-art and state-of-the-practice review. *Construction and Building Materials* 183:544–553.

Sustainable Buildings and Structures: Building a Sustainable Tomorrow – Papadikis et al. (Eds)
© 2020 Taylor & Francis Group, London, ISBN 978-0-367-43019-1

Effect of basalt fibre addition to cementitious mortar at ambient and elevated temperatures

N. Revanna, Charles K.S. Moy, T.D. Krevaikas & C.S. Chin
Department of Civil Engineering, Xi'an Jiaotong-Liverpool University, Suzhou, China

S. Jones
School of Engineering, University of Liverpool, Liverpool, UK

ABSTRACT: A study on the mechanical characteristics of cementitious mortar reinforced with basalt fibres at ambient and elevated temperatures is carried out. Chopped basalt fibres with varying percentages, 0.15%, 0.2%, 0.5%, 1.0% are added to the cement mortar. All the specimens are heated to 200°C, 500°C, 900°C using a muffle furnace. Flexural strength, compressive strength and moisture loss are measured to evaluate the performance of cementitious mortars at elevated temperatures. From the study, it is clear that basalt fibres can be used to reinforce mortar as the fibres remain unaffected up to 500°C, however, the contribution of basalt fibres to the flexural strength, compressive strength development is minimal at both ambient and elevated temperatures.

1 INTRODUCTION

Fire is one of the most frequent causes of damage to buildings and other structures. Exposure to fire leads to loss of moisture in the structural members, weakening of cementitious byproducts resulting in strength deterioration that cause the bound aggregates to loosen resulting in brittleness. Although, cementitious byproducts in the matrix can sustain heat, prolonged exposure can cause serious distress. The main parameters on which the performance of concrete subjected to fire depends are concrete strength, moisture content, concrete density, heating rate, specimen shape and dimension, loading conditions, fibre reinforcement, aggregate type, reinforcement layout and configuration (Kodur et al. 2003).

As basalt fibre is a relatively new fibre, fewer studies have been carried out using basalt fibres in cementitious composites and even less at elevated temperatures. Basalt fibre is an inorganic fibre manufactured from basalt rock that is environment friendly, inert to the alkaline environment in the cement matrix and non-combustible (Borhan 2012). The aim of the study was to test the effectiveness of basalt fibre addition into the cement mortars at high temperatures as basalt fibres have very good thermal resistance (Nováková et al. 2017).

2 METHODOLOGY

2.1 Materials

A 42.5 grade cement was used to cast the samples and the mortar mix was selected from a previous study on textile reinforced concrete (TRC) with slight change in the mix (Barhum & Mechtcherine 2012). The cement mortar composition can be seen in detail in Table 1. Fine sand passing through 600µ was used to cast mortar. The basalt fibres used were 40mm in length. The properties of the fibres is shown in the Table 2.

Table 1. Mortar composition.

Composition	Quantity (kg/m³)
Cement	554
Sand	832
Water	330

Table 2. Properties of basalt fibre.

Length	40 mm
Tensile strength	1050-1250 MPa
Modulus of elasticity	34 GPa

2.2 *Experimental programme*

2.2.1 *Sample preparation*
The size of the prism and cube cast were 160mm x 40mm x 40mm and 50mm x 50mm, respectively as per BS EN196 (2005) and ASTM C109. A total of 120 specimens with four percentages of basalt fibres were chosen viz., 0.15%, 0.2%, 0.5%, 1% calculated as percentage of total volume of mixture added into the cement mortar with a set of three unreinforced samples to serve as reference were cast. Four temperature range were chosen to test the mortar specimen: room temperature (~20 °C), 200°C, 500°C, 900°C. Three specimens per test were cast for each condition. An electrical mortar mixer was used to mix the mortar contents. All the specimens were demoulded after 24hours and then transferred to curing tanks for a period of 28 days. After curing, the specimens were heated in a muffle furnace. The temperature inside the furnace raised at a rate of 1°C every 6 seconds.

2.2.2 *Test*
All the specimens were heated for a period of one hour to maintain steady state thermal condition and taken out from the furnace soon after an hour of heating was completed. All the specimens were allowed to cool down naturally in air and later on subjected to flexural tests based on the recommendations of BS EN196 (2005) and compression tests based on ASTM C109. As the former standard estimates the compressive strength on fractured prisms it was not used to determine the compressive strength as it resulted in less strength compared to the specimen tested as per ASTM C109.

3 RESULTS AND DISCUSSION

3.1 *Flexural strength*

Flexural strength tests were conducted on the heat exposed specimens after they were slowly cooled to ambient temperature. Therefore, the results obtained would represent the residual strength of the samples after the heat treatment. The results of the specimen are plotted in the Figure 1 below.

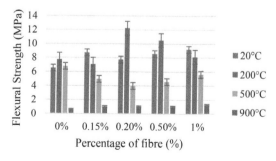

Figure 1. Flexural strength v/s percentage of basalt fiber at various temperatures.

There was only a marginal increase in strength at high temperatures, which implies that the usage of fibres will not be beneficial at high temperatures. However, at 200°C, the specimens resulted in maximum flexural strength. The reduction in strength for higher temperatures could be attributed to the degradation of the cement matrix as the elevated temperatures degrades the essential strength contributing cementitious byproducts. The specimen at 200°C does not show any strength reduction which shows that the cement matrix is able to withstand this temperature.

From Figure 1, it can be seen that unheated samples with 0.15% fibre resulted in more strength compared to 0.2% fibre with a reduction of 11.5% in strength reduction. 0.5% fibre content shows a slight increase in strength compared to 0.2% but it is almost equal to the strength of 0.15% fibre specimen. At higher temperature i.e., at 200°C, neat specimens (without fibres) showed slightly better strength than the 0.15% fibre specimens. However, the 0.2% fibre reinforced specimen showed a sharp increase in strength amounting to 63% compared to neat specimens and the rest of the specimens. The higher percentage of fibre i.e. 0.5% reinforced mortar specimens showed strength reduction of 15% compared to 0.2% fibre reinforced specimen. Finally, 1% fibre reinforced mortar showed a 44% strength decrease compared to 0.2% fibre reinforced specimen. The strength improvement in 1% fibre reinforced specimen could be seen on a marginal scale as it showed the highest strength amongst 0.15%,0.2%,0.5% fibre reinforced specimen but less than the non-reinforced specimen.

The specimens at 900°C, showed maximum strength at 1% fibre reinforcement. The rest of the samples show almost the same strength indicating that the addition of fibre had no effects on their strength. The non-reinforced specimens showed the least strength with a decrease of almost 50% strength compared to the 1% fibre reinforced specimen.

3.2 *Compressive strength*

Specimens were designed for 40 MPa. The compressive strength was calculated from ASTM C109 with a loading rate of 1.8kN/s. The results of non- reinforced specimens confirmed that the strength adhered to the designed compressive strength showing 44.17 MPa. Figure 2 shows the variation of compressive strength with respect to temperature. The compressive strength reduced with the increase in the percentage of the basalt fibres indicating that the use of basalt fibres do not contribute to compressive strength gain.

It can be seen that the compressive strength gradually decreased as the temperature exposure was increased with the highest strength at room temperature and least strength at 900°C. The same trend could be observed for 0.15%, 0.2%, 0.5% and 1% basalt fibre reinforced specimen. The results from Figure 2 show that there is no trend in the compressive strength as the temperature increases with respect to the basalt fibre percentage. There is a sudden drop in the compressive strength at 0.5% fibre reinforcement at room temperature, 200°C, 500°C, whereas the 1% fibre reinforced specimens show almost the same compressive strength at 0%, 0.15%,0.2%,0.5% with a slight increase at 900°C.

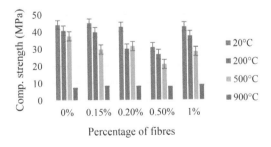

Figure 2.   Compressive Strength v/s percentage of fibre.

3.3 Loss of moisture

A rough estimate of moisture loss was performed by weighing the specimens after exposure to heat. As the temperature increases, the trapped moisture will try to escape from the specimen and will eventually lead to moisture loss making the specimen lighter. Water droplets collected on top of the furnace door as well as near the legs and the rear side of the furnace could be seen as the entrapped moisture was released from the specimens between 300°C-450°C. It was noted that there was gradual loss of moisture irrespective of the percentages of basalt fibre content in the prism or cube specimen. On average, 50% weight reduction was noted from 200°C to 900°C.

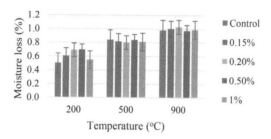

Figure 3.   Moisture loss in prism.

4   BEHAVIOUR OF BASALT FIBRE REINFORCED MORTAR AT HIGH TEMPERATURE

For the specimens at high temperature, the first difference that could be seen was the colour change. The colour of the specimen gradually changed as the specimen was heated. A change from greyish white, light pink to white color differences could be seen as the specimen were heated from room temperature to 900°C. The specimens at room temperature resembled the ideal colour of cement particles. As the specimens were heated to 200°C, the colour of the specimens changed to slightly greyish white which showed that heat treatment was effective enough on the samples as the trapped surface moisture was completely removed. The specimens turned to light pink when they were heated to 500°C, which might result from the decomposition of silica aggregates (Hager 2014). A further analysis of the heat subjected specimens should be made to understand the changes in colour and the reasons associated behind it. At 900°C, the color of the specimen turned out to be whitish grey as a layer of white powder could be seen on the surface of the specimen after letting the specimen to cool down with complete strength loss.

The fibres showed no degradation or colour change upto 500°C, whereas at 900°C, the fibres could be seen changing colour to black as it could not sustain the temperature rise since the fibre was completely degraded. The fibres were so brittle that even touching the fractured surfaces with bare hands could make them spall off from the cross section as seen in Figure 4.

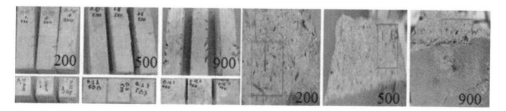

Figure 4.   Specimen colour change at elevated temperature and fibre rupture at 200°C, 500°C & 900°C.

The cross section of the tested specimen showed some of the ruptured fibres having lengthy tails whereas, the specimens at high temperature did not show any of these signs as they fractured by rupturing off from the cross section. The specimens heated to 200°C, 500°C did not show any signs of major cracks whereas the temperature effects could be seen in the form of fine cracks on the surface of the specimen as the temperature was raised to 900°C which could be due to surface shrinkage as the moisture was completely drained out from the surface layers of the specimens.

4.1 *Microstructure analysis of basalt fibre reinforced mortar at elevated temperatures*

Microscopic analysis was conducted on basalt fibre reinforced mortar specimens subjected to elevated temperatures to assess the microstructural changes in the matrix as well as basalt fibres on the fractured surfaces using a Hitachi TM3000, a tabletop scanning electron microscope (SEM). From the findings, it is obvious that there was degradation in the matrix and fibres at high temperatures. One of the main findings from the SEM is that the basalt fibres are very brittle which can rupture in any plane of its cross section leading to brittle filament failure. This could be attributed to the strength loss that was registered on all the heated and non-heated specimens.

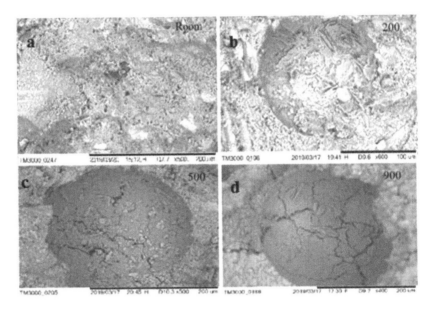

Figure 5.   Basalt fibre matrix at room (a) Room (~20°C) (b) 200°C (c) 500°C (d) 900°C.

5 CONCLUSION

In this study, the behaviour of cementitious mortar samples at high temperatures of 200°C, 500°C, 900°C, with four percentages i.e., 0.15%, 0.2%. 0.5%. 1% of basalt fibre content were studied. The aim of the study was to test the effectiveness of basalt fibre addition into the cement mortars at high temperatures as basalt fibres have very good thermal resistance. Unfortunately, the basalt fibres cannot sustain very high temperature (i.e. above 500 °C) that might have led to their inability in contributing to the strength development. However, to get a fair idea of the strength increment resulting from the addition of basalt fibres, future studies could be taken up by increasing the percentage of basalt above 1% fibres and also by varying the length of basalt fibre.

From the current study, it could be noted that there is marginal flexural strength development as basalt fibres did not contribute to the strength development especially at high temperatures. This could be attributed to the length of the basalt fibre as some of the fibres could be found curled up as 42 mm length basalt fibre was used in this study.

The study confirms that there is no appreciable amount of compressive strength increase with the addition of basalt fibre at high temperatures. Despite having a coating on the surface of the fibre each of the fibre bundles, filaments could possibly get dispersed in the matrix while mixing and casting the mortar leading to loss of strength. The degradation of the basalt fibre at high temperatures is another reason for poor performance. The ruptured fibres at low temperatures were seen to have lengthy fractured remains whereas the fibres at high temperature were simply cut off from the face of the cross section. However, there were no signs of basalt fibre degradation upto 500°C. The degradation of the matrix could also be another main reason for poor performance of cement mortars, as the cement by products would completely degrade at high temperatures. The moisture loss was also obvious from the previous studies making the specimens light in weight and brittle. The spalling effect was not seen as the specimen remained undamaged throughout the course of the test at high temperatures. This shows that the current mortar designed can be subjected to high temperatures as high strength mortar could lead to serious spalling.

ACKNOWLEDGEMENTS

The authors would like to express their gratitude for providing the financial support from Xi'an Jiaotong – Liverpool University (RDF-16-01-17).

REFERENCES

ASTM C109/C109M - 16a. *Standard test method for compressive strength of hydraulic cement mortars.*

Barhum, R. & Mechtcherine, V. 2012. Effect of short, dispersed glass and carbon fibres on the behaviour of textile-reinforced concrete under tensile loading. *Engineering Fracture Mechanics* 92:56-71.

Borhan, T. M. 2012. Properties of glass concrete reinforced with short basalt fibre. *Materials & Design* 42: 265-271.

BS EN 196-2. 2013. British Standards Institution. *Method of testing cement. Chemical analysis of cement.*

Girgin, Z.C. & Yildrim, M.T. 2016. Usability of basalt fibres in fibre reinforced cement composites. *Materials and Structures* 49:3309-3319.

Hager, I. 2014. Colour change in heated concrete. *Fire Technology* 50:945–958.

Kodur, V. K. R. Cheng, F.P. Wang, T.C. & Sultan, M.A. 2003. Effect of strength and fiber reinforcement on fire resistance of high-strength concrete columns. *Journal of Structural Engineering* 129: 253-259.

Nováková, I. Þórhallsson, E. & Bodnárová, L. 2017. Behaviour of basalt fibre reinforced concrete exposed to elevated temperatures. *5th International Conference on concrete spalling due to fire exposure, Boras, Sweden.*

Ralegaonkar, R. V. Aswath, P. B. & Abolmaali, A. 2017. Design investigations of basalt fibre-reinforced mortar. *Proceedings of the Institution of Civil Engineers - Construction Materials* 0:1-9.

Ralegaonkar,R. Gavali, H. Aswath, P. & Abolmaali, S. 2018. Application of chopped basalt fibers in reinforced mortar: A review. *Construction and Building Materials* 164: 589-602.

Shaikh, F. & Haque, S. 2018. Behaviour of carbon and basalt fibres reinforced fly ash geopolymer at elevated temperatures. *International Journal of Concrete Structures and Materials* 12:35.

Sim, J. Park, C. & Moon, D.Y. 2005. Characteristics of basalt fiber as a strengthening material for concrete structures. *Composites Part B: Engineering* 36: 504-512.

Sustainable design in built environment

Sustainable Buildings and Structures: Building a Sustainable Tomorrow – Papadikis et al. (Eds)
© 2020 Taylor & Francis Group, London, ISBN 978-0-367-43019-1

Investigation of efficient self-powered sensors for vibrations in building floors

L. Di Sarno, T.E. Dada & A. Mannis
Department of Civil Engineering and Industrial Design, University of Liverpool, Liverpool, UK

F.A. Fazzolari & P. Paoletti
Department of Mechanical, Materials and Aerospace Engineering, University of Liverpool, Liverpool, UK

ABSTRACT: In the quest for the realisation of safe and smart structures, recent advances in structural health monitoring have rated the application and efficiency of self-powered sensors high, as it drives sustainable development. This paper reviews the state-of-the-art of the self-powered sensors for monitoring vibrations of building floors. It describes the conventional ways of monitoring structures and the existing methods of harnessing energy from vibrations. This study focuses on the application of piezoelectric technologies and in particular the Lead Zirconate Titanate (PZT) technology. It considers essential criteria such as the geometry and support conditions affecting the power generated. Several evaluations of piezoelectric technologies were made based on their features and on the quantity of energy generated using patch optimisation technique. Finally, the cantilever supported PZT of the piezoelectric transduction mechanism was identified to be the most suitable mechanism for application on sensors in building floors due to its efficiency in energy conversion.

1 INTRODUCTION

1.1 Background

The optimization of structural elements is a useful technique in the design of structures in the advent of sustainability and lean construction. This method, coupled with other technological improvements, has paved the way for the possibility of light structures. However, light structures are susceptible to structural vibrations, and these vibrations need therefore to be considered in structural analysis and design. Vibration is a design criterion based on serviceability limit state and, therefore, it could be critical for flexible structures. In specific structures where vibrations above a particular threshold hinder the safe usage of the structure even at safe ultimate limit state, the serviceability vibration is considered as the governing critical design criteria as human-induced vibrations may cause perturbation in the users (Chen et al. 2016). The presence of these potentially dangerous vibrations, among many other criteria, thus requires health monitoring of structures and their elements, a typical example being the building floors.

1.2 Structural health monitoring and sensors

In structural health monitoring (SHM), the strain has been widely employed as the measured quantities of operation when assessing loading, stresses and fatigues in structural engineering to ascertain safe structures (Chew et al. 2017). To measure strain, traditional approaches relied on metal foils and wire gauges. These methods required wired connections between the sensor networks and the receiver, which was proven inefficient due to its significant

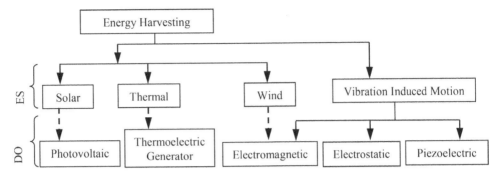

Figure 1. Alternative renewable energy sources and device options. ES: Energy Sources DO: Device Options.

consumption of electric power (Lynch 2006). This inefficiency stimulated the use of wireless sensor networks, as they are cheaper, easy to install and consume less energy compared to a wired system (Liu et al. 2015). Lynch (2006) further stated in his study that there had been an increase in the adoption of wireless technology in SHM applications. Galchev et al. (2011) added that wireless SHM is attractive due to the absence of wires which are vulnerable to damage.

At the start of the millennium, batteries were used to power many wireless sensor nodes, and they operated on tight energy budget as each battery replacement operation attracts a significant cost; therefore there is a need to find sustainable energy alternatives to power the sensors. Galchev et al. (2011) highlighted that the use of some alternative renewable energy sources to power sensors in building floors is also faced with challenges. Firstly, wind energy varies with the location, as the energy that can be obtained is dependent on the prevailing wind speed in each location, posing a challenge to its usage. Also, the wind vanes to be used for harnessing energy from wind are significant in size, and that poses a great challenge to its application in building floors. Secondly, the use of solar energy could also be a potential source for powering the wireless sensors; however, the availability of sunlight and the accumulation of debris on the solar panel pose a significant challenge to its feasibility. Thirdly, thermal energy was also evaluated; however, an excellent thermal connection to the building floor is a challenge. These sources are shown in Figure 1.

Harvesting renewable energies available around the structures intended to be monitored will reduce the cost associated with SHM. Gupta et al. (2016) opined that there had been a recent urgent need for portable and wireless electronic devices with significant life span in the field of harvesting energy. With regards to these needs, attempts have been made to utilise energy harvesting methods that transform available mechanical energy from vibrations into usable electricity as a viable solution for powering Wireless Sensing Nodes (WSN) (Jung et al. 2018). WSNs are electronic devices used to monitor environmental characteristics and signals, such as pressure, humidity, temperature, among others (Khan & Ahmad 2016).

1.3 *Energy harvesting methods for wireless sensors*

The primary methods used to harvest energy from vibrations are evaluated in this section. A review by Beeby et al. (2006) on the possible use of energy from vibration in powering wireless and self-powered micro-systems showed that the transduction mechanism plays a key role in the performance of such devices. The transduction mechanism helps in the conversion of energy from one form to another, and in particular, for SHM it is needed to transform energy from mechanical vibrations to electrical energy. Three transduction mechanisms were evaluated, namely electromagnetic, electrostatic and piezoelectric mechanisms. The electromagnetic transduction generates electricity through the relative motion between a conductor and a magnetic flux gradient. Similarly, the electrostatic transduction generates electricity through the relative motion between electrically isolated charge capacitor plates with energy being harvested by the work done against electrostatic force between the plates. The last transduction mechanism considered was the piezoelectric transduction, and it generates electricity from mechanical strain. In all cases, the relative motions induced by vibration of the structure to be monitored is used as energy input.

The piezoelectric based energy harvester has received much attention with regards to application in wireless sensors. This attention is due to its high power density and ease of installation (Yoon et al. 2016). Gupta et al. (2016) attributed the extensive usage of piezo-electric to the low complexity of its analysis and fabrication while Giuliano & Zhu (2014) ascribed the extensive use of piezoelectric transducers to their simple structure and the capability of generating significant power output. Furthermore, Zhang et al. (2015) revealed that there had been an extensive study of piezoelectric materials and that piezo-electric based devices occupy less space compared to other alternatives when used as a unit in powering sensors. Gupta et al. (2016) opined that the effect of piezoelectricity is best described to have a linear relationship between the mechanical and electrical forms of energy indicating an increasing voltage with increasing vibration induced deformations. This effect is reversible in that mechanical stresses and pressures can also be generated from the piezoelectric material by the application of electrical energy and vice versa. The works reviewed in the literature in tandem with the paper of Wu et al. (2010) therefore suggests the use of piezoelectric materials to generate power for wireless sensors in built environments, thanks to the abundance of vibrations in engineering structures.

2 PIEZOELECTRIC TECHNOLOGY

A brief description of the piezoelectric transduction mechanisms in the previous section sum-marises the piezoelectric technology. This section considers the types of existing piezoelectric technology and their relative outputs in optimised patches.

2.1 Types of piezoelectric technology

A range of piezoelectric types and their uses based on the work of Elhalwagy et al. (2017) are pre-sented in Table 1 and the corresponding layout output in Table 2. The patches cases are shown in Figure 2, and the classification according to power output in patches is presented in Table 3. This classification provides the information needed for a selection procedure during usage.

The output of different patch case for different piezoelectric is presented in Table 3. The floor considered has a dimension of 1 m by 1 m with the patches width in 0.1 m intervals in each case.

In summary, the various types of piezoelectric technology can be classified into three bands, as presented in Table 3.

2.2 Major efficient piezoelectric technology in sensor application

As one of the requirements driving sustainable development, the lean management of oper-ations and materials are important. This method helps in the selection of piezoelectric technol-ogy in the moderate output range for application in sensors. These technologies are the Lead Zirconate Titanate (PZT) and the Polyvinylidene Fluoride (PVDF) piezoelectrics.

The PZT piezoelectric is an inorganic chemical compound (PbZrTiO3)) with a perovskite structure containing Lead (Pb), Zirconium (Zr) and Titanium (Ti). The perovskite structure enables the piezoelectric effect – the transformation of mechanical strain to electricity and vice versa (Beeby et al. 2006). They can be broadly classified into hard and soft categories, and that is dependent on the mechanical and electrical characteristic (Butt & Pasha 2016). PZT exhibits brittle behaviour with a high conversion rate up to 80%, hence it fragile but efficient (Richard et al. 2004). Due to this brittleness and stiffness, it is suited for vibration-based energy harvesting, as the frequency can resonate up to 100Hz under bending (Balouchi 2013). However, the acceptable design frequency is usually slightly higher than the natural

Table 1. Major piezoelectric technologies and their uses (Elhalwagy et al. 2017).

S/NO	Technology	Product size (m)	Features and uses
1	Electro-active Polymers (EAP)	Sheets	- Generally associated with high voltage - Operates on Sensor Network Technology and Sensor Matrix
2	PZT Ceramic (Lead Zirconate Titanate)	Generally manufactured in small sizes	- Brittle in nature - It has up to 80 percent efficiency in energy conversion - It has higher piezoelectric constant than PVDF - It has higher voltage conversion than PVDF - It is more expensive than PVDF
3	Parquet PVDF layers	0.05×0.40 with varying thickness	- It requires simple manufacturing process. - It is considerably ductile compared to PZT - It is very robust and can be fabricated in any shape - Energy generated increases with multiplication of layers - It is usually produced in big size foils - Energy harvested is sufficient for small electric loads
4	Power Leap PZT	0.61×0.61 tile	- The system uses 50 mm by 25 mm PZT plates with brass reinforcement covered in nickel electrode for low current leakage - Generated power can be stored in batteries as direct current power
5	PZT Nanofibre	Nano size	- Drive a single nanowire-based UV sensor to build a self-powered system
6	PVDF Nanofibre	Nano size	- It is a self-powered wireless nanodevices and systems - It powers a pH sensor and an UV sensor. - It can easily synthesize into the required sizes and shapes
7	ZnO Nanowire	Nano size	- Most recent material to be reported for piezoelectric harvesting

Table 2. Major piezoelectric technologies and patches output.

S/NO	Technology	Energy output (mW/m^2)	Patch case 1 (mW)	Patch case 2 (mW)	Patch case 3 (mW)
1	Electro-active polymers (EAP)	1000.000	1000.000	500.000	250.000
2	PZT Ceramic (Lead Zirconate Titanate)	8.400	8.400	4.200	2.100
3	Parquet PVDF layers	0.333	0.333	0.165	0.083
4	Power Leap PZT	1.344	1.344	0.672	0.336
5	PZT Nanofibre	0.030	0.030	0.015	0.008
6	PVDF Nanofibre	0.007	0.007	0.004	0.002
7	ZnO Nanowire	0.005	0.005	0.003	0.001

frequencies of building floors, which ranges between 1HZ for critical working areas like hospitals and 8Hz in offices. Ensuring a higher frequency is necessary because the natural frequency of the human body resonates with the natural frequencies of building floors, causing discomforts to the occupants (Steel for Life & BCSA 2019, Brownjohn & Zheng 2001).

Similarly, the PVDF piezoelectric is a non-reactive thermoplastic fluoropolymer, which is a product of vinylidene difluoride polymerization, with polymeric and flexible characteristics. It was noted that PVDF is more robust than PZT but has conversion efficiency as low as 14.39% (Zhu et al. 2013). Due to its flexibility and higher damping, it is not efficient for use in

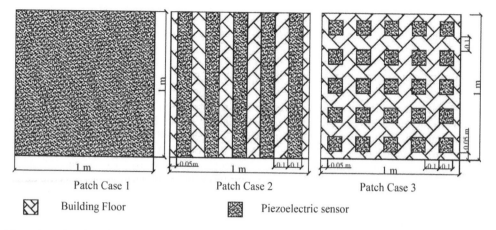

Patch Case 1   Patch Case 2   Patch Case 3

Building Floor   Piezoelectric sensor

Figure 2. The piezoelectric layout in patches.

Table 3. Power classification of major piezoelectric technologies.

S/NO	Power Classification	Energy output (mW/m²)	Patch case 1 (mW)	Patch case 2 (mW)	Patch case 3 (mW)
1	High	1000.000	1000.000	500.000	250.000
2 – 4	Moderate	0.333 - 8.400	0.333 - 8.400	0.165-4.200	0.083-2.100
5 – 7	Low	0.005 -0.03	0.005 – 0.03	0.003 – 0.15	0.001 – 0.008

the vibration-based conversion. However, it is suitable for use in impact based energy conversion devices (Balouchi 2013).

3 FACTORS AFFECTING THE EFFICIENCY OF PIEZOELECTRIC DEVICES

Several factors affect the efficiency of piezoelectric devices, including the support condition and the geometry of the piezoelectric.

The typical support conditions in use are the simple support and the cantilever types of support similar to the standard support conditions in structural mechanics. The simply supported piezoelectric is associated with a low amplitude of displacement coupled with high frequency. This type is inefficient with regards to harvestable energy and therefore feasible for usage only in systems with frequencies exceeding 50 Hz (Li et al. 2016). This minimum operating frequency of 50Hz is far greater than the natural frequencies of buildings floors ranging from 1Hz to 8Hz, highlighted by Steel for Life & BCSA (2019), thereby, distorting resonance.

Conversely, the cantilever system is more efficient when compared to the simply supported system. This efficiency is due to its flexibility and suitability for low-frequency condition, which is prevalent in low self-vibration frequency. Recently, Tao et al. (2018) stated that low frequency is dependent on the increased mass and low stiffness of a structure and that the mode shapes of eigen-function are independent of the external loads. The associated low frequency implies high eigen-mode amplitudes, and it ultimately indicates that more energy can be harvested using the cantilever type piezoelectric model. The cantilever piezoelectric model had been studied extensively by several researchers, and that has also made it a choice model of considerations. However, the cantilever model needs its frequency to resonate with the natural frequency of the system of application to attain the optimum electricity output (Zhaoa et al. 2018).

The geometry of the piezoelectric also affects its efficiency with the shape and dimensions influencing the flexibility and consequently, the frequencies, displacements and electric output. Furthermore, the number of layers influences the generated voltage as Song et al. (2009) pointed out that voltage increases with increasing number of layers.

4 APPLICATIONS OF PIEZOELECTRIC DEVICES

Several relevant studies have been successful in the application of piezoelectric sensors. With regards to the monitoring of bridges in the US, Galchev et al. (2011) studied the use of piezoelectric materials to harvest energy from the vibrations generated by traffic on the bridges, then used in turn to power the sensors used in SHM of the bridges. Also, Sazonov et al. (2009) earlier carried out a study on the use of electromagnetic generators to harvest energy from the vibration due to the passing traffic on bridges, but this is not as efficient as piezoelectric devices due to associated wears. The successes in bridges are based on the associated displacements sufficient to be detected by sensors and harvested through the transduction mechanism adopted. The displacements due to vehicular and human traffic are also large enough to produce voltages that can power sensors and transmits relevant data. In addition, Fazzolari & Violi (2019) studied the active vibration control of structural components by using piezoelectric layers.

Significant success has also been recorded in the application of piezoelectric devices in buildings, specifically on dance floors, where considerable impact is expected (Gupta et al. 2016). Elhalwagy et al. (2017) also carried out a feasibility study on the utilisation of piezoelectric energy harvesting floor in buildings using the footsteps pressure. They highlighted the capability of piezoelectric devices to trap a significant quantity of energy in public spaces with significant traffic, which can power electrical lighting devices and screens. In their report, they opined that there had been limited studies on harnessing the energy from vibrations in building floors especially in private offices and residential spaces noting that the technology had been used in a limited number of building floor projects.

5 CONCLUSIONS AND FUTURE DEVELOPMENT

For the realisation of safe and smart structures, the application of SHM is of paramount importance. However, smart structures may consume a significant amount of energy and therefore needs alternative renewable sources and efficient usage to enhance sustainable development. Self-powered sensors for monitoring building floors have been considered only in a limited number of studies, and therefore, there is a need for adequate research.

Considering the existing methods used in harvesting energy from vibration, the piezoelectric transduction mechanism is the most efficient. The PZT piezoelectric material has also been identified to be the most suitable type for application in sensors due to its efficiency. The cantilever support mode is also found to be the superior support condition in building floors as it is suitable for low frequency-band structures, and it is associated with higher displacement, which gives higher voltage output. The influence of the geometry is also an influencing factor with an increase in the number of layers increasing the amount of voltage generated.

However, the sufficiency of the vibration amplitudes and the corresponding level of energy harvestable from the floors for practical use have not been extensively studied and thus are yet to be evaluated. Future research to address these gaps will be of great importance to the engineering industry. Questions concerning why the application of self-powered sensors in SHM has only been successful in structures associated with high vibrations amplitude like bridges and dance floors and what could be hindering the use of energy harvested in building floors, from powering the sensors used in the monitoring of these buildings, need answers. In order to answer these salient questions, a numerical and experimental investigation must be made into the efficiency of self-powered sensors for vibrations monitoring of building floors.

REFERENCES

Balouchi, F. 2013. Footfall energy harvesting : Footfall energy harvesting conversion mechanisms.

Beeby, S.P. Tudor, M.J. & White, N.M. 2006. Energy harvesting vibration sources for microsystems applications. *Measurement Science and Technology* 17: 175-195.

Brownjohn, J.M.W. & Zheng, X. 2001. Discussion of human resonant frequency. *Proceeding of SPIE-The International Society for Optical Engineering* 4317: 469-474.

Butt, Z. & Pasha, R.A. 2016. Effect of temperature and loading on output voltage of lead zirconate titanate (PZT-5A) piezoelectric energy harvester. *IOP Conference Series: Materials Science and Engineering* 146(1).

Chen, J. Li,G & Racic, V. 2016. Acceleration response spectrum for predicting floor vibration due to occupants jumping. *Engineering Structures* 112:71-80.

Chew, Z. J. Ruan, T. Zhu, M. Bafleur, M. & Dilhac J. 2017. Single piezo-electric transducer as strain sensor and energy harvester using time-multiplexing operation. *IEEE Transactions on Industrial Electronics, Institute of Electrical and Electronics Engineers* 64(12): 9646-9656.

Elhalwagy, A.M. Ghoneem, M.Y.M. & Elhadidi, M. 2017. Feasibility study for using piezoelectric energy harvesting floor in buildings' interior spaces. *International Conference - Alternative and Renewable Energy Quest, AREQ 2017, 1-3 February 2017, Spain, Energy Procedia* 115: 114-126.

Fazzolari, F. A. & Violi, L. 2019. Dynamic response of functionally graded carbon-nanotube reinforced piezoelectric composites with active control. *Second International Conference on Mechanics of Advanced Materials and Structures, Nanjing* 19-22.

Galchev, T.V. McCullagh, J. Peterson, R. L. & Najafi, K. 2011. Harvesting traffic-induced vibrations for structural health monitoring of bridges. *Journal of Micromechanics and Microengineering* 21(10).

Giuliano, A. & Zhu, M. 2014. A passive impedance matching interface using a PC permalloy coil for practically enhanced piezoelectric energy harvester performance at low frequency. *IEEE Sensors Journal* 14(8): 2773-2781.

Gupta, A. Imran, M. Agarwal, R. Yadav, R. Jangir, P. & Poonia, R. 2016. Energy harvesting through dance floor using piezoelectric device. *International Journal of Engineering and Management Research* 6(2): 36-39.

Jung, B.C. Huh, Y.C. & Park, J.W. 2018. A Self-Powered, Threshold-Based Wireless Sensor for the Detection of Floor Vibrations. *Sensors* 18(12).

Khan, F. U. & Ahmad, I. 2016. Review of Energy Harvesters Utilizing Bridge Vibrations. *Hindawi Shock and Vibration*.

Li, Z. Zhou, G. Zhu, Z. & Li, W. 2016. A study on the power generation capacity of piezoelectric energy harvesters with different fixation modes and adjustment methods. *Energies* 9(2).

Liu, C. Teng, J. & Wu, N. 2015. A wireless strain sensor network for structural health monitoring. *Shock and Vibration* 13.

Lynch, J. P. 2006. An overview of wireless structural health monitoring for civil structures. *Philosophical Transactions of the Royal Society A: Mathematical, Physical and Engineering Sciences*.

Richard, C.D. Anderson, M.J. Bahr, D.F. & Richard, R.F. 2003. Efficiency of energy conversion for devices containing a piezoelectric component. *Journal of Micromechanics and Microengineering* 14: 717-721.

Song, H.C. Kim, H.C. Kang, C.Y. Kim, H.J. Yoon, S.J. & Jeong, D.Y. 2009. Multilayer piezoelectric energy scavenger for large current generation. *Journal of Electroceramics* 23:301-304.

Steel for Life & BCSA. 2019. 'Floor Vibration'. Steel construction.

Tao, Z.L. Chen, G.H. & Bai, K.X. 2019. Approximate frequency–amplitude relationship for a singular oscillator. *Journal of Low Frequency Noise, Vibration and Active Control* 1-5.

Wu, T. Yao, W. Wang, S. & Tsai, M. 2010. Analysis of high efficiency piezoelectric floor on intelligent buildings. *SICE Annual Conference 2010, Proceedings of IEEE* 1777–1780.

Yoon, H. Youn, B.D. & Kim, H.S. 2016. Kirchhoff plate theory based electromechanically-coupled analytical model considering inertia and stiffness effects of a surface-bonded piezoelectric patch. *Smart Materials and Structures* 25(2).

Zhang, S. Yan, B. Luo, Y. Miao, W. & Xu, M. 2015. An enhanced piezoelectric vibration energy harvesting system with macro fiber composite. *Shock and Vibration* 1-7.

Zhaoa, Q. Liua, Q. Wangb,L. Yanga, H. & Caod, D. 2018. Design method for piezoelectric cantilever beam structure under low frequency condition. *International Journal of Pavement Research and Technology* 11:153-159.

Zhu, L. Pi, Z. Zhang, W. & Wu, D. 2013. Simulation analysis of transient piezoelectric properties of PvDF structures for energy conversion applications. *The 8th Annual IEEE International Conference on Nano/Micro Engineered and Molecular Systems, 7-10 April 2013, China* 1030-1033.

Numerical evaluation of reinforced concrete frames with corroded steel reinforcement under seismic loading: A case study

F. Pugliese, L. Di Sarno & A. Mannis
University of Liverpool, Liverpool, UK

ABSTRACT: This paper presents a numerical evaluation of reinforced concrete (RC) framed structures exposed to different levels of corrosion. A non-linear finite element fibre-based approach is used. The numerical model utilized in the numerical simulations was first compared and validated against a set of experimental test results from the literature. The framed model is then assessed with respect to the European Standards in order to evaluate the impact of corrosion on the damage associated to different Limit States. The results showed that the corroded RC columns are subjected to a significant reduction, especially for high levels of corrosion and total exposure of the column, both of shear strength and ductility. Finally, a typical existing RC building located in Italy was analyzed through linear and non-linear analyses to evaluate the response of an actual building when exposed to aggressive environments. The results showed that corrosion reduces both the base shear capacity and ductility of the sample framed building. Particularly, while corroded columns decrease mainly the base shear capacity, corroded beams reduce significantly the ductility of the aged building. Moreover, time-history analyses showed that earthquakes with a Peak Ground Acceleration (PGA) of greater than 0.23 g increased the inter-storey displacements of the corroded RC structure, while earthquakes with a PGA of less than 0.23 had a hardly noticeable influence.

1 INTRODUCTION

Many reinforced concrete (RC) structures built in the 60s and 70s tend to be in poor condition and functionally obsolete due to ageing and degradation phenomena, which are related to the exposure to chemical and physical aggressive agents, either in seismic prone-zone (Bhide 1999) or non-seismic prone-zone (Di Sarno & Pugliese 2019). The durability of RC structures, which is strictly related to the environmental exposure of the structure, is the capacity of the concrete to protect steel reinforcements from corrosion processes due to the attack of aggressive agents present in the air, water and soil. As a result, to guarantee the integrity of the RC structures, it is essential to study the concrete composition in terms of consistency and environmental exposure. Standards and codes (EN 1992-1-1 & D.M. 14-01-2008) provide some important measures to ensure the durability of RC structures over the years, especially for damage due to carbonation-induced and chloride-induced corrosion. However, many RC structures show significant deficient conditions due to high levels of corrosion. As a result, the reduction of the mechanical properties of both concrete and steel rebars are among the main consequences of the corrosion impact (Francois et al. 2018, Coronelli & Gambarova 2004, Zhang et al. 1995, Andrade et al. 1991). A few studies have been conducted on the seismic performance of RC sample framed buildings exposed over the years (Zhang et al. 2018, Biondini et al. 2011), but many gaps remain to bridge. This paper aims to analyse the seismic response of typical RC frames exposed to different levels of corrosion. Additionally, a case study representing an ordinary RC building located in Italy is presented. Pushover and time-history analyses were conducted to seismic performance of a common RC structure when exposed to corrosion.

2 MECHANICAL PROPERTIES OF CONCRETE AND STEEL REINFORCEMENT

2.1 Concrete

Due to the effect of corrosion, the increase in volume of the rust, micro-cracking in the core and cover spoiling-off are the main consequences of the concrete deterioration. Generally, concrete is affected by two different exposures: direct exposure, such as carbonate-based and indirect exposure, such as chloride-based. As a result of these two exposures, the reduction of the concrete compressive strength is the leading cause of the damaged concrete. Although, Coronelli & Gambarova (2014) proposed a method to account for the impact of corrosion on the concrete compressive strength based on the numerical evaluation of corroded RC beams, this relationship does underestimate the ultimate capacity of corroded RC columns.

$$\beta = f_c^*/f_c = \frac{1}{\left[1 + \left(K\frac{2\pi X n_{bars}}{b\varepsilon_{c2}}\right)\right]} \qquad (1)$$

where f_c^* represents the corroded compressive strength, f_c the uncorroded compressive strength, K a constant equal to 0.1 for medium rebar, X the corrosion penetration, b the width of the cross-section, ε_{c2} strain at the peak and n_{bars} the number of steel reinforcement in the compressive zone. Therefore, a new method, which consists in splitting the concrete cross-section into three different blocks whilst accounting for the impact of corrosion on the uneffective and effective concrete core, has been proposed by Di Sarno & Pugliese (2019). This has been validated against experimental results providing efficient and accurate results and is herein used.

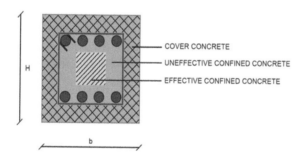

Figure 1. Concrete blocks.

Accordingly:

$$f_c^* = \frac{\beta f_c A_{CC} + \beta f_c A_{UCC} + f_c A_{ECC}}{A_{CC} + A_{UCC} + A_{ECC}} \qquad (2)$$

where f_{cc} is the confined compressive strength.

2.2 Steel reinforcement

Many experimental campaigns have been conducted to evaluate the impact of corrosion on both bare and embedded into concrete steel reinforcements (more details can be found in Di Sarno & Pugliese (2019). As a result, different relationships for yielding stress, ultimate stress and ultimate strain related to the rate of corrosion have been proposed. A comprehensive review of the models for corroded steel rebars present in the literature has been carried out by Di Sarno & Pugliese (2019) who provided an accurate relationship for use in the numerical evaluation of corroded RC components. The general equations have the same form, i.e.

$$f_y^* = fy[1 - \beta CR[\%]] \tag{3}$$

where f_y^* is the corroded yielding stress, f_y, the un-corroded yielding stress, β, the experimental coefficient and $CR[\%] = (M_0-M_C)/M_0 CR \% = M0-MCM0$, the mass loss based on mass before (M_0) and after corrosion (M_C). The relationship proposed by Imperatore et al. (2017) is herein used to take into account the different rate of corrosion.

3   NUMERICAL EVALUATION OF A CORRODED RC COLUMN

3.1   *Reference test*

As a reference test, the corroded RC column under cycling loading tested by Meda et al. (2014), which represents a typical column of a RC structure built in Italy in the 1960s was used. The column had a cross-section of 300 x 300 mm² with a concrete compressive strength of 20 MPa and four Φ16 mm longitudinal ribbed steel reinforcements with a yielding stress of 521 MPa and hardening ratio of 0.005. The transverse reinforcements consist of Φ8 mm stirrups with a spacing of 300 mm. One column was kept as a non-corroded reference, while the second was subjected to a theoretical 20% corrosion on the longitudinal reinforcement. The results were shown in the load-drift ratio plot both for un-corroded and corroded column.

3.2   *Numerical modelling*

The RC column has been implemented in an advanced Finite Element Software [SeismoStruct] through a non-linear Fibre-frame element by using five Gauss-Lobatto points of integrations. The concrete has been modelled using the stress-strain relationship given by Chang & Mander (1994). This concrete model is able to simulate the behavior of both core and cover concrete by modifying the peak strain and the compressive strength as the shape of the constitutive model remains the same. The steel rebars were modelled by using the constitutive model of Monti & Nuti (1992) with the mechanical properties provided by Meda et al. (2014). Hence, the concrete and steel models were modified through the relationships given by Di Sarno & Pugliese (2019) and Imperatore et al. (2017) respectively. Finally, monotonic pushover analyses were performed along x and y, in positive and negative directions, and the outcomes were validated against experimental results (Figure 2).

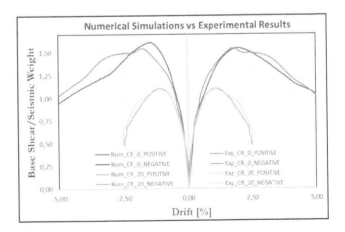

Figure 2.   Monotonic negative-positive pushover.

The results related to both corroded and un-corroded RC column are summarised in the load-drift diagram. The model is able to predict with an excellent accuracy the response of the corroded RC column exposed to a monotonic loading with an error less than 3% in terms of shear strength.

4 CASE STUDY: THREE-STOREY RC STRUCTURE

4.1 *Description*

A three-storey RC ordinary building was considered as testbed for this study. The building is situated in Ortona (Italy), near the sea and consists of 9 columns per floor with a cross-section of 400x900 mm² reinforced with Φ18mm and Φ16mm longitudinally and Φ8mm with spacing 300mm transversally. The beams have different cross-sections and longitudinal ribbed reinforcements composed mostly by Φ16 mm and Φ14 mm. The concrete compressive strength was 20 MPa both for columns and beam, while the steel reinforcement had a yielding stress of 444 MPa. The decks have been implemented through rigid-diaphragms so that they have infinite in-plane stiffness properties, and exhibit neither membrane deformation not report the associated forces, while all the joints were connected through fully-supported-rigid-connection (All degrees of freedom are restrained) to the ground. An accurate loading analysis was conducted and the masses implemented into each node (loading-range (2.38 tons; 45.54 tons). Corrosion has been applied to different scenarios: only columns, only beams and, both columns and beams. Potentially, this procedure allows the evaluation of the impact of corrosion on different RC elements. Two different analyses have been conducted: Non-Linear Static Analysis (Pushover Analysis) and non-linear time-history analysis. The latter time-history analyses have been conducted through real-ground motions using the so-called spectrum-compatibility analysis. Basically, the spectrum-compatibility analysis allows the user to consider all signals that match the elastic spectrum provided by the Italian Code (D.M. 14-01-2008). A reliable software called REXEL (Iervolino et al. 2009) has been utilised for generating the spectrum-compatibility signals (Figure 4). The model of the ordinary RC structures is given in Figure 3.

4.2 *Pushover analysis*

The non-linear Static Analysis (most commonly known as Pushover Analysis) is the most common method used in Structural Engineering to evaluate the response of a RC structure to a lateral load pattern. The non-linearity was assigned through the Fibre-based element frame that spreads the non-linearity through the cross-section. The Push-over analyses were performed in both x and y directions considering the mass distribution of the modal shapes of the RC structure (Adaptive Pushover Analysis) five levels of corrosion rate (CR [%] = [0, 5, 10, 15, 20]).

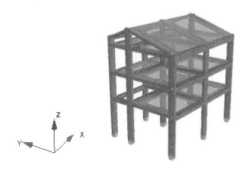

Figure 3. Finite model of the sample structure implemented in SeismoStruct.

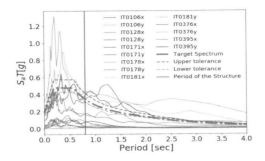

Figure 4. Spectrum-compatible accelerograms (Rexel Output).

The Pushover analyses were performed in both x and y directions considering the mass distribution of the modal shapes of the RC structure (Adaptive Pushover Analysis) five levels of corrosion rate (CR [%] = [0, 5, 10, 15, 20]). The results have shown that the base shear along the x-axis is reduced by 17% as the corrosion rate increases, while the ductility remains essentially the same (Figure 5a). On the other hand, the capacity curves along the y-axis showed both a reduction of the base shear (15%) and ductility, which becomes critical when the corrosion rate reaches 15% and 20% (Figure 5b). Moreover, the impact of corrosion was also considered for the beams to evaluate the response of a RC structure under different corrosion expositions (Figure 6a and Figure 6b). The results, as expected, showed a reduction of the base shear by 15% along x-axis and 12% along y-axis as the corrosion rate goes up. However, ductility is more significant in the case of all beams exposed to corrosion. Particularly, the ductility is reduced by half when the corrosion rate is between 15% and 20%.

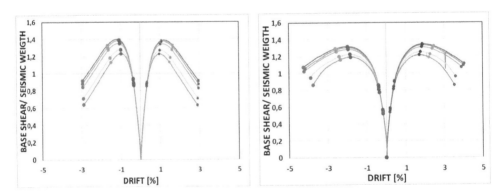

Figure 5. Capacity curve: a) X-axis b) Y-axis (only columns exposed).

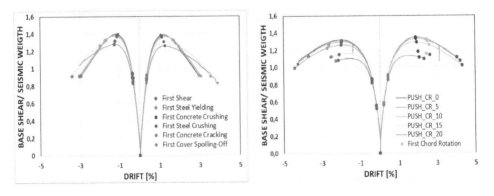

Figure 6. Capacity curve: a) X-axis b) Y-axis (only beams exposed).

4.3   Time-history analyses

Dynamic non-linear analysis is commonly used to predict the nonlinear response of a structure subjected to earthquake ground motions. The results are herein presented in terms of Mean-Relative Storey-Displacements (Figure 8, Figure 9a and Figure 9b) Maximum Base Shear (Figure 11 and 12) and Maximum Displacement at the top of the building (Figure 7a and Figure 7b). All the storey-displacements have been combined using the following formulation:

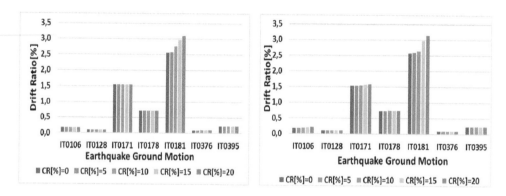

Figure 7.   Top displacement; a) columns corroded b) beams corroded.

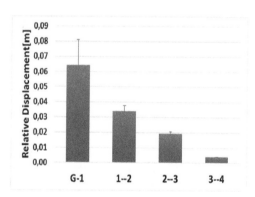

Figure 8.   Mean relative storey-displacements (uncorroded structure).

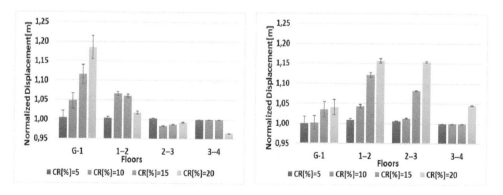

Figure 9.   Mean-relative storey-displacements. a) Corroded columns b) corroded beams.

117

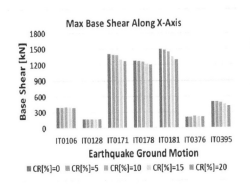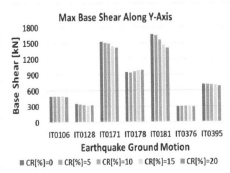

Figure 10. Max base shear for all the signals (only columns corroded).

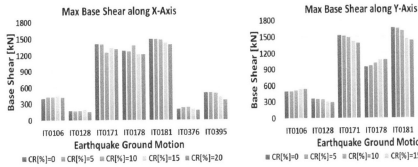

Figure 11. Max base shear for all the signals (only beams corroded).

$$Disp = \sqrt{Dx^2 + Dy^2} \qquad (4)$$

The time-history analyses provided an essential response of the corroded structure by means of Base Shear, Max Displacement at the top of the building and Storey-Displacements. In this study, the records were chosen using the spectrum-compatibility analysis, which consists of retrieving all the real accelerograms compatible with the elastic spectrum defined by using the Italian Code during the Limit State of the Collapse. Storey displacements, which are considered as the most useful response for time-history analysis, were obtained along the height of the RC building and presented in the Storey Displacement-Time plots. The plotted curves (Figure 7a, 7b,9a and 9b) clearly show an increase in the Storey displacement where the structure is less stiff over time. The maximum relative displacement Ground-Roof floor (Figure 7a and 7b) is mainly affected when the earthquake has a PGA more than 0.23g (IT0181 Event). As a result, the corroded RC structure can reach a collapse condition earlier than the un-corroded building. Finally, Figure 10 and Figure 11 showed how the base shear was influenced by the corrosion impact for the different ground motions herein considered.

5 CONCLUSIONS

The objective of the present study was to evaluate the numerical response of a RC building affected by different values of the corrosion rate. The results of the Pushover analyses have shown that the base shear decreases with the increase of the corrosion rate when both beams and columns are exposed to aggressive environments. In particular, the maximum reduction of the base shear when only the columns of the RC building are exposed to corrosion is 17%, while the beams showed a smaller reduction equal to 15%. In addition to this, the ductility was examined. The results showed that the ductility was mainly affected by the impact of the corrosion on the beams. In particular, the ductility was reduced by half when the corrosion rate was 15% or 20%, especially along y-axis where the structure is less stiff. On the other hand, smaller variations of ductility, which can be seen especially along y-axis, were obtained when the case of corroded columns was considered. Moreover, time-history analyses were performed using the Spectrum-Compatibility analyses and seven ground motions were accounted. The main results were plotted using the Storey-displacements and the base shear. The Storey-Displacements were combined using the equation (4). The increase of the corrosion up to 20% had significant results only when the PGA of the ground motion was more than 2.3%, while signals with a PGA less than 2.3 had an irrelevant influence. In particular, the ground motion IT0181 showed an increase of the Storey-Displacement of 19% as the corrosion goes up until 20%. The results also showed the increase of the mean interstorey displacements, which are higher (15% and 18%) for corroded columns compared with corroded beams (12% and 15%), when the structure is exposed to a significant corrosion rate. Finally, the base shear of the corroded RC building showed a relevant decrease as the corrosion increased, especially for corrosion rates of 10%,15% and 20%, i.e. moving from 1600 kN to 1400 kN in the IT0181 Earthquake. The main consequences of these results can be summarized in earlier failure condition compared with an uncorroded RC building, and thus a significant decrease of the safety and the lifetime of a RC Structure.

ACKNOWLEDGEMENTS

The authors would like to acknowledge the gracious support of this work through the EPSRC and ESRC Centre for Doctoral Training on Quantification and Management of Risk and Uncertainty in Complex Systems Environments Grant No. (EP/L015927/1)

REFERENCES

Andrade, C. Alonso, C. Garcia, F. & Rodriguez, J. 1991. Remaining Lifetime of Reinforced Concrete Structures: Effect of Corrosion in the Mechanical Properties of the Steel. *Life Prediction of Corrodible Structures*, NACE, Cambridge, UK. 12/1-12/11

Bhide, S. 1999. *Material usage and condition of existing bridges in the (U.S)*. Skokie, Ill.

Biondini, F. Palermo, A. & Toniolo, G. 2011. Seismic performance of concrete structures exposed to corrosion: case studies of low-rise precast buildings. *Structure and Infrastructure Engineering* 7(1–2): 109–111

Chang, G.A. & Mander, J.B. (1994). Seismic energy-based fatigue damage analysis of bridge columns: Part 1 – Evaluation of seismic capacity, *NCEER Technical Report No. NCEER-94-0006*, State University of New York, Buffalo, N.Y.

Coronelli, D. & Gambarova, P. 2004. Structural assessment of corroded reinforced concrete beams: Modeling guidelines. *Journal of Structural Engineering* 130(8): 1214-1224.

D.M. 14.01. 2008. Norme Tecniche per le Costruzioni e Circolare 2. 02.09, N.617: *Istruzioni per l'applicazione delle nuove Norme Tecniche per le Costruzioni*. (in Italian)

Di Sarno, L. & Pugliese, F. 2019. Critical review of models for the assessment of the degradation of reinforced concrete structures exposed to corrosion. *Conference SECED 2019*, Earthquake Risk and Engineering towards a Resilient World

EN 1992-1-1. 2005. Eurocode 2: Design of concrete structures - *Part 1-1: General rules and rules for buildings E. for Standardization*. Brussels, EN, CEN.

François, R. Laurens, F. & Deby, F. 2018. Corrosion and its Consequences for Reinforced Concrete Structures. *ISTE Press* ISBN 978-1-78548-234-2

Iervolino, I. Galasso C. & Cosenza, E. 2009. REXEL: computer aided record selection for code-based seismic structural analysis. *Bulletin of Earthquake Engineering* 8:339-362.

Imperatore, S. Rinaldi, Z. & Drago C. 2017. Degradation relationships for the mechanical properties of corroded steel rebars. *Construction and Building Materials* 148: 219-230

Meda, A. Mostosi, S. & Rinaldi, Z. 2014. Experimental evaluation of the corrosion influence on the cyclic behaviour of RC columns. *Engineering Structures* 76: 112-123. https://www.sciencedirect.com/science/article/pii/S014102961400399X - !

Monti, G. & Nuti, C. 1992. Nonlinear cyclic behaviour of reinforcing bars including buckling. *Journal of Structural Engineering* 118(12): 3268-3284.

SeismoStruct, FEM Software, *SeismoSoft, Earthquake Engineering Software Solutions*, Piazza Castello 19, 27100 Pavia – Italy, info@seismosoft.com

Zhang, M. Liu, R. Li, Y. & Zhao, G. 2018. Seismic performance of a corroded reinforce concrete frame structure using pushover method. *Advances in Civil Engineering.*

Zhang, P. S. Lu, M. & Li, X. 1995. The mechanical behaviour of corroded bar. *Journal of Industrial Buildings* 25:41-44.

Sustainable Buildings and Structures: Building a Sustainable Tomorrow – Papadikis et al. (Eds)
© 2020 Taylor & Francis Group, London, ISBN 978-0-367-43019-1

Nonlinear dynamic analysis on progressive collapse resistance of a multi-story reinforced concrete building with slab

S. Nyunn, J. Yang, F.L. Wang & Q.F. Liu

State Key Laboratory of Ocean Engineering, Shanghai Jiao Tong University, Shanghai, China
School of Naval Architecture, Ocean and Civil Engineering, Shanghai Jiao Tong University, Shanghai, China
Collaborative Innovation Centre for Advanced Ship and Deep-Sea Exploration (CISSE), Shanghai, China

ABSTRACT: In a reinforced concrete (RC) frame structure, beam-slab floor system can significantly increase the resistance of progressive collapse under damage of a critical load bearing element. In this research, an eight-story RC building is investigated by using nonlinear dynamic analysis in order to study the contribution of slabs in progressive collapse and to evaluate the robustness of the RC structure under column damage at different locations. Two reinforced concrete frame models, one of which is simulated with slabs while the other without slabs, are analyzed. As per GSA regulation, the potential of progressive collapse is assessed by considering damaged column at three different locations, namely, corner, exterior and interior. Analysis results show that slabs help in transferring the load and enhance the structural resistance after losing the column. In the model without slabs, hinge formation is more severe and the displacement response is considerably increased in comparison with beam-slab system.

1 INTRODUCTION

Progressive collapse is the situation in which initial local damage results in disproportionate collapse of the whole structure or a major part of it. Researchers and design engineers began to notice the risk of progressive collapse after the partial failure of the Ronan Point building in London in 1968 (Byfiel et al. 2014). In addition to this, other serious collapses have also been found in the past such as the U.S. Marine Barracks (Beirut 1983), the A.P. Murrah Federal Building (Oklahoma 1995) and World Trade Center (New York 2001). These instances have demonstrated that conventional designed buildings may not be robust enough to withstand local failure in order to mitigate structural collapse (Adam et al. 2018). Therefore, several design methods for progressive collapse resistance have been proposed in different design codes and guidelines such as ASCE (ASCE 2014), Eurocode (BS EN 1991-1-7: 2006), the General Service Administration (GSA 2013) and Department of Defense (DOD 2016).

In recent years, numerous studies of experimental and numerical works have been done in the research field of progressive collapse. Generally, experimental tests have mainly focused on the sub-assemblages and numerical methods have been performed for investigation of multi-story buildings or sometimes for validation of experimental studies. Tsai and Lin studied the evaluation of progressive collapse for an eleven-story reinforced concrete structure by using nonlinear static and nonlinear dynamic analyses (Tsai & Lin 2008). Likewise, Zahrai & Ezoddin (2014) conducted four analysis procedures (linear static, linear dynamic, nonlinear static and nonlinear dynamic) and they concluded that dynamic analysis method was more accurate and easier to carry out than other methods (Zahrai & Ezoddin 2014). Besides reinforced concrete buildings, progressive collapse analyses for multi-story steel structures were also performed by Kwasniewski and Tavakoli (Kwasniewski 2010) (Tavakoli & Rashidi Alashti 2013).

Although progressive collapse analysis of frame models had been widely studied in recent years, the evaluation of slabs contribution under column damage situations is still limited, especially for high rise buildings. It is noticed that some experimental studies of beam-slab substructures have been performed in the literature. Ren et al. (2016) and Lim et al. (2017) tested RC sub-assemblage specimens with and without slabs and studied the influence of slabs in the behavior of collapse. It was investigated from their researches that beam-slab system can significantly improve the load-carrying capacity of the specimens. In this present research, progressive collapse resistance of an 8-study reinforced concrete (RC) structure is studied by performing non-linear dynamic analysis based on GSA guideline. The building is analyzed under three column damage scenarios which include corner, exterior and interior column damage conditions.

2 MODELING OF RC FRAME STRUCTURE

Two eight-story reinforced concrete (RC) buildings are simulated by using finite element commercial software SAP2000 (SAP2000® 2019) and designed in accordance with ACI 318 specifications (ACI Committee 318 2014). The building configuration and member details of the two buildings are identical. The only difference between the two models is that there are slabs in the first model (Model 1), while the frame is simulated as without slabs in the second model (Model 2). The buildings are symmetric structures with four bays in both longitudinal and transverse directions as shown in Figure 1. Each span is 3.66 m long in both directions. The first floor is 3.66 m in height and the other floors are 3.05 m in height. The structures are considered as fixed supports on the ground. All the columns and beams are modeled as frame elements. The design details of all the columns and beams are shown in Table 1. All the slabs have 100 mm thickness and are modeled as shell elements in Model 1 and there is no slabs in Model 2.

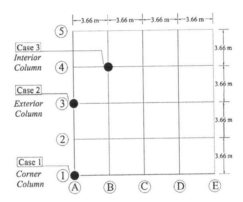

Figure 1. Floor plan and damage cases.

The compressive strength of concrete (f_c') is 17236 kN/mm² and yield strength of reinforcing steels (f_y') is 275790 kN/mm² for both main bars and stirrups. In addition to the own weight, dead load of 1 kN/m² and live load of 2 kN/m² are applied on the slabs in Model 1 and these loads are directly applied on the beams in Model 2. The exterior brick walls and interior partitions of 115 mm thickness is assumed and applied as distributed loads on the beams. In this present research, three critical column damage scenarios at the first floor are studied based on GSA guideline: corner removal (Case 1) and exterior removal (Case 2) and interior removal (Case 3) as shown in Figure 1.

Table 1. Details of beams and columns for Model 1 and 2.

	Floor	Dimension (mm)	Main bar	Stirrup
Beam	1F-8F	230x305	4: 16 mm	8 mm @ 152 mm
Coloumn	1F-2F	457x457	6: 22 mm	8 mm @ 152 mm
Coloumn	3F-4F	406x406	6: 22 mm	8 mm @ 152 mm
Coloumn	5F-6F	355x355	4: 22 mm	8 mm @ 152 mm
Coloumn	7F-8F	305x305	4: 18 mm	8 mm @ 152 mm

3 PROGRESSIVE COLLAPSE ANALYSIS

The nonlinear dynamic analysis is the most accurate method compared to other analysis proced-ures as it considers both inelastic and dynamic mechanism during the analysis. In this study, nonlinear time history analysis is carried out to predict the structural behavior when the column is destroyed quasi-instantaneously. Based on GSA guideline, the loading is initiated from zero and then the gravity load is constantly increased on the intact structure until equilibrium is achieved. After reaching the equilibrium condition, the column is instantaneously removed by assigning an appropriate time function. The time function of sudden column loss is illustrated in Figure 7. The period of 1.5 seconds is taken to ensure the initial equilibrium condition and then the column is suddenly removed. The column removal time has to be less than one tenth of the period for vertical motion as per GSA and Chinese Code also mandates that the time for removal must be smaller than 0.005 seconds (Standard CN 2017). In this analysis, it is taken as 0.001 seconds for all the models. Rayleigh damping of 5% for reinforced concrete buildings is assumed. It has been mentioned in the GSA guideline to apply the load combination of full dead load and 0.25 times live load for column failure condition. Therefore, load combination for this dynamic analysis is equal to dead load plus 0.25 times live load.

In this work, plastic hinge models are applied in both models by adopting default hinge proper-ties of ASCE 41-13 guidelines (ASCE 2014). Under the loss of a load bearing column, structural

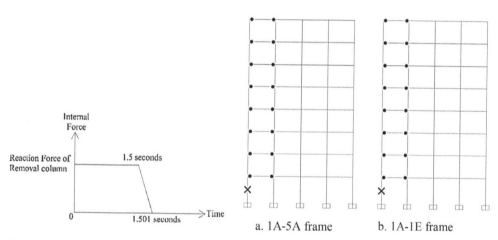

Figure 2. Simulation of column removal time. Figure 3. Hinge formation of Model 1 for Case 1.

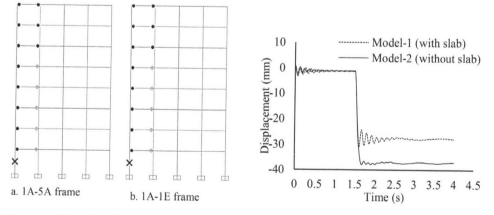

Figure 4. Hinge formation of Model 2 for Case 1. Figure 5. Vertical displacement vs. time for Case 1.

123

damage is dominated by flexural mechanism of beams. Therefore, only flexural plastic hinges are assigned at both ends of the beams and columns are considered to be within elastic region (Tsai & Lin 2008). The damage level of the elements is assessed by various stages of hinge formation after each analysis. Immediate occupancy (IO) hinges is illustrated as black circle and life safety (LS) hinges are illustrated as circle with crossing inside. Furthermore, the vertical displacement at the point of column removal, the most critical location, is also inspected as shown in Figure 5, 8 and 11.

4 DISCUSSION OF RESULTS

4.1 Corner column removal

The plastic hinge formation of Case 1 is illustrated in Figures 3 and 4 for Model 1 and Model 2 respectively. The first plastic hinge occurs at the time of 1.519 s and 1.515 s, respectively, for Model 1 and Model 2 which are activated on the beams connected to the point of damage column. After that, as the time increases, the formation of hinges spread gradually throughout the upper levels in the span of the removed column. It is observed that all the hinges are within the state of Immediate Occupancy (IO) in the building with slabs as displayed in Figure 3. But, in the case of bare frame model, plastic hinges of life safety (LS) level are activated as shown in Figure 4. The time history of displacement at the column removal point is plotted in Figure 5. The maximum deflection is 31.07 mm for Model 1 and 37.91 mm for Model 2. It is computed that displacement is increased 22% in the model with no slabs. The amplitudes are highest at the times equal to 1.602 s for Model 1 and 1.655 s for Model 2. These peak amplitudes are occurred immediately after the damage of column and the vibration rate is steadily decreased due to 5% viscous damping.

4.2 Exterior column removal

Figures 6 and 7 show the formation of plastic hinges for Model 1 and Model 2 under exterior column removal scenario. Two plastic hinges of IO state are simultaneously formed at the time of 1.515 s on the beams linked to the removed column in both models. After formation of initial hinges, the IO hinges are gradually activated to the higher floor levels in the bays of the lost column. It is investigated that only IO level hinges are occurred in Model 1 as shown in Figure 6. In Model 2, at the time of 1.524 s, IO hinges are found on the beams of all the floor level. Then, LS level hinges are activated at the locations of first plastic hinge when the time is reached to 1.604 s. The final distribution of plastic hinges for Model 2 is shown in Figure 7. It can be seen that LS hinges are found in the lower five levels. In the frame with

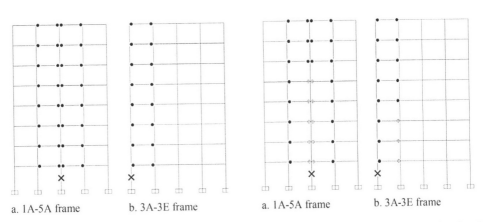

| a. 1A-5A frame | b. 3A-3E frame | a. 1A-5A frame | b. 3A-3E frame |

Figure 6.   Hinge formation of Model 1 for Case 2   Figure 7.   Hinge formation of Model 2 for Case 2

slabs, the maximum deflection is 23.04 mm at 1.577 s which is 25.84 % smaller than Case 1. The maximum deflection is 33.88 mm at 1.642 s for the model without slabs and it is 10.63% smaller than Case 1. Furthermore, it is calculated that 32% displacement is increased due to loss of slabs in this exterior column removal condition.

4.3 Interior column removal

For interior column removal case, plastic hinge distribution of Model 1 and 2 are displayed in Figures 9 and 10. It is observed that more plastic hinges are found in interior column damage case than corner and exterior column removal cases. This is because more beams are connected to the column removal point in interior column damage scenario. Initial hinges are activated immediately after the column lost at the time of 1.515 s and 1.512 s for Model 1 and Model 2 respectively. As observed in Case 1 and 2, hinges are within IO state in Model 1 and LS level hinges are occurred in Model 2. In Model 2, at the time of 1.581 s, hinges are reached to LS state in the lowest level. Then, LS level plastic hinges are gradually distributed to the higher floor levels. The displacement of the structure with slabs is 25.13 mm which is 19% smaller than Case 1 and is nearly equal to Case 2. In the structure without slabs, the displacement is 52.87 mm which is 28.3% higher than Case 1 and 35.92% higher than Case 2. To this point, it is important to notice that the effect of

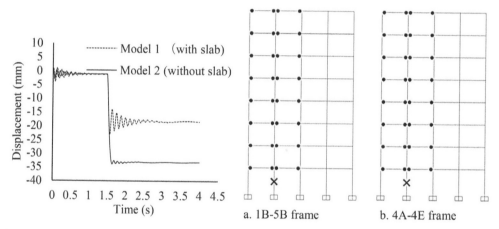

Figure 8. Vertical displacement vs. time for Case 2.   Figure 9. Hinge formation of Model 1 for Case 3.

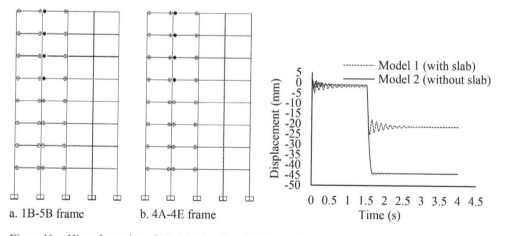

Figure 10. Hinge formation of Model 2 for Case 3. Figure 11. Vertical displacement vs. time for Case 3.

125

slabs is more significant in Case 3 than Case 1 and 2. The displacement is increased approximately 50% in Model 2 in comparison with Model 1 due to the loss of slabs in the structure.

5 SUMMARY AND CONCLUSION

This research work presents nonlinear dynamic analysis of reinforced concrete building to evaluate the potential of progressive collapse with and without the presence of slabs. The above analysis results reveal that the corner column removal is the most vulnerable among the three column damage scenarios in the structure with slabs. However, in the structure without slabs, the displacement is highest and the plastic hinge formation is the most severe in the interior column removal case. So, the risk of progressive is highest in the interior column removal scenario than the other two column removal scenarios analyzed in Model 2. Out of the three cases considered, slabs contribution is the most significant in Case 3 which results in large displacements. Hinge formation is within IO state in Model 1 which implies that the potential of progressive collapse is very low in the model with slabs; however, LS level hinges are formed in the model without slabs. Hence, Model 2 is more susceptible to progressive than Model 1 as expected. Therefore, it is concluded from this research that the building is more susceptible to progressive collapse when the slabs are not included in the structure.

REFERENCES

ACI COMMITTEE 318. 2014. *Building code requirements for structural concrete and commentary*, Farmington Hills, Michigan, American Concrete Institute.

Adam, J.M. Parisi, F. Sagaseta, J. & Lu, X. 2018. Research and practice on progressive collapse and robustness of building structures in the 21st century. *Engineering Structures* 173: 122-149.

ASCE 2014. *Seismic Evaluation and Retrofit of Existing Buildings*, Reston, Virginia, American Society of Civil Engineers.

BS EN 1991-1-7:2006 2006. *Eurocode 1 - Actions on structures - Part 1-7: General actions -Accidental actions*, London, UK, BSI.

Byfiel, M. Mudalige, W. Morison,C. & Stoddart, E. 2014. A review of PC research and regulations. *Structures and Buildings* 167: 447-455.

DOD 2016. *Design of buildings to resist progressive collapse. Washington*, Washington, D.C, Unified Facilities Criteria (UFC).

GSA 2013. *Progressive collapse analysis and design guidelines for new federal office buildings and major modernization projects*, Washington, D.C, General Service Administration.

Kwasniewski, L. 2010. Nonlinear dynamic simulations of progressive collapse for a multistory building. 32(5): 1223-1235.

Lim, N.S. Tan, K.H. & Lee, C.K. 2017. Experimental studies of 3D RC substructures under exterior and corner column removal scenarios. *Engineering Structures* 150(1): 409-427.

Ren, P. Li, Y. Lu, X. Guan, H. & Zhou, Y. 2016. Experimental investigation of progressive collapse resistance of one-way reinforced concrete beam–slab substructures under a middle-column-removal scenario. *Engineering Structures* 118(1): 28-40.

SAP2000® Version 21.0.2 2019. *Linear and nonlinear static and dynamic analysis and design of three-dimensional structures*, Berkeley, California, USA.

Standard CN. 2017. *Technical standard for assembled buildings with concrete structure*.

Tavakoli, H. R. & Rashidi Alashti, A. 2013. Evaluation of progressive collapse potential of multi-story moment resisting steel frame buildings under lateral loading. *Scientia Iranica* 20(1): 77-86.

Tsai, M.H. & Lin, B.H. 2008. Investigation of progressive collapse resistance and inelastic response for an earthquake-resistant RC building subjected to column failure. *Engineering Structures* 30(12): 3619-3628.

Zahrai, S.M. & Ezoddin, A.R. 2014. Numerical study of progressive collapse in intermediate moment resisting reinforced concrete frame due to column removal. *Civil Engineering Infrastructures Journal* 47(1): 71-88.

Sustainable Buildings and Structures: Building a Sustainable Tomorrow – Papadikis et al. (Eds)
© 2020 Taylor & Francis Group, London, ISBN 978-0-367-43019-1

Numerical modelling of structural behaviour of transverse beam under interior column removal scenario

S. Shah & F. Wang
Shanghai Jiao Tong University, Shanghai, China

J. Yang
Shanghai Jiao Tong University, Shanghai, China
University of Birmingham, Birmingham UK

ABSTRACT: This paper presents a numerical investigation on structural behaviour of two specimens: one without transverse beam and another with transverse beam, under interior column removal scenario subjected to progressive collapse. Finite element models are developed for these two cases and initial stiffness, first peak load, ultimate load capacity and transverse beam effects are analyzed for both. Developed finite element models are validated by comparing the results with lab tests in literature. Results show that the transverse beam increases the first peak load by 72.2%. Furthermore, parametric studies were carried out to examine the effects of beam longitudinal reinforcement ratio, beam section depth and concrete compressive strength. Results under parametric studies show that increase in beam longitudinal reinforcement ratio increases initial stiffness, flexural and ultimate capacity and while increase in beam section depth and concrete compression strength increases the capacity only during the beam flexural stage.

1 INTRODUCTION

Progressive collapse is a catastrophic event, and these losses are initiated by gas explosions, accidental fire or terrorist attacks. In such a situation a failure of whole or small part of structure occurs by failure of small part of structure such as column or masonry wall. When such structures are lost, loads are supposed to safely pass through the adjacent structure to prevent progressive collapse. Government bodies such as, General Service Administration (GSA 2003) and Department of Defense (DoD 2009) published guidelines to provide necessary requirements for design of multistory building to prevent progressive collapse. International Building Code (ICC 2009) and American Concrete Institute Building Code (318 2008) introduced a minimum level of structural integrity requirement of multistory buildings.

For reinforced concrete structures, academic researchers such as Yu & Tan (2013) tested six RC beam-column sub assemblages consisting of two singly bay beams, two end column studs and middle joint quasi-statically under middle column removal scenario and found out with adequate axial restraints, both compressive arch action(CAA) and catenary action(CA) could be mobilized significantly increasing the structural resistance beyond the beam flexural capacity. Furthermore, (Alogla et al. 2016) performed tests on three RC beam-column sub assemblages' specimens quasi statically under middle column removal scenario with addition longitudinal beam longitudinal rebar at different depths throughout the beam to propose economical scheme to increase the progressive collapse capacity. (Lim et al. 2017) designed a test set up to for the complete measurement of support reactions under exterior and corner column removal scenario of 3D RC structures to properly elucidate the interaction between beam and column and isolate and quantify the slab contributions by comparing skeletal frame and frame-slab specimens. (Weng et al. 2017) proposed a new modelling procedure to analyze the progressive collapse scenario different from pushdown analysis which incorporates the effect of service load before the column removal by specially designed searching algorithm for member and substructure removal based on combined actions of flexural/shear/axial failures. For pre-cast structures, (Tohidi et al. 2014) modelled a pulled out behavior of strands in the keyways of precast concrete blocks and a detailed 3D non-linear behavior of precast concrete floor joints under underlying wall support removal scenario to study the ductility behavior of floor joints.

The main objective of this study is to investigate the effects of transverse beam effects on simple beam column sub-assemblage on resisting mechanism of progressive collapse at first peak load and at ultimate capacity. For this purpose two series of finite element models are developed, one without transverse beam and another with transverse beam and the mechanism of progressive collapse resistance under interior span point loading is analyzed. Furthermore, parametric studies are carried on both models varying concrete compressive strength, beam longitudinal reinforcement ratios and beam section depth.

2 FINITE ELEMENT MODELLING

In order to investigate the effects of transverse beam under interior column removal scenario in response to Progressive Collapse, non-linear finite element analysis is performed using ABAQUS/STANDARD 2017. As shown in Figure 1 b) same geometric, material non-linear properties and boundary conditions were considered to properly stimulate the behavior of Progressive Collapse considered in the experiment performed by (Qian et al. 2015).

2.1 Material model for steel reinforcement

The stress strain curves of different bar types used in this model are shown in Figure 1 a). The tensile stress-strain behavior of steel bars was assumed to be elastic and for plastic phase was modelled using bilinear behavior.

2.2 Material model for concrete

Among various constitutive models available in ABAQUS the Concrete Damage Plasticity Model (CDPM) is selected and introduced to simulate the behavior of concrete to the

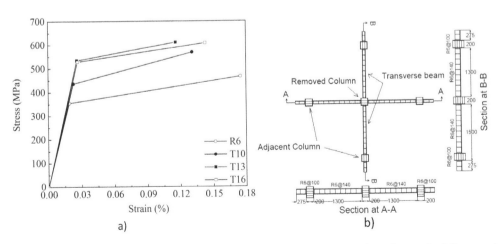

Figure 1.   a) Stress-strain curve of different diameters reinforcing bars & b) dimensions and reinforcement details of P2 and T2, units in mm (Qian et al. 2015).

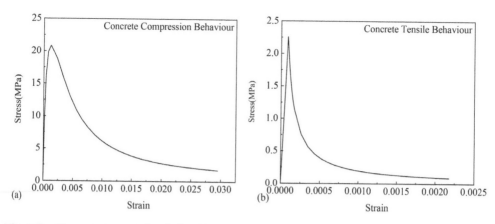

(a)

(b)

Figure 2. Concrete stress strain relationship.

numerical model (Abaqus-6.13 2013). The stress-strain compressive curve of concrete was derived using Popovics model (Popovics 1973) as shown in Figure 2 a) and tensile curve based on the model presented by Belarbi et al. (Belarbi et al. 1996) as shown in Figure 2 b).

CDPM requires other parameters such as dilation angle, eccentricity, and biaxial loading ratio, the coefficient K and viscosity parameter to be defined in the model and these values are 20, 0.1, 1.16, 0.667 and 0.001 respectively.

2.3 Analysis approach

Finite Element Models were developed on **ABAQUS/STANDARD** 2019, i.e., Newton Rapson approach was employed in solving non-linear system of equations. In this study, embedded model is used. The concrete block of the test specimen is idealized as homogenous material and modelled as 3D cubic element with eight nodes (brick) elements identified as (C3D8R) element in ABAQUS. 3D truss element with two nodes (T3D2) are used for finite element modeling the steel bars. For this study the FE model mesh are selected as 30mm via several trials with different mesh size and comparing the outcomes with experimental results.

3 MODEL VALIDATION

In order to validate the developed FE model applied load versus interior joint displacement was compared with the experimental results as described below.

3.1 Applied load versus joint displacement

The load displacement curve depicts the overall progressive collapse behaviour i.e., three stages of load resisting capacity subjected to progressive collapse; a) flexural stage, initial linear ascending portion in the curve b) transition stage, descending portion in the curve and c) catenary stage, finally again ascending portion in the curve till ultimate (see Figure 3). These all the stages are in good agreement as shown in Figure 3 a) and 3 b) and compares the applied load versus interior joint displacement curve of P2 and S2 specimen with simulated FEM curve. The FEM curve can satisfactorily simulate initial stiffness, first peak load, ultimate load capacity and the general trend at large displacement (See Table 1). The FEM curve could not simulate the vertical drops since the embedded model could not capture the rebar fracture as in test.

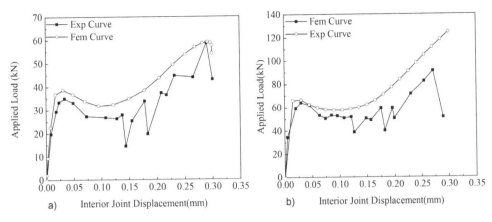

Figure 3.   Applied load vs middle joint displacement a) P2 specimen & b) T2 specimen.

Table 1.   FEM Results.

Specimens	Initial Stiffness (kN/mm)	First Peak Load (FPL) (kN)	Ultimate Load Capacity (ULC) (kN)
P2	1.7	38.6	54.2
T2	3.7	66.5	124.

3.2 Comparison between FEM P2 and T2 specimens

As shown in Figure 3 b) shows that the due to the transverse beam the first peak load capacity is increased by 72.2%. Also, it can be seen that the presence of transverse beam increases the initial stiffness RC structure.

4 PARAMETRIC STUDY

Parametric studies includes the effect on each parameters discussed in two aspects: a) effect on initial stiffness, first peak load (FPL) and ultimate load capacity (ULC) and b) effect of transverse beam on the same. It should be noted that during each parametric analysis only one parameter was changed and all others were kept constant in the developed FEM model.

4.1 Effect of beam longitudinal ratio (BLR)

Figure 4 a) and b) shows the applied load versus interior joint displacement of without transverse beam effect (P2) and with transverse beam specimen (T2) respectively under varying degree of beam longitudinal reinforcement. The beam longitudinal reinforcement ratio was varied by changing the diameters of longitudinal beam rebar as 12 mm and 8 mm against 10 mm with FEM P2 and T2 respectively. In both the specimens increasing the BLR does not only increases the initial stiffness and yield strength but also increases the FPL and ULC and vice-versa.

4.2 Effect of beam depth

Figure 5 a) and b) shows the applied load versus interior joint displacement of without transverse beam (P2) and with transverse beam specimen (T2) respectively under varying degree of

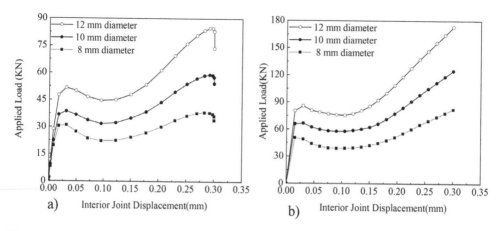

Figure 4. Applied load vs interior joint displacement a) P2 specimen & b) T2 specimen.

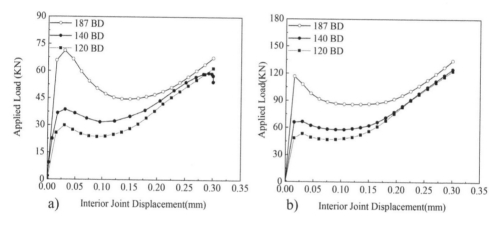

Figure 5. Applied load vs interior joint displacement a) P2 specimen & b) T2 specimen.

beam section depth. For parametric analysis of beam section depth one was increased to 187mm and another was decreased to 120mm against 140mm FEM P2 and T2 respectively. Result shows that increase and decrease in beam section depth increases and decreases the initial stiffness and FPL but does not affect the ULC in both the specimens. That means, higher section depth has higher flexural capacity and vice-versa.

4.3 Effect of concrete compression strength

Figure 6 a) and b) shows applied load versus interior joint displacement of without transverse beam (P2) and with transverse beam (T2) specimens under varying degree of compressive strength of concrete. In both the specimens, concrete compressive strength was increased from 20 MPa to 25 MPa and then to 30 MPa. It was observed that increasing strength of the concrete increases both initial stiffness and FPL but the does not affect the ULC in both the specimens

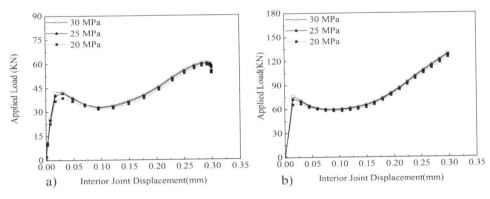

Figure 6.   Applied load vs interior joint displacement a) P2 specimen & b) T2 specimen.

5   CONCLUSIONS

- A simplified FE model was developed in order to investigate the structural behavior of RC beam-column sub-frame with transverse beam under interior column removal scenario and its reliability was verified by the experimental results in literature. Therefore, the proposed model is able to predict the collapse resistance capacity at first peak load and ultimate load with satisfactory accuracy.
- Result shows that the transverse beam increases the first peak load by 72.2 %.
- With increase in beam longitudinal reinforcement ratio increases the initial stiffness, first peak load and ultimate load capacity under both scenario i.e., with transverse and without transverse beam.
- Increase in beam section depth and compressive strength of concrete increases the initial stiffness, FPL while ULC remains constant. The rate of increase in capacity is significant with increasing beam section depth than to compressive strength of concrete.

ACKNOWLEDGEMENT

The corresponding author would like to thank the supports from the National Key Research and Development Program of China [Grant No. 2016YFC0701400], Science and Technology Program of Guangzhou, China [Grant No. 201704030057] and the Opening Project of State Key Laboratory of Green Building Materials.

REFERENCES

318. A.C. 2008. *ACI 318-08 Building Code Requirements for Structural Concrete (ACI 318-08) and Commentary.*

Abaqus-6.13. 2013. *ABAQUS-6.13 2013. Analysis User's Manual*, SIMULIA.

Alogla, K. Weekes, L. & Augusthus-Nelson, L. 2016. A new mitigation scheme to resist progressive collapse of RC structures. *Construction and Building Materials* 125:533-545.

Belarbi, A. Zhang, L. & Hsu, T. 1996. Constitutive laws of reinforced concrete membrane elements. *Eleventh World Conference of Earthquake.*

DoD. 2009. Unified Facilities Criteria (UFC). *Design of buildings to resist progressive collapse.*

GSA. 2003. *Progressive collapse analysis and design guidelines for new federal office buildings and major modernization projects.*

ICC. 2009. International building code.

Lim, N.S. Tan, K.H. & Lee, C.K. 2017. Experimental studies of 3D RC substructures under exterior and corner column removal scenarios. *Engineering Structures* 150: 409-427.

Popovics, S. 1973. A numerical approach to the complete stress-strain curve of concrete. *Cement and Concrete Research* 3:583-599.

Qian, K. Li, B. & Ma, J.X. 2015. Load-carrying mechanism to resist progressive collapse of RC buildings. *Journal of Structural Engineering* 141:04014107.

Tohidi, M. Yang,J. & Baniotopoulos, C. 2014. Numerical evaluations of codified design methods for progressive collapse resistance of precast concrete cross wall structures. *Engineering Structures* 76: 177-186.

Weng, J. Tan, K. H. & Lee, C. K. 2017. Modeling progressive collapse of 2D reinforced concrete frames subject to column removal scenario. *Engineering Structures* 141:126-143.

Yu, J. & Tan, K.H. 2013. Structural Behavior of RC Beam-Column Subassemblages under a Middle Column Removal Scenario. *Journal of Structural Engineering* 139: 233-250.

Sustainable Buildings and Structures: Building a Sustainable Tomorrow – Papadikis et al. (Eds)
© 2020 Taylor & Francis Group, London, ISBN 978-0-367-43019-1

Compressive behavior of FRP-confined steel-reinforced recycled aggregate concrete columns: An experimental study

Z. Xu
School of Civil and Transportation Engineering, Guangdong University of Technology, Guangzhou, China

G.M. Chen
School of Civil Engineering & Transportation, South China University of Technology, Guangzhou, China

M.X. Xiong
Protective Structures Centre, School of Civil Engineering, Guangzhou University, Guangzhou, China

ABSTRACT: This paper presents an experimental study on axial compressive behavior of FRP-confined steel-reinforced recycled aggregate concrete (FCSRAC). A series of tests have been carried out to evaluate its compressive behavior, with the effects of key parameters such as replacement ratio of recycled coarse aggregate (RCA) and FRP confining stiffness being investigated. The failure modes and failure mechanism, load and deformation capacities, confinement and composite effects were discussed in this paper. It has been revealed that the use of RAC led to higher axial and lateral deformation capacities and in turn higher confinement ratio for the FCSRAC columns as compared to FRP-confined steel-reinforced concrete (FCSRC). Favorable composite effect was found for the combination of FRP, steel and concrete.

1 INTRODUCTION

An important way to make use of the waste concrete is to crush it into recycled concrete aggregates (RCAs), and then replace part of or all the natural concrete aggregates (NCAs) in concrete to produce recycled aggregate concrete (RAC). The RCAs often suffer some damages, i.e. defects or micro-cracks, due to the stresses in service and crushing process. Meanwhile, there are also weak interfacial zones (ITZs) between original aggregates and old mortar, old mortar and new mortar matrix, original aggregates and new mortar matrix (Xiao et al. 2013). The damages and weak interfacial strengths generally lead to lower strength and stiffness (Kisku et al. 2017, Soares et al. 2014), poorer water permeability and resistance to chloride and sulfate ions (Kou & Poon 2012, Amorim et al. 2012), and larger shrinkage and creep under long-term loading (Xiao et al. 2015) when compared to the natural aggregate concrete (NAC) with the same mix proportion. The above differences are more obvious with the increase of replacement ratio of the RCAs. To overcome these adverse effects, the RCAs may not be pre-wetted to improve the strength of RAC (Xu et al. 2017). Alternatively, the RAC could be filled into the FRP tube to form concrete-filled FRP tube (CFFT). This is because the FRP tubes can improve both strength and deformation capacities of the RAC due to the confinement effect and enhance the corrosion resistance of RAC by isolating it from the corrosive environment. In addition, the FRP tubes can also be used as permanent formwork for concrete casting, which enables elimination of additional formwork and thus fast track construction. However, the bending and tension performance of CFFT is inferior. Under such circumstances, pure steel column could be encased in the CFFT to form a hybrid FRP-confined steel-reinforced concrete (FCSRC) column with excellent performance (e.g., compression, bending and tension). On the other hand, encasing steel section in the CFFT enhances both the local and global buckling resistance of the steel section, as shown in recent studies associated with FCSRC (Karimi et al. 2011, Feng et al. 2015, Ozbakkaloglu 2015, Yu et al. 2017, Huang et al. 2017, Karagah et al. 2018). However, few studies have examined the mechanical characteristics of FCSRAC columns. Against this background, an experimental study on the axial compressive behavior of FCSRAC stub columns were carried out. Three replacement ratios (0%, 50%, and 100%) were adopted. Due to space limit, only the test results of failure modes, axial load and deformation capacities, load versus deformation curves, confinement and composite effects were presented and discussed in this paper, more complete test results can be found elsewhere (Xiong et al. 2019).

2   EXPERIMENTAL PROGRAM

Figure 1 shows the cross-sectional sizes of specimens, where "CC", "SRC", "CFFT" and "FCSRC" respectively denotes the concrete column, steel reinforced concrete column, concrete-filled FRP tube, and the FRP-confined steel-reinforced concrete column. A total of 24 specimens were tested under axial compression. All specimens had a nominal diameter of 200 mm and a height of 400 mm.

After crushing without wetting, the RCAs were sieved and divided into two groups based on the particle size (i.e., 5~10 mm and 10~20 mm), respectively. Then, they were mixed together according to a proportion of 2:3 (i.e., smaller size RCAs: larger size RCAs). The mix proportions of NAC and RAC are detailed in Xiong et al. (2019). The compressive strength of R0, R50 and R100 on 28 days 45.6 MPa, 41.6 MPa and 40 MPa, respectively, which is 48.2 MPa, 45.9 MPa and 44.3 MPa on the days of column test. The yield strength, elastic modulus, and Poisson's ratio of steel section were 249 MPa, 198 GPa and 0.28, respectively. The cruciform steel columns with a height of 400 mm were also tested under axial compression. The solid GFRP tubes were fabricated in a filament-wound process with fiber winding angles of ±80° to the longitudinal axis of the tube. Two types of GFRP tube, 4-ply and 8-ply, were prepared, having a thickness of 1.96 mm and 3.69 mm, respectively. The axial strength of 4-ply and 8-ply is 77.7 MPa and 83.4 MPa, respectively. The compressive strength and elastic modulus of the 8-ply tubes were higher than those of the 4-ply tubes, but the Poisson's ratio was lower. The hoop strength of 4-ply and 8-ply is 40.3 MPa and 44.2 MPa, respectively. The 8-ply tubes also exhibited higher elastic modulus and lower Poisson's ratio, when compared with the 4-ply tubes.

3   TEST RESULTS AND DISCUSSIONS

3.1   *Test observations and failure modes*

According to EN 1993-1-1 (CEN 2005), the webs and flanges of the cruciform steel column can be classified as Class 1 and Class 2, respectively. For Class 1 sections, the local buckling would not occur in both elastic and plastic ranges, but for Class 2 sections, the local buckling could appear in the plastic stage. When the steel column was encased in concrete, the global buckling was effectively restrained. The local buckling was still be found, however, it tended to be less severe when the steel column was encased in RAC or confined by the thicker GFRP tube. All CFFT and FCSRC specimens failed by rupture of the GFRP tubes.

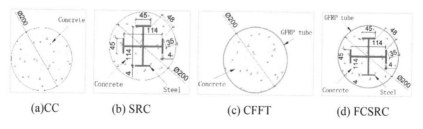

(a)CC   (b) SRC   (c) CFFT   (d) FCSRC

Figure 1.   Cross-sectional sizes of tested specimens.

3.2 Axial load – strain responses

The axial load-strain curves considering the effects of replacement ratio of RCA are shown in Figure 2. Loads of the specimens without GFRP dropped after their peaks; but for the specimens confined by the GFRP tubes, such load drop was not observed, and the post-yield hardening continued until the GFRP tubes ruptured. Moreover, the axial stiffness in terms of the slope decreased with the increase of replacement ratio of RCA, however, the lateral stiffness in both elastic and plastic stages showed little variation with the said ratio. Furthermore, the encasement of steel column increased the load capacity and the axial stiffness especially in the post-yield stage, but had only slight effect on the lateral stiffness and the deformation capacity in terms of the failure strains. It is also of much interest to know if the addition of RCA would increase the deformation capacity of CFFT or FCSRC columns. By observing Figure 2 (b) and (c), it can be found that the axial deformation capacity has been improved as the CFFT and FCSRC specimens failed at a larger axial strain for a higher replacement ratio of RCA, meanwhile the lateral deformation capacity was only slightly improved according to the ultimate lateral strains.

3.3 Confinement effect

The confinement level from the GFRP tube to the core concrete can be interpreted by the confinement ratio (CR) defined in Eq.(1), where A_f and A_c are the cross-sectional area of GFRP tube and concrete, respectively; and $\sigma_{h,rup}$ is the hoop stress of GFRP tube at rupture. The GFRP tube is subjected to both hoop stress $\sigma_{h,rup}$ and axial stress $\sigma_{a,cu}$ at failure. They can be calculated according to Eq.(2) and (3) (Yazdanbakhsh & Bank 2006), where $\varepsilon_{a,cu}$ and $\varepsilon_{h,rup}$ are the axial and hoop strains at failure, respectively. The CR values for the CFFT and FCSRC specimens are provided in Xiong et al. (2019). Basically, the CR values were less than 1.0 for all specimens confined by the 4-ply GFRP tubes, but greater than unity for those confined by the 8-ply GFRP tubes. Besides, it is found that the CR value generally increased with the increase in the replacement ratio of RCA, due to the slightly improved lateral deformation capacity as mentioned above.

$$CR = \frac{A_f \sigma_{h,rup}}{A_c f'_{co}} \tag{1}$$

$$\sigma_{h,rup} = \frac{\nu_a E_h}{1 - \nu_h \nu_a} \varepsilon_{a,cu} + \frac{E_h}{1 - \nu_h \nu_a} \varepsilon_{h,rup} \tag{2}$$

$$\sigma_{a,cu} = \frac{E_a}{1 - \nu_h \nu_a} \varepsilon_{a,cu} + \frac{\nu_h E_a}{1 - \nu_h \nu_a} \varepsilon_{h,rup} \tag{3}$$

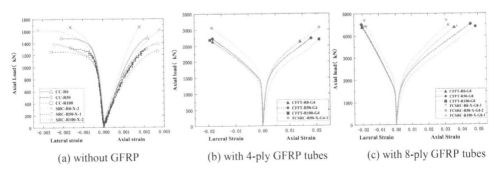

(a) without GFRP   (b) with 4-ply GFRP tubes   (c) with 8-ply GFRP tubes

Figure 2. Effects of replacement ratio of RCA on axial load-strain relationship (Notes: X=steel of cross section; the last number represents the serial number of specimen).

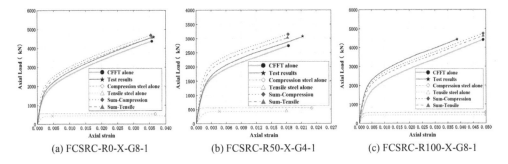

| (a) FCSRC-R0-X-G8-1 | (b) FCSRC-R50-X-G4-1 | (c) FCSRC-R100-X-G8-1 |

Figure 3.   Composite effect of constituent materials.

3.4   *Composite effect*

The composite effect is deemed to be exerted when the composite load capacity N_{cu} is greater than the superimposed load capacity N_0 from the individual load capacities of the constituent materials of the composite section. The N_{cu} and N_0 values for the SRC and FCSRC specimens are detailed in Xiong et al. (2019). For better illustration, the individual, superimposed, and composite load – axial strain curves of the FCSRC specimens are shown in Figure 3. To avoid double counting the steel cross-sectional area in the superimposition, the load capacity of CFFT column was multiplied by a reduction factor of $(A_c-A_s)/A_c$, where A_c is the gross cross-sectional area inside the GFRP tube. Besides, it should be mentioned that the load capacity of steel column to calculate the superimposed load capacity (N_0) was based on either the results of tension coupon tests of steel samples or compression tests of steel tubes; it can be seen that the difference in the load capacities determined using the two methods is insignificant. For SRC specimens, the composite load capacity was generally smaller than the superimposed load capacity based on the CC specimen and pure steel column. This could be attributed to the following two facts: 1) the global bending and local buckling of the encased steel section had caused more micro-cracks of the concrete than the plain concrete column (i.e., CC specimen), and 2) the loads carried by concrete and steel do not peak simultaneously in SRC specimens (see Xiong et al. 2019 for detailed discussions). Therefore, a reduction factor of 0.85 should be applied for the load capacity of the concrete section (i.e. $A_c f_{co}'$) to calculate the characteristic axial load capacity of SRC column according to the American code (ANSI/AISC 2016) or the Eurocode 4 (CEN 2004). For the FCSRC specimens, the composite capacity was closer to the superimposed load capacity based on the CFFT specimen and pure steel column, but replacement ratio of RCAs has a slight effect on the composite effect: for NAC (R=0%) and RAC with R=50%, the composite load at an axial strain in the late loading stage is slightly higher than the test value, but for RAC with R=100%, the former is slightly lower than the latter. This might be due to the local buckling of the steel section in the case of R=100% due to the larger lateral dilation/deformation of RAC with high replacement ratio in the late loading stage. However, in terms of load capacity, the composite effect was not affected by the addition of RCA in notable manner.

4   CONCLUSIONS

(1) For FRP-confined steel-reinforced recycled aggregate concrete (FCSRAC), the replacement of RCA does not alter the failure mode (FRP rupture) of concrete under compression but increase the possibility of local buckling of encased steel to a certain extent. Increasing the thickness of GFRP tube is also an effective way to suppress the local buckling.

(2) Increasing replacement ratio of RCA reduces the axial stiffness of the FCSRAC column but improves its axial deformation capacity in terms of the increased axial strain at failure. The said increase also slightly improves the lateral deformation capacity but has little

effect on its lateral stiffness. The confinement of GFRP tube improves both the load and deformation capacities as well as the stiffness of the FCSRAC column especially in the post-yield stage.

(3) Favorable composite effect generally exists in the combination of FRP tube, concrete and steel section in the FCSRAC proposed in this study, especially in terms of load-carrying capacity.

ACKNOWLEDGEMENT

The authors gratefully acknowledge the financial support provided by National Natural Science Foundation of China (Project No. 51678161) and Natural Science Foundation of Guangdong Province (Project No. 2018A030313752).

REFERENCES

Amorim, P. De Brito, J. & Evangelista, L. 2012. Concrete made with coarse concrete aggregate: influence of curing on durability. *ACI Materials Journal* 44:195-204.

ANSI/AISC. 2016. *Specification for structural steel buildings.* ANSI/AISC 360-16, Chicago, Illinois.

China Metallurgical Construction Association. 2010. *Technical code for infrastructure application of FRP composites. GB 50608-2010, Beijing, in Chinese.*

European Committee for Standardization (CEN). 2004. *Eurocode 4: Design of composite steel and concrete structures – Part 1-1: General rules and rules for buildings. EN 1994-1-1, Brussels.*

European Committee for Standardization (CEN). 2005. *Eurocode 3: Design of steel structures – Part 1-1: General rules and rules for buildings. EN 1993-1-1, Brussels.*

Feng, P. Chen, S. Bai, Y. & Ye, L.P. 2015. Mechanical behavior of concrete-filled square steel tube with FRP-confined concrete core subjected to axial compression. *Composite Structures* 123: 312-324.

Huang, L. Yu, T. Zhang, S.S. & Wang, Z.Y. 2017. FRP-confined concrete-encased cross-shaped steel columns: Concept and behavior. *Engineering Structures* 152: 348-358.

Karagah, H. Dawood, M. & Belarbi, A. 2018. Experimental study of full-scale corroded steel bridge piles repaired underwater with grout-filled fiber-reinforced polymer jackets. *Journal of Composites for Construction* 04018008.

Karimi, K. El-Dakhakhni, W.W. & Tait, M.J. 2011. Performance enhancement of steel columns using concrete-filled composite jackets. *Journal of Performance of Constructed Facilities* 25(3):189-201.

Kisku,N. Joshi, H. Ansari, M. & Panda, S.K. 2017. A critical review and assessment for usage of recycled aggregate as sustainable construction material. *Construction and Building Materials* 131:721-740.

Kou, S.C. & Poon, C.S. 2012. Enhancing the durability properties of concrete prepared with coarse recycled aggregate. *Construction and Building Materials* 35: 69-76.

Ozbakkaloglu, T. 2015. A novel FRP–dual-grade concrete–steel composite column system. *Thin-Walled Structures* 96: 295-306.

Soares, D. Brito, J.D. Ferreira, J. & Pacheco, J. 2014. Use of coarse recycled aggregates from precast concrete rejects: Mechanical and durability performance.*Construction and Building Materials* 71: 263-272.

Xiao, J.Z. Li, W.G. Sun, Z.H. Lange, D.A. & Shahm, S.P. 2013. Properties of interfacial transition zones in recycled aggregate concrete tested by nanoindentation. *Cement and Concrete Composites* 37: 276-292.

Xiao, J.Z. Fan, Y.H. & Tam, V.W.Y. 2015. On creep characteristics of cement paste, mortar and recycled aggregate concrete. *European Journal of Environmental and Civil Engineering* 19: 1234-1252.

Xiong, M.X. Xu, Z. Chen, G.M. & Lan, Z.H. 2019. FRP-confined steel-reinforced recycled aggregate concrete columns: Concept and compressive behavior. Under reparation.

Xu, J.J. Chen, Z.P. Xiao, Y. Demartino, C. & Wang, J.H. 2017. Recycled Aggregate Concrete in FRP-confined columns: A review of experimental results. *Composite Structures* 174: 277-291.

Yu, T. Chan, C.W. Teh,L. & Teng, J.G. 2017. Hybrid FRP-concrete-steel multitube concrete columns: Concept and behavior. *Journal of Composites for Construction* 04017044.

Sustainable Buildings and Structures: Building a Sustainable Tomorrow – Papadikis et al. (Eds)
© 2020 Taylor & Francis Group, London, ISBN 978-0-367-43019-1

Dynamic analysis of short-circuit effects on a flexible conductor system

P. Yang & G. Gong

Department of Civil Engineering, Xi'an Jiaotong-Liverpool University, Suzhou, P.R. China

ABSTRACT: An open-air electric transmission system with cable conductors belongs to a type of substation structures and it is subjected to electromagnetic loadings in addition to the common loadings from a structural engineering viewpoint. There are two methods commonly used to analyze substation structures, i.e. the simplified method according to IEC865-1 and advanced method with simulation software. It is found that for flexible cable conductor systems with short-circuit loadings, if the conductor's initial sag is larger than 4% of the span length, the simplified method will lead to non-conservative results in terms of cable tension and therefore the advanced method with simulation must be used. Advanced method, such as using ADINA, is also recommended for simulating complex cases, for modelling dynamic response of flexible conductor under short-circuit current.

1 INTRODUCTION

The substation structure comprises rigid or flexible conductor, supporting structure, and other circuit components, such as insulator, jumper and dropper, etc. The analysis of a substation structure with flexible conductors is much more complicated than that with rigid conductor, because the former one experiences significantly large displacements and this results in a highly geometry-nonlinear behaviour (Gong & Jiang 2015). The linearity may not lead to a conservative solution for geometry-nonlinear problems, which is not in case of material-nonlinear problems, according to Cook et al. (2001).

There is a simplified method introduced by IEC865-1 (IEC 1993), which adopts simplified analytical representations of short-circuit effects which can be further checked by hand calculations. The simplified method requires only general data, for example span length, static tension, distance between conductors in different phases, supporting structure stiffness, short-circuit current and duration (Yang 2019). However, the simplified method typically gives maximum values about conductor tension and no information about the time history or evolution of the short-circuit phenomena. The simplified calculation method based on IEC865-1 is feasible and expedient for design purposes in most practical situations.

After 1970s, researchers began to develop a more reliable method to analyze the conductor response during short-circuit current (CIGRE 1996), for example using Finite Element Method (FEM) to calculate the dynamic response of flexible busbar systems under short-circuits. As a result, several methods for the calculation of short-circuit effects in flexible conductors have been developed.

This paper aims to use ADINA to simulate the dynamic behavior of flexible conductor during three phase balanced short-circuit current of a substation structure, and also investigate the influence of flexible conductor sag on conductor tension by comparing the results from ADINA simulation and from the simplified method according to IEC865-1.

2 THEORETICAL BACKGROUND

It is well known from electromagnetic theory (Schmitt 2002) that attractive forces will be induced between the conductors with electric currents flowing in the same direction, while with opposing currents repelling forces will be induced. In normal electric power system, the current is small and the induced electromagnetic force is negligible for conductors compared with the effects caused by other loadings such as gravity. However, during a short-circuit current condition, the current magnitude is large enough to introduce significant conductor tension and deflection. According to CIGRE (1996), the maximum initial symmetrical (to be explained later in this section) short-circuit currents I_k'' (rms) can be estimated according to the voltage levels and the data are listed in the following table.

Table 1. Expected maximum initial symmetrical short-circuit currents I_k'' (rms).

Voltage level	I_k'' (rms)
123 - 170 kV	40 kA
245 - 300 kV	40 kA
362 - 525 kV	80 kA

The flexible conductors will experience large deflections under forces induced by short-circuit currents. Experimental investigate of electric fields were performed by Charalambos et al. (2012). For two conductors in different phases, a conductor carrying current $i_1(t)$ is in a magnetic field created by another conductor carrying current $i_2(t)$ and undergoes an electromagnetic force defined for the differential element (ADINA R&D 2011).

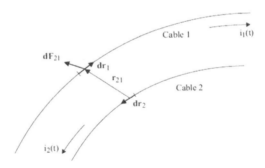

Figure 1. Elementary force d^2F_{21} created on flexible conductor 1 by conductor 2.

$$d^2F_{21} = \frac{\mu_0}{4\pi} i_1(t) i_2(t) \frac{dr_1 \times (dr_2 \times r_{21})}{r_{21}^3} \tag{1}$$

where, μ_0 is the magnetic permeability and r_{21} is the direction vector from dr_2 to dr_1, r_{21} refers to the distance of vector r_{21}.

The electromagnetic force is dependent on the short-circuit currents and distance between elements. The flexible conductor system will experience a non-linear response under dynamic short-circuit current loadings.

Three phase alternating current short-circuit current is the current in abnormal connection between phase and phase or phase and ground (or neutral line) in power systems. There are two categories, symmetrical and unsymmetrical short-circuit currents which are further

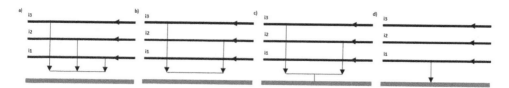

Figure 2. Characterization of short-circuit currents.

divided into four types according to CIGRE (1996), as shown in Figure 2, where the direction of current arrows is chosen arbitrarily.

In Figure 2, there are four kinds of short-circuit currents, (a) is balanced three-phase short-circuit current, (b) is line-to-line short-circuit without earth connection, (c) is line-to-line short-circuit with earth connection (d) is line-to earth short-circuit. The symmetrical short-circuit current are cases (a) and (b), since there is no earth connection. Cases (c) and (d) are two types of unsymmetrical short-circuit current.

The short-circuit current may cause mechanical effects on flexible conductors, supporting structure and other circuit components. According to CIGRE (1996), the lowest significant mechanical frequency of flexible conductor is less than the electrical frequency of the system. In flexible conductor configurated bus system, the mechanisms may be caused by short-circuit effect from three aspects. The tensile force can damage the flexible conductors including other circuit components, such as the jumper. Forces in the insulators and the supporting structures can cause yielding or insulator breakage. Thermal expansion of flexible conductors during short-circuit can be the cause of secondary short-circuits.

In flexible conductor configurated substations, conductors are in geometry of catenary curve under gravity. The static tension of flexible conductors under gravity depends on the conductor's sag value. Considering the highly non-linear phenomenon in addition to the fact the mechanical loadings and electromagnetic loadings coexist, the analysis of flexible busbar system is complicated. Therefore, there is very limited literature available in this related research area.

The flexible conductor's initial sag is a critical design parameter to determine the static tension and deflection of conductor. With the decreasing of conductor sag, the static tension will be increased and the flexible conductor will be more vulnerable to vibrations (Mohammadi et al. 2012). The design of overhead conductors against fatigue has been controlled by the level of the conductor sag (Kalombo et al. 2017). According to CIGRE (1996), a sag of about 8% of the span length will usually lead to acceptable results. Otherwise, if there is no other requirement, a static tension of 5-10 N/mm² is recommended as an initial design value. The final and optimal sag should be based on further calculations with the consideration of short-circuit current. Similar condition in determining sag of suspension bridge is in range of 1/9~1/11 of span length (JTG 2015).

The simplified method provided by IEC865-1 is a traditional method for calculating stresses in conductors under short-circuit current. However, the simplified method is mainly used for rigid conductor system and the systems of equations for calculating the stress is linear according to CIGRE (1996). Using this approach for analyzing flexible conductor systems may oversimplify the scenario. An advanced method like FEM can be used for solving differential equations, including the electromagnetic field equation. The computer simulation according to FEM is one kind of advanced analysis method and can be used for displacement and stress analysis in this investigation.

3 NUMERICAL SIMULATION

3.1 General description and simulation parameters

This research was based on a practical National Grid project in UK: Blyth 275kV Substation Emergency Return to Service Connection. The short-circuit current is in type of three phase balanced short-circuit current with duration of 0.3s. The short-circuit current magnitude is 31.5kA (RMS value) which is recommended in this real project.

Table 2. Three phase balanced short-circuit current properties.

RMS of Current, I	31500A
Angular Frequency, ω	18000 (deg/time)
Phase Angle, ϕ	-15, -135, -255 (degrees)
Impedance Angle, α	0 (degrees)
Constant, τ	0.081
Magnetic Permeability, μ_0	$1.26 * 10^{-6}$(Volt. second/meter. Ampere)

The time function of the short-circuit current uses 'Short-circuit (Type 2)' case. The formula is

$$f(t) = f^*(t)\sqrt{\frac{\mu_0}{4\pi}} \times \sqrt{2} \times I \times \left[\sin(\omega t - \phi + \alpha) + \exp\left(\frac{-t}{\tau}\right) \times \sin(\phi - \alpha)\right] \qquad (2)$$

where, $f^*(t)$ is interpolated from the time function curve according to input data and the resulting function $f(t)$ is the actual short-circuit current. ω is the angular velocity in degrees per unit time, ϕ is the phase angle in degrees, τ is time constant, α is impedance angle in degrees, and μ_0 is the magnetic permeability in (Volt. second/meter. Ampere). The input data is listed in Table 2.

For the flexible conductor, the simulation procedure is the same with the preliminary cable design in suspension bridge. The equation of cable in suspension bridge under gravity is in geometry of catenary curve, and can be obtained from a parabolic curve with further deformation under gravity loading (Meng et al. 2016). A typical parabolic curve can be described by

$$y = y_0 - \frac{4f}{l^2}x(l - x) \qquad (3)$$

where, y is the elevation of any point in the conductor and y_0 is the elevation of two end points, f is the initial (input) sag, i.e. the vertical distance from the mid-point of flexible conductor to the straight line connecting end points of conductor.

The flexible conductor and jumper are made of copper with density of 8900 kg/m³, Poisson's ratio of 0.3 and Young's modulus of 1.05 x 1011 N/m². The C30 concrete is used for the supporting structure with a density of 2500 kg/m³, Poisson's ratio of 0.2 and Young's modulus of 3.0 x 1010 N/m². Elastic materials are used for both concrete and conductor cables in the simulations. It should be noted that the flexible conductor can only take tension since otherwise buckling will easily occur under compression.

The supporting structures adopts a standard HS43 substation structure (Layton 2015) with the height of 13m and the conductor span length is 60 m. The cross section of the beam is in rectangular shape with dimensions of 0.381m x 0.533 m. The columns are in rectangular shape with dimensions of 0.457m x 0.457 m.

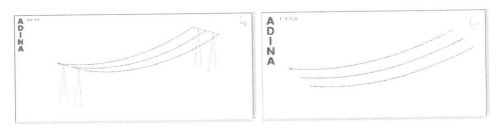

Figure 3. ADINA simulation with PI structure and pinned support cases.

The flexible conductor is modelled with the kinematic property of large deflection and small strains. Beams and columns are simulated using Euler-Bernoulli beams. The boundary condition at the bottom of columns are treated as fixed support (foundation was not included in consideration). A dynamic analysis (dynamic implicit) has been conducted where Rayleigh damping has been used. The gravity is applied at the beginning of analysis, and the short-circuit current is applied after the flexible conductor is stable under gravity.

3.2 Simulation results

The supporting structure will deform under various kinds of loadings, which will influence the conductor's dynamic response, such as reducing the conductor tension and increasing conductor deflection. In order to investigate the dynamic response only from flexible conductors, the support structure is excluded for consideration and the conductors are treated as pinned supported at ends. The conductor tension at the mid-point of flexible conductor in different supporting conditions and various sags is investigated. Figures 4 and 5 show the conductor tension in cases of initial sag is 8% of span length in pinned supported and PI structure supported.

From Figures 4 and 5, the tendency of tension in a specified conductor is nearly the same and the effects of support structures are negligible. In order to compare the maximum tension of each conductor in these two cases, the cable tension is listed in Table 3 as follows, where the conductor initial sag is taken as 8% of the span length, F_{st} refers to tension at mid-point of flexible conductor after the conductor is stable under gravity, $F_{t,ADINA}$ refers to the maximum conductor tension at mid-point of a conductor with short-circuit current and gravity from ADINA.

From Table 3, the PI supporting structure has minor influences on static tension and maximum conductor tension during short-circuit current, since the simulated PI structure is rigid

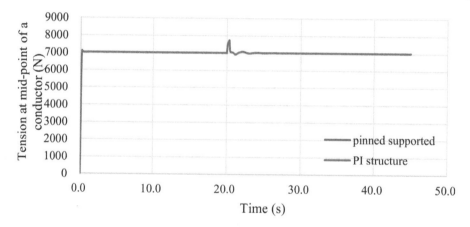

Figure 4.   Flexible conductor tension.

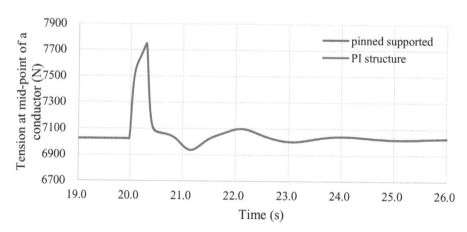

Figure 5.   Flexible conductor tension within time period of 19.0-26.0s.

Table 3. ADINA simulation results of support structure and pinned support cases.

| | with supporting structure | | with pinned support | |
	F_{st} (N)	$F_{t,ADINA}$ (N)	F_{st} (N)	$F_{t,ADINA}$ (N)
Conductor 1	7196.64	7907.96	7198.33	7910.23
Conductor 2	7196.64	7695.14	7198.33	7695.17
Conductor 3	7196.64	7931.20	7198.33	7932.82

enough. The maximum conductor tension with pinned support is more critical than that with the PI structure properties.

With the increasing of flexible conductor sag, the conductor will become more slack and the conductor tension will also be influenced. Table 4 shows the conductor tension at the mid-point of the conductor calculated from ADINA and the simplified method (According to IEC865-1) with the variances of flexible conductor sag. $F_{t,IEC}$ is the tension force during short circuit caused by swing out (short circuit tensile force) calculated according to IEC865-1.

he conductor tension will decrease with the increasing of conductor sag. The $F_{t,ADINA}$ is calculated from ADINA and is regarded as accurate results, since this kind of advanced method is more suitable for complex cases (CIGRE 1996). However, $F_{t,IEC}$ is the hand calculation result according to IEC865-1 without considering non-linear behaviors. According to Table 4, for cases of conductor sag is less than 4% of span length, the simplified method could give a typical value since $F_{t,IEC} > F_{t,ADINA}$. For initial sag larger than 4% of span length, the simplified method

Table 4. Flexible conductor tension with various sags.

Initial sag (%)	F_{st} (kN)	$F_{t,ADINA}$ (kN)	$F_{t, IEC}$ (kN)	$\left(\frac{F_{t,ADINA}}{F_{st}} - 1\right)\%$	$\left(\frac{F_{t,IEC}}{F_{t, ADINA}} - 1\right)\%$
1.00	35.37	37.66	40.50	+6.47	+7.56
2.00	24.84	26.66	27.72	+7.35	+3.98
3.00	17.88	19.26	19.51	+7.71	+1.30
4.00	13.73	14.80	14.74	+7.80	-0.40
5.00	11.09	11.99	11.78	+8.15	-1.83
6.00	9.29	10.10	9.79	+8.70	-2.81
7.00	8.00	8.75	8.37	+9.35	-3.12
8.00	7.02	7.73	7.32	+10.03	-4.28
9.00	6.26	6.94	6.50	+10.75	-5.32
10.00	5.66	6.31	5.85	+11.49	-6.29
11.00	5.16	5.79	5.33	+12.25	-7.19

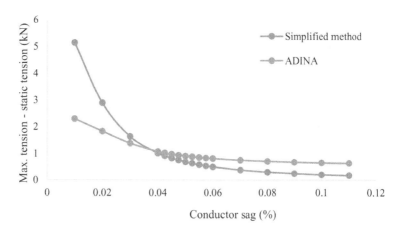

Figure 6. Differences of maximum conductor tension and static tension with various sags.

144

is not suitable to analyse the highly nonlinear condition. The short circuit will be more critical compared to static tension F_{st} in slack flexible conductor cases. With the increasing of conductor sag, the differences between maximum conductor tension and static tension are plotted in Figure 6.

4 CONCLUDING REMARKS

Simulations and hand calculations have been performed for substation structures with flexible conductors with various initial sags. The supporting structure of flexible conductor can be ignored by comparing the results with and without support structures. A pinned support condition without support structures at the ends will give a more critical results in terms of conductor tension for flexible conductor systems. Geometry nonlinearity with large deformation considering electromagnetic force are taken into consideration and can be reasonably modelled in ADINA. It is found that for flexible cable conductor systems with short-circuit loadings, if the conductor's initial sag is larger than 4% of the span length, the simplified method will lead to non-conservative results in terms of cable tension and therefore the advanced method with simulation must be considered.

ACKNOWLEDGEMENTS

We appreciate the funding supported by Xi'an Jiaotong - Liverpool University (RDF-15-01-38, RDF-14-02-44 and RDF-18-01-23). Also, the funding supported by Key Program Special Fund in XJTLU (Grant No: KSF-E-19) is greatly appreciated.

REFERENCES

ADINA R&D. 2011. ADINA theory and modeling guide report, Watertown. ADINA.
Charalambos, P. Antonis, P. Papadakisb, P.A. Ra-zisa, G. & Kyriacouc, S. 2012. Experimental measurement, analysis and prediction of electric and magnetic fields in open type air substations. *Electric Power Systems Research* 90: 42-54.
CIGRE Study Committee 23 (CIGRE). 1996. The mechanical effects of short-circuit currents in open air substations (rigid and flexible bus-bars). *Paris: CIGRE.*
Cook, R.D. Malkus, D.S. Plesha, M.E. & Witt, R.J. 2001. Concepts and applications of finite element analysis. *4th Edition, Wiley.*
Gong, G. & Jiang, W. 2015. Dynamical analysis of a substation structure under short-circuit current loadings. The first International Conference on Sustainable Buildings and Structures (ICSBS), Suzhou, China. *Sustainable Buildings and Structures* 149-152.
International Electrotechnical Commission (IEC). 1993. Short-circuit currents-calculation of effects. *Part 1: Definitions and calculation methods (IEC865-1:1993). London: BSI Standard Publication.*
International Electrotechnical Commission (IEC). 1994. Short-circuit currents – Calculation of effects. *Part 2: Examples of calculation (IEC 865-2:1994). London: BSI Standard Publication.*
JTG. 2015. Specifications for design of highway suspension bridge (JTG/T D65-05-2015). *China: China Communications Press.*
Kalombo, R.B. Pestana, M.S. Ferreira J.L.A. da Silva, C.R.M. & Araujo, J.A. 2017. Influence of the catenary parameter (H/w) on the fatigue life of overhead conductors. *Tribology International* 108:141–149.
Layton, L. 2015. Substation Design Volume II: Physical layout, USA: PDH Center.
Meng, C.F. Wang R.G. & Xu G.P. 2016. *Suspension Bridge.* China: China Communications Press.
Mohammadi, R.K. Akrami, V. & Nikfar, F. 2012. Dynamic properties of substation support structures. *Journal of Constructional Steel Research* 78:173–182.
Schmitt R. 2002. Electromagnetics Explained, 1st Edition. Newnes.
Yang, P. 2019. Simulation of short-circular effect on irregular configurated flexible conductors on substation structure. *Final-year-project (FYP) thesis, Department of Civil Engineering, Xi'an Jiaotong-Liverpool University (XJTLU).*

Sustainable Buildings and Structures: Building a Sustainable Tomorrow – Papadikis et al. (Eds)
© 2020 Taylor & Francis Group, London, ISBN 978-0-367-43019-1

A comparative study of concrete filled steel tubes using FEM & DEM

M. Zhu
Department of Civil Engineering, Xi'an Jiaotong - Liverpool University, Suzhou, P.R. China
Department of Civil Engineering, University College London, London, UK

G. Gong
Department of Civil Engineering, Xi'an Jiaotong - Liverpool University, Suzhou, P.R. China

ABSTRACT: A comparative study on the bearing capacity of the concrete filled steel tube (CFST) columns under axial compression is carried out using FEM and DEM. Three groups of specimens are simulated in this research using the commercial software ANSYS and PFC^{3D} respectively. The overall results demonstrate that both the FEM and DEM are capable to reproduce the mechanical behavior of CFSTs qualitatively and quantitatively. The results of the ultimate bearing capacities show that the DEM simulation (-8.25% to -15.58%) has a larger deviation in comparison with FEM (-1.69% to 2.64%), which illustrates that the micro-parameters in DEM should still be calibrated and modified in order to achieve better results. Furthermore, by comparing the two methods with each other, it is noted that the discrete element analysis is capable but not very efficient to investigate large objects with lots of elements/particles.

1 INTRODUCTION

Concrete-filled steel tubes (CFSTs) have been increasingly used in the construction of factories, high-rise buildings and arch bridges in recent years. This type of structure is a composite structure generally manufactured by filling the hollow steel tubes with fresh concrete, and simultaneously combines the positive characteristics of steel and concrete materials, including the high compressive or tensile strength, high ductility, high stiffness and effective seismic resistance (Dundu 2012). Essentially, the steel tube provides tensile strength for the concrete core and enhances its compressive strength and ductility, as well as acting as a permanent formwork. Meanwhile, the filled-in concrete specifically delays local buckling of the steel tube, which means that local buckling will happen at higher stress compared with pure steel tubes (Chen & Wang 2009).

Finite element method (FEM) is a computational technique used to obtain the approximate solutions of boundary value problems in engineering by introducing specific boundaries and variables of a field and solving with interpolation functions (Reddy 1993). Finite element analysis has been frequently used in previous researches to predict the behavior of CFST using available commercial finite element software such as ANSYS (Zhong 2006, Abed et al. 2013). During the analysis, a three-dimensional non-linear numerical model of CFST columns is normally proposed, with concrete core and steel tube simulated separately from each other to model the bonds between them by simulating the composite actions. Meanwhile, the appropriate elastic-plastic material constitutive models are applied for concrete core and steel tube, based on a wide range of experimental data and verifications selected from the literature (Gupta et al. 2014).

Discrete element method (DEM) was originally developed by Cundall (1971) for the analysis of rock mechanics problems. It is a micromechanics-based numerical method, which could explicitly capture the behavior and characteristics of granular materials via tracing contact forces, particle orientations and velocities, further providing a better understanding of mechanism leading to the macroscopic behaviour. In this method, there can even be no boundary conditions by introducing a periodic cell, where stresses can be measured over any volume and variables can be controlled effectively. Cundall & Hart (1993) has demonstrated that DEM is more valid in analyzing a discontinuous media than other numerical methods such as FEM and BEM (boundary element method). Cundall (2001) also predicted that the difficulty to apply such particle method to large-scale problems could be improved with the development of computer technology. Nowadays, the DEM analysis is applied in many fields, such as geotechnical engineering, wave propagation and massively parallel computing. The particle flow code (PFC^{3D}) is a common commercial software to explore the macroscopic and microscopic behavior of granular materials.

The present work is intended to study the axial capacity of CFST columns. The finite element program ANSYS and the discrete element program PFC^{3D} are used to conduct the uni-axial compressive test of CFST samples in the comparative analyses.

2 SIMULATION DETAILS

2.1 Specimen selections

Three groups of specimens are selected and simulated in this investigation. The uniaxial compressive tests will be conducted for these specimens numerically, using ANSYS and PFC. The detailed information of these specimens is listed in Table 1. The steel ratio is a specific value between the cross-sectional area of steel (A_s) and concrete (A_c).

2.2 Finite element modelling

In finite element modelling, the 'Solid 185' element is used for steel modelling in ANSYS. This element adopts the bilinear kinematic hardening model which adopts the Von Mises yield criterion in this case. The plasticity, creeping, stress hardening and large strain analysis of steel can be well performed using this model. For simplification, the steel is assumed as an ideal elasto-plastic material in ANSYS with the tangent modulus of 0. The yield strength is 310 MPa, which happens at the strain of approximately 0.15%. The curve is linear with a certain elastic modulus 206 GPa before yielding, but horizontal after the yield point. According to Mechanical Handbook (Wen 2010), the density of steel is taken as 7830 kg/m^3.

Furthermore, the 'Solid 65' element is provided for concrete to simulate its behavior under the triaxial compressive stress state. Zhong (1994) proposed a unified and rational theory for the constitutive relationship of concrete under such state, following the hardening stress-strain development pattern. To perform this behavior, the multi-linear kinematic hardening model and the Willam-Warnke yield criterion are adopted in this element, which focuses on predicting the occurrence of failure in cohesive-frictional materials. To implement this model appropriately, the several key parameters of concrete are chosen from the Mechanical Handbook (Wen 2010), the Chinese Standard GB 50164-2011 (2011) and the Chinese Standard GB 50010-2010 (2010), as listed in Table 2. The constitutive relationship of concrete based on the multi-linear kinematic hardening model is chosen from the Engineering Construction Standard (2003), as depicted in Figure 1(a). The meshing pattern for sample A1-SC is displayed in Figure 1(b) and the other samples are meshed similarly. The specimens are loaded by applying displacement on top face.

Table 1. Dimensions of specimens.

Group	Serial No.	Cross Section	Dimension $D \times t \times h$ mm × mm × mm	Steel Ratio (A_s/A_c)
1	A1-SC	Solid-Circular	700 × 12 × 1500	0.0723
2	A2-SC	Solid-Circular	165 × 3.54 × 480	0.0917
3	A3-HC	Hollow-Circular	700 × 12 × 1500	0.0867

* D - diameter of the section; t - thickness of steel tube; h - height of specimens.

147

Table 2. Parameters of concrete model in FEM.

Parameters	Value	Unit
Poisson's ratio	0.2	-
Young's modulus	34.5	GPa
Density	2420	kg/m^3
Open shear transfer coefficient	0.35	-
Closed shear transfer coefficient	0.6	-
Uniaxial cracking stress	2.64	MPa
Uniaxial crushing stress coefficient	-1	-

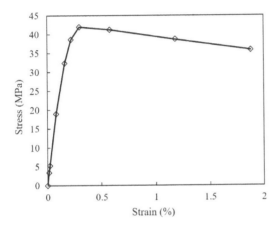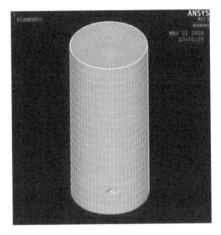

Figure 1. (a) The multilinear kinematic hardening model of concrete. (b) The meshing pattern in ANSYS for sample A1-SC.

2.3 Discrete element modelling

In discrete element modelling, the selection of contact constitutive models is significant since they define the relationship between external excitations and the response of a material. In this research on CFSTs, the parallel bond model will be attempted which provides the mechanical behavior of a finite-sized piece of cement-like material deposited between the two contacting particles and can transmit both force and moment. This model provides an elastic interaction between the two notional surfaces of particles, and the rigid interaction will be removed when the bond breaks. The local microscopic parameters should be assigned to these particles to represent the core concrete and the steel tube, respectively. The microscopic parameters are totally different from the macroscopic parameters. Normally, the interactions between two particles are established by means of the normal stiffness (k_n) and tangent stiffness (k_s) to characterize the elastic behavior of the materials in microscopic perspective. Rousseau et al. (2010) proposed the 'micro-macro' relationships using the Young's modulus (E) and the Poisson's ratio (v), expressed as:

$$\begin{cases} k_n = E \frac{S_{int}}{D_{init}} \frac{1+\alpha}{\beta(1+v)+\gamma(1-\alpha v)} \\ k_s = k_n \frac{1-\alpha v}{1+v} \end{cases} \quad (1)$$

where S_{int} is the area of the interaction surface, D_{init} is the initial distance between particles, and α, β, γ are the parameters identified by quasi-static and elastic tests.

Furthermore, a modified Mohr-Coulomb criterion associated with material softening is applied to model the non-linear behavior of CFSTs. Microscopic parameters, such as local tensile strength (T), cohesion strength (C_o) and softening factor (ζ) are identified from macroscopic parameters such as compressive and tensile strengths corresponding to those used in

Table 3. Microscopic parameters in DEM.

Parameters	Concrete/Interface	Steel	Unit
Poisson's ratio	0.2	0.3	-
Young's modulus	34.5	206	GPa
Density	2420	7830	kg/m³
Local tensile strength	5.7	33.4	MPa
Cohesion strength	17.2	310	MPa
Friction angle	20	0	degree

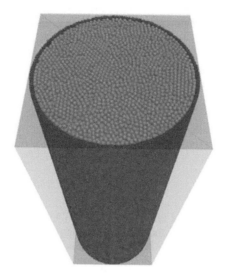

Figure 2. The particle distribution in PFC³ᴰ for sample A1-SC.

finite element models. A classical friction constitutive behavior law proposed by Rousseau et al. (2008) is used for the contact models, including concrete to concrete, steel to steel and concrete to steel interface contacts. Through calibration, the input parameters for the uniaxial compressive test of CFST are identified in Table 3. The final appearance of sample A1-SC in PFC³ᴰ is shown in Figure 2, where the blue balls (191594 particles) represent concrete and the red balls (988079 particles) represent steel. The specimens are loaded by moving the top wall down and the bottom wall up.

3 RESULTS AND DISCUSSION

3.1 Theoretical expectation

Many researchers (Zhong 1994, Han & Yang 2007) have proposed the formulas to estimate the bearing capacity of CFST members for design based on their experiments and concepts. Considering the non-linear working mechanism of concrete and the complexity of the whole performance of CFSTs, the ultimate bearing capacity N_u of CFST columns can be determined as (Zhong 1994):

$$N_u = f_{ck}A_c + kf_yA_s \qquad (2)$$

where f_{ck}, f_y are the characteristic strength of concrete and the yield strength of steel, A_c, A_s are the cross-sectional areas of concrete and steel, respectively, and k is an enhanced coefficient of strength. However, many researchers (Knowles & Park 1970) consider that the

Table 4. Theoretical bearing capacity of specimens.

Group	Serial No.	Cross Section	Steel strength MPa	Concrete strength MPa	Theoretical capacity kN
1	A1-SC	Solid-Circular	310	32.4	23008.59
2	A2-SC	Solid-Circular	310	32.4	1857.29
3	A3-HC	Hollow-Circular	310	32.4	20888.34

assumption of this formula does not confirm to the actual situations because experimental results have demonstrated that the longitudinal stress of the steel tube will not be zero and its hoop stress will not reach the yield point at ultimate load. In other words, the hardening stage of the steel tube is ignored in the formula. Chinese code (Han & Yang 2007) summarized an improved design formula considering the confinement effects which is given by:

$$N_u = (1.14 + 1.02\xi)f_{ck} \qquad (3)$$

where ξ is the confining factor ($\xi = \alpha f_y/f_{ck}$) and α is the steel ratio of the CFST column ($\alpha = A_s/A_c$). In order to obtain unified results for the three groups of specimens, this formula is adopted to determine the theoretical bearing capacities as a reference. The results are listed in Table 4.

3.2 Comparative analysis between FEM and DEM

To study the validation of the FEM and DEM on CFST members, the results of ultimate bearing capacities of the three specimens for both simulation methods are summarized in Table 5.

The deviations between the theoretical and simulated results of the ultimate compressive capacities for the three groups of specimens demonstrate the validity and feasibility of the applications of both FEM and DEM on the CFST members. The maximum deviation in ANSYS (2.64%) is much smaller than that in PFC^{3D} (-15.58%). The results in FEM are very close to the theoretical values probably due to the rational meshing of the specimens and the implementation of suitable constitutive relationships. Moreover, the study to conduct the non-linear analysis on CFSTs using FEM has been done frequently by many researchers, and hence this study is precedented to avoid detours. For the DEM, the results have a relatively larger deviation but are acceptable since this is the first attempt of the DEM on CFSTs. Nevertheless, the large deviations still indicate that the model needs to be calibrated, especially for the microscopic parameters of the two materials, a better consideration of the contact models and the configuration of the particles.

Considering the axial load against displacement diagrams of FEM and DEM for one specific specimen A2-SC, it is indicated that the tendency of the diagrams is quite different in Figure 3.

For the FEM result in Figure 3(a), the axial load against displacement relationships at initial stage are generally linear before the strain of approximately 0.15% which corresponds to the elastic working stage of CFSTs. The elasto-plastic stage approximately ranges from strain

Table 5. FEM and DEM simulation results of ultimate bearing capacities.

Group	Serial No.	FEM simulated bearing capacity kN	FEM deviation %	DEM simulated bearing capacity kN	DEM deviation %
1	A1-SC	23375.00	1.59	19423.64	-15.58
2	A2-SC	1825.91	-1.69	1704.09	-8.25
3	A3-HC	21440.51	2.64	18501.58	-11.43

150

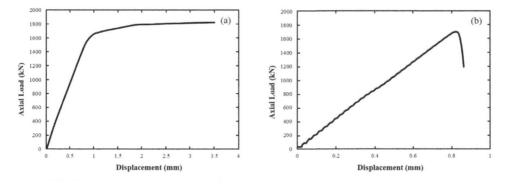

Figure 3.   Axial load against displacement for specimen A2-SC using (a) FEM and (b) DEM.

0.15% to 0.40%. During this stage, the elastic modulus of the steel tube significantly decreases, while the one of the core concretes scarcely decreases. This phenomenon raises the variation of the distribution ratio of axial load between the steel and concrete, and thus the load-displacement curve deviates from the straight line, and the transition zone is generated. There is a slight rising trend after the elasto-plastic stage, which demonstrates the appearance of the hardening stage.

For the DEM results in Figure 3(b), the elastic stage is similar to that obtained in FEM results before the strain of 0.15%. There is a small vibration during this stage which tends to be stable at a strain of approximately 0.075%. This phenomenon may be attributed to the fact that the initial loads between particles are difficult to be set to zero due to the large number of particles generated for the specimens. Moreover, the elasto-plastic stage in DEM is relatively short comparing to that in FEM. Once it has reached the peak load bearing capacities, the curve immediately declines with a steep slope. This may be due to the fact that the stiffness of the model is very high but with low softening. The possible solution in future works to retard the yielding of specimens is to restrict the rotations of the particles or try alternative contact models.

The other two specimens also show similar results on the tendency of the axial load against displacement diagrams. From the simulation processes of the results, it is indicated that the discrete element analysis is capable but not very efficient to investigate massive objects. The number of spheres used to build the numerical model for specimen A2-SC is 1179673 unlike other researchers who mainly restrict the number of particles below 5000 (Nitka & Tejchman 2015). Such fact may increase the complexity and instability of the validation of the numerical results. Comparing with the DEM, the FEM is much more capable on the efficiency of calculation and analysis due to the good optimization in ANSYS.

4   CONCLUSIONS

This paper reports the investigation of the axial capacity of CFST columns using FEM and DEM. The results show good correlation with the theoretical expectations, where the maximum deviations are 2.64% and -15.58% for FEM and DEM simulations, respectively. Both methods are capable to reproduce the working behavior of CFSTs qualitatively and quantitatively. For FEM, the accuracy is highly dependent on the suitable constitutive models and fine meshing and the results tend to be consistent. For DEM, the accuracy is related to the configuration of particle distribution, the contact models and the calibration of micro-parameters. The DEM simulation of large objects may require large computational efforts. The future study will focus on the optimization of DEM simulation on CFSTs, especially for the selection of contact models and micro-parameters.

ACKNOWLEDGEMENTS

The first author would like to thank Mr. Shiva Prashanth Kumar Kodicherla at Xi'an Jiao-tong-Liverpool University for his valuable suggestions and encouragement and the second author would like to thank the funding supported by Xi'an Jiaotong-Liverpool University (RDF-15-01-38, RDF-14-02-44 and RDF-18-01-23). Also, the funding supported by Key Program Special Fund in XJTLU (Grant No: KSF-E-19) is greatly appreciated.

REFERENCES

Abed, F. Alhamaydeh, M. & Abdalla, S. 2013. Experimental and numerical investigations of the compressive behavior of concrete filled steel tubes (CFSTs). *Journal of Constructional Steel Research* 80:429-439.

Chen, B. & Wang, T. 2009. Overview of Concrete Filled Steel Tube Arch Bridges in China. *Practice Periodical on Structural Design and Construction* 14: 70-80.

Cundall, P.A. & Hart, R.D. 1993. Numerical modeling of discontinua. *Analysis and Design Methods* 9: 231-243.

Cundall, P.A. 1971. A computer model for simulating progressive large-scale movements in blocky rock systems. *Proc. Symp. Int. Soc. Rock Mech* 2: 2-8.

Cundall, P.A. 2001. A discontinuous future for numerical modelling in geomechanics. *Proc. of the Institution of Civil Engineers-Geotechnical Engineering* 149(1): 41-47.

Dundu, M. 2012. Compressive strength of circular concrete filled steel tube columns. *Thin-Walled Structures* 56: 62-70.

Engineering Construction Standard (DBJ 13- 51-2003). 2003. Technical Specification for Concrete-filled steel tube. *Beijing: China Building Industry Press.*

Gupta, P.K. Khaudhair, Z.A. & Ahuja, A.K. 2014. Modelling, verification and investigation of behaviour of circular CFST columns. *Structural Concrete* 15(3):340-349.

Han, L. & Yang, Y. 2007. *Modern concrete filled steel tubular structural technology.* Beijing: China Build-ing Industry Press.

Knowles, R.B. & Park, R. 1970. Axial load design for concrete filled steel tubes. *Journal of Structural Division* 96(10): 2125-2153.

National Standard of the People's Republic of China (GB 50010-2010). 2010. Standard for design of concrete structures. *Beijing: China Architecture & Building Press.*

National Standard of the People's Republic of China (GB 50164-2011). 2011. Standard for quality control of concrete. *Beijing: China Architecture & Building Press.*

Nitka, M. & Tejchman, J. 2015. Modelling of concrete behaviour in uniaxial compression and tension with DEM. *Granular Matter* 17(1):145-164.

Reddy, J.N. 1993. *An introduction to the finite element method.* New York: McGraw-Hill.

Rousseau, J. Frangin, E. Marin, E. & Daudeville, L. 2008. Damage prediction in the vicinity of an impact on a concrete structure: a combined FEM/DEM approach. *Computers and Concrete* 5(4): 343-358.

Rousseau, J. Marin, P. Daudeville, L. & Potapov, S. 2010. A discrete element/shell finite element coupling for simulating impacts on reinforced concrete structures. *European Journal of Computational Mechanics* 19:153-164.

Wen, B. 2010. *Mechanical Handbook.* Beijing: China Machine Press.

Zhong, S. 1994. *Concrete Filled Steel Tubular Structures.* Heilongjiang: Science & Technology press.

Zhong, S. 2006. *Research and Application Achievement of Concrete-Filled Steel Tubular (CFST) Structures.* Beijing: Tsinghua University Press.

Experimental and numerical study on seismic performance of the SRC beams to inclined circular CFST columns connection

B.L. Jiang & Y.M. Li
School of Civil Engineering, Chongqing University, Chongqing, China

S.Y. Ji
School of Construction Management and Real Estate, Chongqing University, Chongqing, China

ABSTRACT: Concrete filled steel tubular (CFST) columns are well known for excellent structural behavior and have been widely used in building structures, especially in high-rise and super high-rise buildings. However, the seismic performance of the steel reinforced concrete (SRC) beams to inclined circular CFST columns connection under cyclic loads was rarely studied before. This paper presents experimental and numerical study on the seismic performance of the connection. Based on the cyclic loading test results, with good ductility and energy dissipation capacity, the tested connection encased by the SRC ring beam showed good seismic performance and satisfied the principle of "strong column & weak beam" and "strong joint & weak member" in seismic design. Finite element model was developed to verify the ultimate strength and hysteretic damage of the connection, and the stress distribution was illustrated accordingly.

1 INTRODUCTION

Concrete filled steel tubular (CFST) columns have been increasingly used in building structures and infrastructures, especially in high-rise and super high-rise buildings, due to its excellent mechanical and seismic performance. In practical super high-rise building structures, CFST columns are often connected with SRC beams, forming complex beam-column connections. Evidence from past earthquakes indicates that connection failure may result in structural failure. The key point of the earthquake resistant design for beam-column connections is to ensure and maintain the energy absorption capacity of plastic hinges of adjacent members, usually the beams, avoiding shear failure in the connection core area. Therefore, the SRC beam to CFST column connections are critical elements and their seismic performance plays an important safety role on structural design.

In previous research, a great effort has been made to study the seismic performance of beam to CFST column connections. However, the pioneering work mainly focused on cyclic loading tests of regular symmetrical reinforced concrete (RC) or steel beam to CFST column connections (Ding et al. 2017, Zhang et al. 2012) and axial loading tests of multi-CFST-column complex connections (Han et al. 2010, Zhen et al. 2017). However, the seismic performance of the SRC beams to inclined circular CFST columns connection under cyclic loads was rarely studied before. Therefore, the seismic performance of the SRC beams to inclined circular CFST columns connection was investigated by experiment and numerical analysis in this paper. Several aspects reflecting the seismic performance of the connection, such as the hysteretic response, the skeleton curves, the ductility and energy dissipation, were analyzed based on the cyclic loading test results. Finite element modeling was also implemented in order to obtain further details, such as stress characteristic, which was difficult to measure experimentally.

2 EXPERIMENTAL PROGRAM

2.1 *Test specimen*

A specimen of SRC beams to circular CFST columns connection with a scale factor of 1:4 to the actual structure was tested. It was assumed that the ends of beams and columns of the specimen corresponded to the infection points in the mid span of beams and the mid height of columns. Four 200 mm-diameter circular CFST columns with 6 mm in thickness of steel tube were connected with two beams ($b \times h \times l$ = 112.5×300×1000 mm, named L-1 and L-2 respectively). The CFST columns were 600 mm, 625 mm, 600 mm and 600 mm in length with inclined angle of 21 degrees, 26 degrees, 13 degrees and 20 degrees, respectively. In order to enhance seismic performance of the connection, the SRC ring beam ($b \times h$ = 80×300 mm) was designed around the joint area of CFST columns. Details of the specimen are shown in Figure 1 where all dimensions were given in mm.

C60 and C30 concretes were used for the columns and beams, respectively. The compressive strength of the concrete was obtained by testing 150 mm-standardized cubes and the compressive strength f_c of C60 and C30 were 40.1 MPa and 24.1 MPa, respectively. Moreover, Q345B and Q235 steel were used for the steel tube of columns and steel plates in beams, respectively, and HRB400 steel was used for the reinforcing bars. The mechanical properties of steel and reinforcing bars are given in Table 1.

2.2 *Test setup*

The setup for the cyclic loading test is shown in Figure 2. In the loading system, two hydraulic jacks were utilized to apply axial load on the top of CFST columns, and the other two

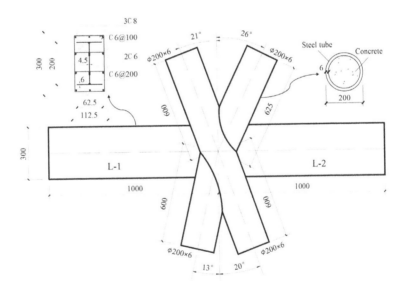

Figure 1. Dimensions and reinforcement.

Table 1. Mechanical properties of steel.

Type	f_y/MPa	f_u/MPa	E_s/GPa
Q345B	383	543	206
Q235	295	402	206
HRB400	437	647	210

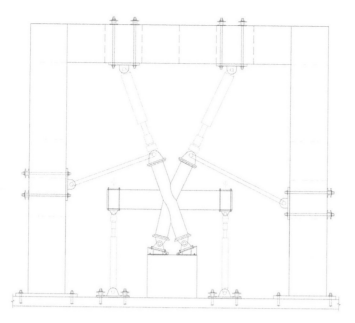

Figure 2.   Test setup.

hydraulic jacks were utilized to apply antisymmetric cyclic vertical load at the beam-ends to simulate the seismic action causing shear and bending moment in beams. Moreover, the bottom of CFST columns were connected to a spherical hinge.

2.3   Loading procedure

At the beginning of the test, axial forces of 752kN and 503kN were applied on the top of the left (west) and right (east) columns, and kept constant during the testing process. Then, the anti-symmetric cyclic vertical load was applied on the left (west) and right (east) beam ends. Based on the Chinese code (JGJ/T 101-2015), the loading history of the vertical load on the end of the beam included a force control stage and a displacement control stage. During the force control stage, the applied peak load of each cycle increased by $0.25P_y$, where P_y was the estimated yield strength of the specimen. After the yield of the specimen, the displacement control stage began

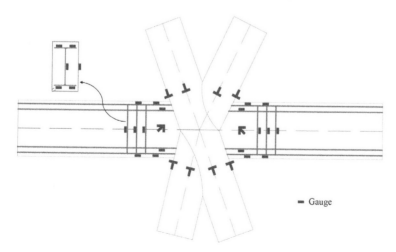

Figure 3.   Arrangement of strain gauge.

with the control displacements of Δ_y, $2\Delta_y$, $3\Delta_y$..., where Δ_y was the yield displacement corresponding to the yield load P_y. The test was terminated when the vertical load on the beam end decreased below 85% of the maximum measured load capacity of the test specimen or the specimen can no longer be sustained.

2.4 *Measurements*

Force sensors were attached to the hydraulic jack to measure the axial loads on the columns and the cyclic loads on beam ends. The displacements at the beam ends were measured by linear variable displacement transducers. Strain gauges were used to measure the strains of the steel tube of columns, steel and reinforcements of beams close to the joint core region. The arrangement of the strain gauges is shown in Figure 3.

3 EXPERIMENTAL RESULTS

3.1 *Failure mode*

The test phenomena before failure can be divided into four stages as the following.

1. In the first stage, the specimen was in elastic state with no crack and damage.
2. In the second stage, the crack firstly occurred on the beam under 15kN of the applying cyclic load and the cracks developed as the increase of load. There was no obvious phenomenon of CFST columns and the load-displacement curve was basically linear, indicating that the specimen was basically in elastic state as a whole.
3. In the third stage, the load reached yield point where the displacement value was the yield displacement Δ_y. With the increase of displacement amplitude, the bending cracks of beams increased gradually, and developed from tension zone to compression zone. Then, the beam cracks near the joint region began to extend to the ring beam of the connection and the beam concrete in plastic hinge zone was divided into many small pieces by cracks. The strain of the longitudinal reinforcement and the flange of section steel in beams already entered the yielding status. Meanwhile, the strain of steel tube at the junction of CFST column and ring beam increased rapidly. The large local deformation of the flange of the steel in the beam caused some concrete outside the flange to crack and fall off.
4. In the fourth stage, after the specimen reached its ultimate bearing capacity, the cracks in the plastic hinge area of the beam were wider and wider, and several main cracks occurred, developing and extending from the tension zone of the beam to the compression zone and to the ring beam of the connection. Then partial concrete of beam was spalling and peeling. At this time, the buckling was partially present at the CFST column below the ring beam. Under the repeated load with increasing displacement, the concrete damage in the plastic hinge area of the beam became more serious with all stirrups yielded and the steel deformed greatly. With the increasing displacement, the applying load at beam end correspondingly reduced. The test was over with the fracture of the steel web and the longitudinal rebar which made the specimen not suitable for continued load bearing.

The final failure mode of the specimen is shown in Figure 4. The final damage was concentrated in the plastic hinge area of the beam and penetrated to the ring beam, but the ring beam was less damaged. The CFST columns were only slightly damaged near the ring beam, while the connection part in the ring beam was not obviously damaged. Therefore it can be observed that the design of the SRC ring beam around CFTS columns in the joint area ensured the performance of the connection, indicating that the connection had a strong capacity meeting the design requirements and the expected seismic performance targets.

3.2 *Hysteretic relationship*

Figure 5 shows the load-displacement hysteretic curves at the beam ends. It can be seen that the curves are not completely symmetrical. The main reason is the asymmetry of the specimen,

Figure 4.   Failure mode.

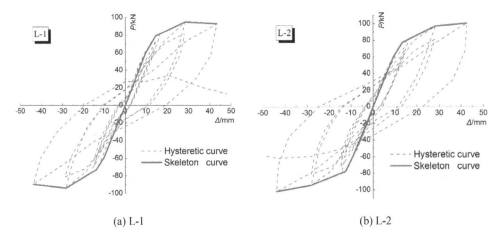

(a) L-1   (b) L-2

Figure 5.   Load-displacement hysteretic curves.

which leads to different calculation length in the tensile and compression side. The specimen response keeps elastic initially and the hysteresis loop is closely linear. With the increase of cyclic loading displacement, the loading stiffness of the specimen gradually degenerated and the unloading stiffness changed little. Meanwhile, the bearing capacity of the specimen increased gradually. After the peak bearing capacity, the beam load decreased slowly with the increasing displacement. The specimen still had a high bearing capacity before destroyed. During the cyclic loading history, the hysteresis loop became plump and the envelope area increased. There was no obvious pinching phenomenon found in the hysteresis curves of the specimen, which indicates that the connection has a good seismic performance with good ductility and energy dissipation capacity.

3.3   Ductility and energy dissipation

The displacement ductility coefficient is used to show the deformation capacity and ductility of the connection specimen. In the test, the displacement ductility coefficient was shown in Table 2. The ductility coefficients of the tested connection are from 2.99 and 3.44, indicating that the connection specimen meets the ductility requirements and has good deformation capacity.

The energy dissipation capacity is an important index to measure the seismic performance of a structure, which is usually measured by the area enclosed by the load-displacement hysteresis curve. In this paper, the equivalent viscous damping coefficient h_e is adopted to evaluate the energy dissipation capacity of the specimen. Figure 6 shows the curve of equivalent viscous damping coefficient calculated from the hysteresis hoops of the specimen. During the

Table 2. Displacement ductility coefficient.

Beam end	Loading direction	Cracking load/kN	Yield displacement/ mm	Yield load/kN	Ultimate displacement/ mm	Ultimate load/kN	Ductility coefficient
L-1	clockwise	20	14.5	79.1	43.4	94.8	2.99
	anticlockwise	15	13.8	73.1	43.5	94	3.15
L-2	clockwise	20	12.4	77.9	42.6	100.8	3.44
	anticlockwise	15	13.4	77.3	43.8	101.8	3.27

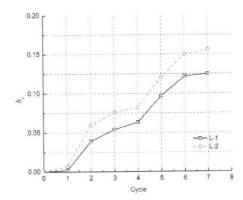

Figure 6. Equivalent viscous damping coefficient.

whole test, the equivalent viscous damping coefficient changes from small to large. In the initial elastic stage, the hysteresis loop is not plump in shape; the equivalent viscous damping coefficient is very small. With the increase of the loading cycles, the plastic hinge at the beam end forms and the hysteresis loop gradually turns into a plump spindle shape. The equivalent viscous damping coefficient increases, and the energy dissipation capacity increases. The equivalent viscous damping coefficient of all specimens increased steadily with the test process and finally reached about 0.15, indicating that the specimen could play a good role in energy dissipation and have stable energy dissipation performance.

4 FINITE ELEMENT MODELING

4.1 General

Besides experimental study, the seismic performance of the test specimen was also simulated. ABAQUS was used to build the finite element model and conduct nonlinear analysis. Eight-node brick elements with reduced integration (C3D8R) were used to model the circular steel tube, profile steel and concrete, while two-node 3D truss elements (T3D2) were used for the reinforcing bars which were assumed to embed in the concrete.

The interaction between steel tube and concrete at the interface was modelled by a surface based contact. By specifying a hard contact property in the direction normal to interface plane, there was no contact pressure unless the two surfaces were in contact. The friction coefficient in terms of contact pressure used in the penalty frictional formulation was defined as 0.6.

The loading procedure and boundary conditions were in accordance with the test. The typical finite element model is shown in Figure 7.

4.2 Material properties

The concrete damaged plasticity (CDP) model was used to simulate the inelastic behavior of concrete. The concrete damaged plasticity model is primarily intended to provide a general

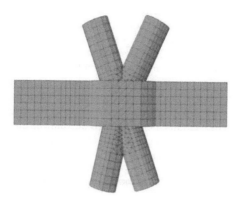

Figure 7. Finite element model.

capability for the analysis of concrete structures under cyclic and/or dynamic loading. The plastic-damage model in ABAQUS is based on the models proposed by Lubliner et al. (1989) and by Lee and Fenves (1998). The adopted values of basic parameters of concrete damaged plasticity model including: dilation angle; eccentricity (flow potential eccentricity); f_{b0}/f_{c0} (the ratio of initial equibiaxial compressive yield stress to initial uniaxial compressive yield stress); K (the ratio of the second stress invariant on the tensile meridian to that on the compressive meridian), and viscosity parameter are 30, 0.1, 1.16, 0.667 and 0.005, respectively. In the concrete damaged plasticity model, concrete behavior or damage of compression and tension are defined according to the Appendix C of Code for design of concrete structures (GB 50010-2010).

An elastic-plastic model, considering Von Mises yielding criteria, Prandtl-Reuss flow rule, and isotropic strain hardening, was used to describe the constitutive behavior of steel. In consideration of the Bauschinger effect under cyclic loading, the mixed hardening model was applied for the steel. The yield stress at zero plastic strain and elastic modulus were set according to the test. The value of Poisson's ratio, Kinematic hard parameter, the change ratio of the back stress, the maximum change of the yield surface and Hardening parameter was defined as 0.3, 7500, 50, $0.5f_y$ and 0.1, respectively.

4.3 Numerical results

4.3.1 Skeleton curve
The comparison of the experimental and FE analysis results is shown in Figure 8. Through comparison and analysis, it can be seen that the overall trend of skeleton curves obtained by

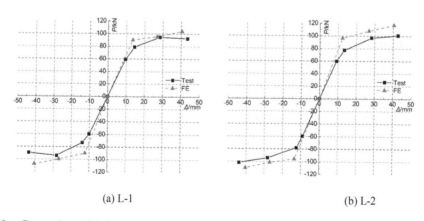

(a) L-1   (b) L-2

Figure 8. Comparison of skeleton curves.

simulation and experiment is basically the same, but the simulation value is slightly higher than the experimental value, and there are some errors. The reasons for the above errors are analyzed as follows: 1) Calculation model error. In this simulation analysis, the reinforced bar element is embedded in the concrete element. The node of the reinforced bar and the surrounding concrete element deforms together without relative slip. Although the tensile hardening introduced in the model can simulate the bond slip of steel bar and concrete to some extent, it is difficult to accurately simulate the serious bond slip of the actual structure at the later stage of loading. In addition, the effect of crack surface cannot be taken into account in the concrete damage plasticity model, thus the local concrete bears high stress state during repeated compression and compression. 2) Experimental errors. It is unavoidable that experimental loading devices and measuring devices will affect the test results. For example, when the bearing is under greater stress in the later stage of loading, slight sliding or disturbance, the gap between beams, columns, hinges, plates and connectors will lead to errors in the experimental results.

4.3.2 *Concrete damage*
Figure 9 shows the distribution of concrete damage factors at various loading stages. The compression damage factor reflects the degree of concrete damage in each stage. It can be seen that in the forward loading (clockwise) to the yield displacement. Firstly, the compressive band is formed at the upper edge of the left beam and the lower edge of the right beam at the junction of the beam segment and the connection ring beam. Then during reverse loading (counterclockwise) to yield displacement, the damage area increases, and the compression strip runs through the entire beam height. At this time, the concrete stress of the ring beam is small, and the damage is not obvious.

4.3.3 *Steel stress*
As can be seen from Figure 10, the most stress-carrying part of the longitudinal rebar appears at the first slide of the connection near the center. The stirrups and longitudinal rebars are far less stress away from the center of the connection. As the displacement value increases from Δ_y to $3\Delta_y$, the maximum part of the longitudinal reinforcement of the beam (red) gradually extends to both sides. The yielding section of the longitudinal rebar is lengthened, and the number of yielding of the stirrup is also increased from the center position near the connection to the distal end of the beam. When comparing clockwise and anticlockwise loading, the stress of each main steel bar is the same as that of the same section of the same layer. It can be seen, during the entire loading process, the stress is concentrated on the connection parts. The steel of beams has uniform stress distribution with no sudden change. As the concrete damage increases, the load shared by the beam-shaped steel increases rapidly. The maximum stress of the steel is concentrated at the connection steel between the beam and the ring beam, and the web stress value is higher than the flange edge.

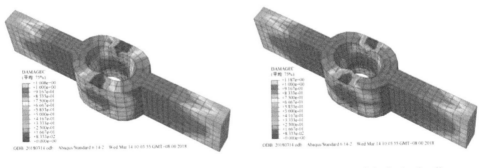

(a) Clockwise loading   (b) Anticlockwise loading

Figure 9.   Compression damage of concrete (40.5mm, $3\Delta_y$).

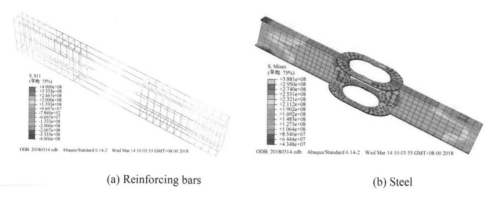

| (a) Reinforcing bars | (b) Steel |

Figure 10.   Mises stress of steel (clockwise loading at 40.5mm, $3\Delta_y$).

5   CONCLUSIONS

Seismic performance of the SRC beams to inclined circular CFST columns connection has been investigated by experimental and numerical study. The following conclusions can be drawn based on the study in this paper:

1. Cyclic loading test was conducted on a scaled specimen. The SRC ring beam around CFTS columns in the joint area ensured the good performance of the connection. The strength and stiffness degradation of the specimen decreases smoothly with the increasing loading displacement. With good ductility and energy dissipation capacity, the proposed connection showed good seismic performance and satisfied the principle of "strong column & weak beam" and "strong joint & weak member" in seismic design.
2. The damage was mainly concentrated in plastic hinge areas located in the junction between the SRC beam and the SRC ring beam around CFST columns, which can be seen from both experimental and numerical study. It is indicated that the proposed connection has a ductile failure mode from the development of the concrete cracks (damage) and residual plastic deformation with the increasing displacement.
3. A finite element model was built to conduct numerical analysis. The finite element model gave efficient simulation of the performance of the connection, which could well verify the ultimate strength and hysteretic damage of the connection. Therefore, it can be used for analyzing this kind of connections and structural design in practice.

ACKNOWLEDGEMENTS

The research in this paper was supported by the National Key R&D Program of China (2016YFC0701902) and the National Natural Science Foundation of China (51638002).

REFERENCES

Ding, F.X. Yin, G.A. Wang, L.P. Hu, D. & Chen, G.Q. 2017. Seismic performance of a non-through-core concrete between concrete-filled steel tubular columns and reinforced concrete beams. *Thin-Walled Structures* 110: 14-26.
GB 50010-2010. 2010. *Code for design of concrete structures.* Beijing: China Architecture & Building Press.
Han, X.L. Huang, C. Fang, C. Wei, H. Ji, J. & Tang, J.M. 2010. Experimental study on spatial inter-secting connections used in obliquely crossing mega lattice of the Guangzhou West Tower. *Journal of Building Structures* 31(1): 63-69.
JGJ/T 101-2015. 2015. *Specification for seismic test of buildings.* Beijing: China Architecture & Building Press.

Lee, J. & Fenves, G.L. 1998. Plastic-damage model for cyclic loading of concrete structure. *Journal of Engineering Mechanics* 124(8): 892-900.

Lubliner, J. Oliver, J. Oller, S. & Oñate, E. 1989. A plastic-damage model for concrete. *International Journal of Solids and Structures* 25(3): 299-326.

Zhang, Y.F. Zhao, J.H. & Cai, C.S. 2012. Seismic behavior of ring beam joints between concrete-filled twin steel tubes columns and reinforced concrete beams. *Engineering Structures* 39: 1-10.

Zhen, W. Yang, Q.S. Tian, C.Y. Sheng, P. & Cheng, T. 2017. Experimental study on intersecting connections used in CFT diagrid frame tube. *Journal of Building Structures* 38(10): 29-37.

Sustainable Buildings and Structures: Building a Sustainable Tomorrow – Papadikis et al. (Eds)
© 2020 Taylor & Francis Group, London, ISBN 978-0-367-43019-1

Smart construction for urban rail transit based on energy-efficient bi-directional vertical alignment optimisation

C. Wu
Department of Electrical and Electronic Engineering, Xi'an Jiaotong-Liverpool University, Suzhou, China
Department of Electrical Engineering and Electronics, University of Liverpool, Liverpool, UK

S. Lu & F. Xue
Department of Electrical and Electronic Engineering, Xi'an Jiaotong-Liverpool University, Suzhou, China

L. Jiang
Department of Electrical Engineering and Electronics, University of Liverpool, Liverpool, UK

G. Gong
Department of Civil Engineering, Xi'an Jiaotong-Liverpool University, Suzhou, China

ABSTRACT: Due to the extensive use of urban rail transit systems in modern cities, the energy consumption of it rises significantly. To help reduce the energy consumption from the planning stage, the paper proposes a linear model to concurrently optimise the vertical alignment and bi-directional train speed profiles based on Mixed Integer Linear Programming (MILP), taking into account the regenerative energy. The paper investigates the scenarios with same station elevation and different station elevation, which clearly illustrates that the optimisation on vertical alignment can reduce net energy consumption by 39.7 % when compared to the reference scenario with vertical alignment not being optimised.

1   INTRODUCTION AND LITERATURE REVIEW

Urban rail transit is regarded as a green and sustainable transportation modal around the world. For improving its energy efficiency, in the past few decades, many academics and engineers have contributed much in energy-saving strategy for urban railway system this field especially at planning and operation stage. Normally, the train needs to accelerate when it departs from the departure station and decelerate when it arrives at the arrival station. The motor needs to consume the energy for acceleration and the braking system needs to dissipate the kinetic energy for deceleration when it arrives. In practical case, the route gradient, also referred to as the "vertical alignment", is generally designed to be sloping but not flat to assist train's acceleration and deceleration, as shown in Figure 1.

Vertical alignment design and optimisation problem has been studied by some researchers since 1990s. Kim designed the symmetric vertical alignment then simulated the train operation on it, showing that introducing the downgrade and upgrade will significantly reduce the traction energy and braking wear (Kim & Schonfeld 1997). The work was extended to simulate train movement, afterwards calculated energy consumption and travel time under various vertical track alignments and train controls on a time-driven basis (Kim & Chien 2010). Inspired by the above studies, asymmetric vertical alignments has been designed (Kim & Schonfeld 2012). The cruising speed was optimized and train operation is simulated to minimize the operational cost for both directions. By taking into account the construction and operating costs, vertical alignment optimisation has been investigated as well (Bababeik & Monajjem 2012). These studies mainly focused on the designing and evaluation of the vertical alignment

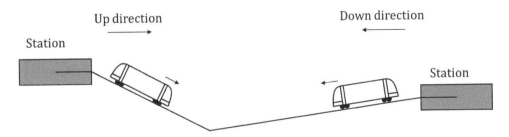

Figure 1. The schematic of the inter-station operation of bi-directional trains.

and the train operation is merely given as the input parameters or simply modelled. In 2018, an integrated model which concurrently optimises one-directional vertical alignment and train operation has been established based on Mixed Integer Linear Programming (MILP) (Wu et al. 2018).

Apart from vertical alignment, finding optimal train speed profile is a traditional way to help save energy consumption. The relevant research have been conducted by many researchers in the field. The optimal speed profile can be obtained through numerical algorithm e.g. Pontryagin maximum principle (PMP) by determining the cruising regimes and connecting two consecutive constant-speed regimes with traction, coasting or braking operation (Khmelnitsky 2000, Albrecht et al. 2016a, 2016b). In recent years, heuristic approach like artificial neural networks (ANN) and genetic algorithms (GA) are applied in finding the optimal speed trajectory (Söylemez & Açıkbaş 2008). Mathematical programming is also used in finding the optimal speed profile. Speed-based model, distance-based model has been proposed to optimise the partial speed trajectory during speed-varying processes such as acceleration and decelerations for recovering more regenerative energy (Lu et al. 2016, Tan et al. 2018, Wu et al. 2018). One key observation is that in these studies about train operation optimization, the gradient information is considered as constraints but not the variables to optimize.

From the literature above, it can be seen that an appropriate vertical alignment design can bring significant energy-saving effect on train operation while the studies on it are still simple e.g. train operation considered as constant parameters or lack of consideration on regenerative energy. To tackle this problem, this paper, based on the previous work (Wu et al. 2018), proposes a more practical mathematical model to optimise both the vertical alignment to minimise bi-directional net energy consumption of train operation, in which the optimal train speed profile is also obtained.

2 MATHMATICAL MODELLING

2.1 Potential energy change analysis

In this paper, the horizontal length between two stations is discretized and the whole journey will be divided into several segments with the same distance Δd, see Figure 2. The amount of the distance segments is $N = L/\Delta d$. At each distance segment, the vertical alignment might

Figure 2. The discretization of one route with vertical segment Δh_i and horizontal segment Δd.

rise, flat or fall thus Δh_i is set to be elevation difference between i^{th} distance segment and $i+1^{th}$ one. Here we set downhill potential energy change $\Delta E_{i,pd}$, uphill potential change $\Delta E_{i,pu}$ as two sets of variables and denote $\Delta E_{i,p}$ as the potential energy change.

Therefore, when the train runs on the downhill, for each distance segment the change of the potential energy of rail vehicles can be expressed in $\Delta E_{i,p} = \Delta E_{i,pd} = Mg\Delta h_i$ (1), where M is the mass of the train and g is the gravitational constant.

$$\Delta E_{i,p} = \Delta E_{i,pd} = Mg\Delta h_i \tag{1}$$

When the train runs on the uphill, for each distance segment the change of the potential energy of rail vehicles can be expressed in $\Delta E_{i,p} = \Delta E_{i,pu} = Mg\Delta h_i$ (2):

$$\Delta E_{i,p} = \Delta E_{i,pu} = Mg\Delta h_i \tag{2}$$

For the downhill and uphill, the gradients of them cannot exceed $\theta = 3\%$ according to the manual (usually 4.5% and here we set it to be 3%) (WMATA 2002), thus we have $\Delta h_i \Delta d \le \theta$ (3):

$$\frac{\Delta h_i}{\Delta d} \le \theta \tag{3}$$

It imposes the corresponding constraints of the potential energy change as shown in $0 \le \Delta E_{i,p} \le Mg\theta\Delta d$ (4):

$$0 \le \Delta E_{i,p} \le Mg\theta\Delta d \tag{4}$$

Since the elevations of the stations are known, thus the potential energy change of train of the whole journey is fixed and needs to satisfy the following constraint as shown in $\sum_{i=1}^{N} \Delta E_{i,p} = Mg(H_1 - H_2)$ (5).

$$\sum_{i=1}^{N} \Delta E_{i,p} = Mg(H_1 - H_2) \tag{5}$$

where H_1 is the elevation of the initial station and H_2 is the elevation of the terminal station.

2.2 Single train motion modelling

For N distance segments, there will be N+1 speed points v_i where i = 1, 2, 3 ... N+1. In the research, the train is assumed to do uniformly accelerated or decelerated motion in each distance segment, which has been commonly recognised by several existing studies (Scheepmaker et al. 2017). Thus we use $-amax \le a_i = \frac{v_{i+1}^2 - v_i^2}{2\Delta d} \le a_{max}$ (6) to indicate that the acceleration or deceleration of each distance segments a_i and its constraints imposed for the comfort and safety of standing passengers.

$$-a_{max} \le a_i = \frac{v_{i+1}^2 - v_i^2}{2\Delta d} \le a_{max} \tag{6}$$

where the a_{max} is the maximum acceleration and deceleration rate.

For ensuring the punctuality, the journey time T of each train is set to be an equation constraint $i = 1N\Delta t_i = \sum_{i=1}^{N} \frac{\Delta d}{v_{i,avg}} = T$ (7).

$$\sum_{i=1}^{N} \Delta t_i = \sum_{i=1}^{N} \frac{\Delta d}{v_{i,avg}} = T \tag{7}$$

where $v_{i,avg}$ is the average speed of each Δd.

During the journey, the drag force for the train can be calculated according to Davis equation $R_i = A + Bv_{i,avg} + Cv_{i,avg}^2$ (8), where A, B and C are Davis coefficients;

$$R_i = A + Bv_{i,avg} + Cv_{i,avg}^2 \tag{8}$$

In order to linearize the nonlinear constraints in $i = 1N\Delta t_i = \sum_{i=1}^{N} \frac{\Delta d}{v_{i,avg}} = T$ (7) and $R_i = A + Bv_{i,avg} + Cv_{i,avg}^2$ (8), a series of ascending key speed points V_1, V_2, ..., V_K are selected to represent the any speed within the range of V_1 and V_K guided by PWL. Hence the decision variables v_i^2 can be expressed by (9). The approximation of the speed v_i' and the average speed $v_{i,ave}'$ can be obtained by (10) and (11). We can further approximate $\frac{1}{v_{i,ave}'}$ and $v_{i,ave}'^2$ as in (12) and (13).

$$v_i^2 = \sum_{k=1}^{K} V_k^2 \cdot \alpha_{i,k} \tag{9}$$

$$v_i' = \sum_{k=1}^{K} V_k \cdot \alpha_{i,k} \tag{10}$$

$$v_{i,ave}' = \frac{v_i' + v_{i+1}'}{2} = \sum_{k=1}^{K} V_k \cdot \beta_{i,k} \tag{11}$$

$$\frac{1}{v_{i,ave}'} = \sum_{k=1}^{K} \frac{1}{V_k} \cdot \beta_{i,k} \tag{12}$$

$$v_{i,ave}'^2 = \sum_{k=1}^{K} V_k^2 \cdot \beta_{i,k} \tag{13}$$

where $\alpha_{i,k}$ and $\beta_{i,k}$ are variables of two sets of SOS2 (Special ordered set of type 2) in each distance segment, in which the integer variables need to be added to formulate these constraints.

2.3 *Energy conversion modelling for bi-directional operation*

In the paper, bi-directional operation and regenerative braking is considered. For modeling the energy conversion of the train operation, according to the law of conservation of energy, for up-line train, the relevant constraint (14) and (15) need to be added.

$$\Delta E_{i,U} \eta_t - \Delta E_{i,p} - \frac{1}{2} M \left(v_{i+1,U}^2 - v_{i,U}^2 \right) - R_{i,U} \Delta d \geq 0 \tag{14}$$

$$\Delta E_{i,U} / \eta_b - \Delta E_{i,p} - \frac{1}{2} M \left(v_{i+1,U}^2 - v_{i,U}^2 \right) - R_{i,U} \Delta d \geq 0 \tag{15}$$

where $\Delta E_{i,U}$ is the energy consumed or regenerated by the train motor in the ith distance segment for up-directional operation, η_t is the conversion efficiency of the traction and η_b is the conversion efficiency of energy regeneration. It is easy to see that when $\Delta E_{i,U} > 0$, train is in traction mode and since $\Delta E_{i,U} \eta_t < \Delta E_{i,U} / \eta_b$, (15) is relaxed; when $\Delta E_{i,U} < 0$, train is in regeneration mode and as $\Delta E_{i,U} \eta_t > \Delta E_{i,U} / \eta_b$, (14) is relaxed. As a result, both constraints can ensure that the energy conversion is reasonable in both traction and regeneration.

Similarly, for down-line train operation, (16) and (17) need to be imposed.

$$\Delta E_{i,D}\eta_t - \Delta E_{N-i+1,p} - \frac{1}{2}M\left(v_{i+1,D}^2 - v_{i,D}^2\right) - R_{i,D}\Delta d \geq 0 \tag{16}$$

$$\Delta E_{i,D}/\eta_b - \Delta E_{N-i+1,p} - \frac{1}{2}M\left(v_{i+1,D}^2 - v_{i,D}^2\right) - R_{i,D}\Delta d \geq 0 \tag{17}$$

It should be noticed is that since the down-line trains run on the opposite direction, along with the direction of movement, the 1st distance segment of the down-line trains is the Nth one for the up-line trains, thus $\Delta E_{1,p}$ is actually $\Delta E_{N,p}$ when the i begins from up-direction.

Considering the characteristics of the train vehicle, the maximum traction/braking force and traction/braking power of the motor should be added as the constraints. Therefore, (18), (19), (20) and (21) are used to ensure the traction/braking force and power that the motor supplies do not exceed their own limits $\overline{F_t}$, and $\overline{P_t}$, $\overline{P_b}$.

$$-\eta_b\overline{F_b}\Delta d \leq \Delta E_{i,U} \leq \frac{\overline{F_t}\Delta d}{\eta_t} \tag{18}$$

$$-\eta_b\overline{P_b}\Delta t_i \leq \Delta E_{i,U} \leq \frac{\overline{p_t}\Delta t_i}{\eta_t} \tag{19}$$

$$-\eta_b\overline{F_b}\Delta d \leq \Delta E_{i,D} \leq \frac{\overline{F_t}\Delta d}{\eta_t} \tag{20}$$

$$-\eta_b\overline{P_b}\Delta t_i \leq \Delta E_{i,D} \leq \frac{\overline{p_t}\Delta t_i}{\eta_t} \tag{21}$$

Both constraints ensure the practicality of the model and make the model extendable to fit different types of rolling stocks.

The objective of the paper is to minimise the net energy consumption of bi-directional train operation. Therefore, the objective function of the integrated optimisation model is shown in (22).

$$\min \sum_{i=1}^{N}(\Delta E_{i,U} + \Delta E_{i,D}) \tag{22}$$

s.t. (1) – (21)

$v_{i,U}^2$ and $v_{i,D}^2$ is the decision variables the model. The objective function shows an integrated optimization process which the optimal vertical alignment can be obtained concurrently when their respective v_i^2 are obtained. Δd is chosen to be 50 m here and this value can be modified according to the precision requirement.

3 CASE STUDIES AND RESULTS DISCUSSION

Based on the linear programming model we propose above, this section will illustrate the case study results, followed by the discussions on the specific scenarios. The parameters used in the optimisation process are tabulated in Table 1 and Table 2. In the first case study, scenario 1 and reference scenario, are demonstrated. The second case study illustrates scenario 2.

3.1 Case 1 - Same station elevations

In this case study, the elevations for both the departure station and arrival station are set to be the same, which is -10 m below the ground level. The height of the tunnel is set to be 2.5 m thus the white area in the soil is the tunnel determined by the optimised vertical

Table 1. Parameters for train operation used in the case studies.

	L	T	V_{max}	a_{max}	A	B	C
Parameter	(m)	(s)	(km/h)	(m/s^2)	(kN)	$(kN \cdot h/km)$	$(kN \cdot h^2/km^2)$
Value	2000	120	80	1.38	2.0895	0.0353	0.0842

Table 2. Parameters for rolling stock used in the case studies.

	M	η_t	η_b	$\overline{F_t}$	$\overline{F_b}$	$\overline{P_t}$	$\overline{P_b}$
Parameter	(t)			(kN)	(kN)	(kW)	(kW)
Value	178	0.8	0.8	200	200	5000	5000

alignment, as shown in Figure 3. The traction force it positive and the braking force is negative in the figure. Figure 1-(a) shows the optimal vertical alignment and its corresponding optimal train speed profiles of both up direction and down direction. It is easily observed that the optimal vertical alignment is symmetric, and so are the train speed profiles. In addition, the cruising operation for both directions only depend on the optimised down slope, which also help train maintain a relatively high speed without traction to guarantee the punctuality. The reason for this is that when the bi-directional operation is taken into account, a compromise of operation on both directions need to be made, leading to a balance between the length of the down slope and up slope for both directions to save total energy consumption.

Figure 3-(b) shows the reference scenario for only optimising train speed profiles when vertical alignment is flat with no gradient change. It should be noted that though the train operation for reference scenario can achieve the minimum net energy consumption by applying optimal speed profile, the net energy consumption of it, which is 50.41 MJ for both directions, is still higher than that of the scenario 1, which is 30.40 MJ for both directions. It implies that making the most of down slope/up slope is able to assist acceleration/deceleration and further reduce net energy consumption. This assistance directly reduces the traction energy consumption and it can be easily observed in Figure 1 that the traction distance and braking distance of reference scenario are longer than that of the scenario 1.

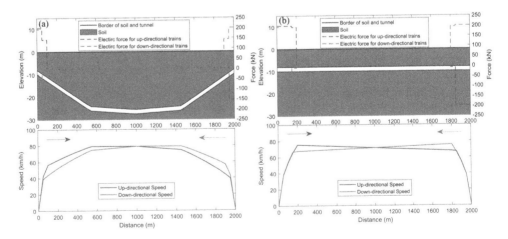

Figure 3. Optimal vertical alignment (a) and reference vertical alignment (b).

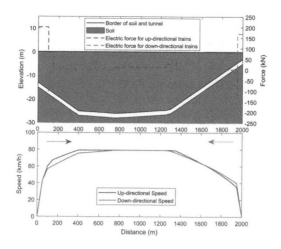

Figure 4. Optimal vertical alignment and train speed profile for different station elevations.

3.2 *Case 2 - Different station elevations*

Normally train stations are located on different elevations, which causes a height difference of the two ends of the route. Relative height difference can be adjusted according to the requirement of different case scenarios. In this scenario, the elevation of the initial station and terminal station is set to be -15 m and -5 m below horizontal line respectively. The optimal vertical alignment and train speed profiles of this scenario is not symmetric, as shown in Figure 2. It is demonstrated clearly that the down slope for down-directional operation is longer that for up-directional operation, resulting in a shorter traction distance and lower traction force for down-direction trains.

Additionally, since for up-directional operations the train will meet with a longer up slope when compared to down-directional operations, it needs to conduct traction in the middle of the journey to maintain a constant speed, to ensure a less radical deceleration on the relatively steep slope followed. The total net energy consumption for both directions is 33.71 MJ, which is a bit higher than that of the optimal result for scenario 1. Theoretically when there is a different height different between two ends of the route, the energy consumption of train operation from higher position to lower position is reduced since the potential energy stored by the height difference helps reduce the traction energy consumption, leading to a less net energy consumption. Thus the net energy consumption of the down-directional operation of scenario 2 is less than 15.20 MJ (the net energy consumption of train operation for either direction from scenario 1). However, it can be deduced that the total net energy consumption of scenario 2 is still higher than that of scenario 1 due to its more frequent traction operations of up-directional trains which raises the traction energy consumption and consequently neutralises the energy-saving effect brought by the potential energy assistance from down-directional operations.

4 CONCLUTION AND FUTURE WORK

Taking into account the bi-directional operation and regenerative energy of railway system, this paper proposes an integrated linear model to concurrently optimise the vertical alignment and train speed profile based on Mixed Integer Linear Programming (MILP). The work is an extension for the previous work and gives more practical solutions by considering more constraints in real operation. The model can solve the vertical alignment optimisation problem with complex constraints such as different station elevations, different rolling stock characteristics and different operational conditions. The paper conducts two case studies, gives two optimised results based on two scenarios, one with same station elevation and another with different station elevation, and compared them to the reference scenario when vertical alignment is flat with not gradient change. The results shows that the optimised vertical alignment can bring substantial energy-saving potential, reducing net energy consumption by 39.7 % compared to the reference scenario.

In the future, the research will be extended further to solve the vertical alignment optimisation problem considering the multiple inter-station operation and civil constraints such as soil conditions and construction cost. The exploration will help reduce the energy cost of the urban railway system from the planning stage, which will bring significant contribution in the field.

REFERENCES

Albrecht, A. Howlett, P. Pudney, P. Vu, X. & Zhou, P. 2016a. The key principles of optimal train control-Part 1: Formulation of the model, strategies of optimal type, evolutionary lines, location of optimal switching points. *Transportation Research Part B: Methodological* 94: 482–508.

Albrecht, A. Howlett, P. Pudney, P. Vu, X. & Zhou, P. 2016b. The key principles of optimal train control-Part 2: Existence of an optimal strategy, the local energy minimization principle, uniqueness, computational techniques. *Transportation Research Part B: Methodological* 94:509–538.

Bababeik, M. & Monajjem, M. S. 2012. Optimizing longitudinal alignment in railway with regard to construction and operating costs. *Journal of Transportation Engineering* 138(11): 1388–1395.

Khmelnitsky, E. 2000. On an optimal control problem of train operation. *IEEE Transactions on Automatic Control* 45(7): 1257–1266.

Kim, D. N. & Schonfeld, P. M. 1997. Benefits of dipped vertical alignments for rail transit routes. *Journal of Transportation Engineering* 123(1): 20–27.

Kim, K. & Chien, S.I.J. 2010. Simulation-based analysis of train controls under various track alignments. *Journal of Transportation Engineering* 136(11): 937–948.

Lu, S. Wang, M.Q. Weston, P. Chen, S. & Yang, J. 2016. Partial train speed trajectory optimization using mixed-integer linear programming. *IEEE Transactions on Intelligent Transportation Systems* 17(10): 2911–2920.

Kim, M. Schonfeld, P. & Kim, E. 2012. Comparison of vertical alignments for rail transit. *Journal of Transportation Engineering* 138(1): 90–97.

Scheepmaker, G. M. Goverde, R.M. P. & Kroon, L. G. 2017. Review of energy-efficient train control and timetabling. *European Journal of Operational Research* 257(2): 355–376.

Söylemez, M. T. & Açıkbaş, S. 2008. Coasting point optimisation for mass rail transit lines using artificial neural networks and genetic algorithms. *IET Electric Power Applications* 2(3): 172–182.

Tan, Z. Lu, S. Bao, K. Zhang, S. Wu, C. Yang, J & Xue, F. 2018. Adaptive partial train speed trajectory optimization. *Energies* 11(12): 3302.

Washington Metropolitan Area Transit Authority (WMATA). 2002. *Manual of design criteria.*

Wu, C. Lu, S. Xue, F. & Juang, L. 2018. Earth potential as the energy storage in rail transit system - on a vertical alignment optimization problem. *21st International Conference on Intelligent Transportation Systems (ITSC), IEEE 2729–2734.*

Wu, C. Zhang, W. Lu, S. Tan, Z. Xue, F. & Yang, J. 2018. Train speed trajectory optimization with on-board energy storage device. IEEE Transactions on Intelligent Transportation Systems, IEEE, PP(Mld) 1–11.

Sustainable Buildings and Structures: Building a Sustainable Tomorrow – Papadikis et al. (Eds)
© 2020 Taylor & Francis Group, London, ISBN 978-0-367-43019-1

Optimization of neutral section location on high-speed railways with consideration of train operations

R. Miao, C. Wu, S. Lu & F. Xue
Department of Electrical and Electronic Engineering, Xi'an Jiaotong-Liverpool University, Suzhou, China

Z. Tian & S. Hillmansen
Department of Electric, Electrical and Systems Engineering, University of Birmingham, Birmingham, UK

ABSTRACT: Neutral Section (NS), also referred to as the "phase separation section", is an important component for electrified railways. NS leads to forced coasting operation and thus its location has an impact on energy consumptions of train operation for both running directions. This paper proposes a Mixed Integer Linear Programming (MILP) model to optimize the NS location and the train speed trajectory concurrently, so as to achieve a minimum energy consumption of train operation. A case study on a 21-km inter-station journey for high-speed railway is done, and the energy consumptions for the optimal NS location have been calculated and compared with the ones for other NS locations. The energy consumptions for scenarios with various NS locations are calculated by applying the optimal train trajectory with pre-determined NS locations. It is found that the minimum energy consumption based on the optimal NS location can achieve a 1.12% energy-saving rate in comparison to the mean value of energy consumptions for scenarios with other NS locations and can save 6.12% energy consumption comparing to the worst case. This is because the forced coasting resulted from NS contributes less to increase of energy consumption within some areas, where more operations of cruising and coasting are likely to be taken and less traction and braking can be replaced.

1 INTRODUCTION

For electrified railways, a neutral section, also referred to as the "phase separation section", is an important component and needs to be designed systematically. Neutral section (NS) as an electrical insulator, is used for separating two electric circuit loops with different electrical phases in railway power supply systems (KieBling et al. 2018). NS can be classified as long NS, short NS and split NS, according to the length of the NS and the distance between pantographs. This is introduced in the standard requirement (British Standard Institution 2017). The location of the neutral section can affect operation safety, running time, and energy consumption. Without power supply in NS, trains run with a forced coasting operation. Based on the theory of train optimal control, such a forced coasting operation upon different locations will systematically affect the total energy consumption in a journey (Albrecht et al. 2016a). This provides room for optimal design of neutral section location to achieve the minimum energy consumption.

As shown in Figure 1, NS is a gap in the catenary between Phase A and Phase B. For a train running across the neutral section, its main circuit of the train is opened before entering. After coasting out of NS, the train is repowered. For both directions, the train undergoes the same process.

In the past studies, the location of the neutral section has been optimized to achieve the minimum delay time and energy consumption. In 2013, (Han et al. 2013) mentioned that the NS location design relied on engineers' experience without designing systematically. This

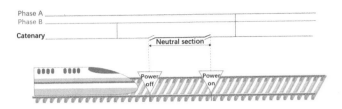

Figure 1. Neutral section demonstration.

paper applied the genetic algorithm on the optimization scheme of neutral section and demonstrated that the proposed method can reduce the energy consumption with less running time compared to the schemes in industry. In 2014, (Song et al. 2014) used a differential evolution algorithm on the optimization of the layout of neutral sections in terms of length and locations. The paper aimed to obtain the minimum total delay time. It was focused on the chasing section and analysis of the relationship between the delay time and different entering speeds as well as different final speeds (the speed when trains exit the neutral section).

Apart from NS location optimization, many approaches have been applied for train energy-efficient operation. Numerical methods based on Pontryagin's Maximum Principle (PMP) are popular indirect methods to obtain continuous solutions (Albrecht et al. 2016a & b). Genetic Algorithm (GA) along with other heuristic algorithms are tools for train energy-efficient operations, as proposed by (Yan et al. 2013), a two-stage model was demonstrated to be effective in locating the optimal solution for urban rail transit line. On the other hand, mathematical programming as a direct method has been applied in the field. In 2017, (Ye & Liu 2017) used nonlinear programming method to do optimal train control based on closed-form expressions. (Tan et al. 2018) compared MILP to PMP method and demonstrated the MILP has a short computational time. In 2018, (Wu et al. 2018) proposed an integrated model combining both train operation and energy storage management based on MILP.

From the above literature review, research on NS location optimization has not been fully studied. The location of the neutral section is usually designed relying on engineers' experience based on safety consideration but not necessarily on energy consumption. In the meantime, most of the papers on train energy-efficient operation has not considered the design of neutral section location. This paper proposes an integrated model based on MILP to concurrently address the optimization problem of neutral section location and train speed trajectory. A case study on the section of a typical high-speed railway has been conducted to demonstrate that the proposed method is able to obtain optimal NS location and train speed trajectory.

The rest of this paper is arranged as follows. Introduction of the model is in Section 2. The case study and discussion are in Section 3 followed by the conclusion and future work in Section 4.

2 METHODOLOGY

2.1 Objective function

In this model, the track (L) is divided into N intervals and the train is supposed to do uniformly accelerating or decelerating in every interval. The train is treated as a particle with its length not being considered here. The objective of this model is to achieve two pieces of speed trajectories for both directions, with such speed trajectory optimization having minimum energy consumption. The distance for every interval is Δd. The energy consumed for the motion on the i_{th} interval is ΔE_i where i is the distance interval index.

In the problem about the neutral section, both up running and down running directions are considered. Therefore, this model considers the minimum energy for both directions. $\Delta E_{i,up}$ and $\Delta E_{i,down}$ denotes the energy supplied for up running and down running on the i_{th} interval, respectively.

172

Its objective function is

$$\min : \Sigma_{i=1}^{N} \Delta E_{i,up} + \Sigma_{i=1}^{N} \Delta E_{i,down} \tag{1}$$

2.2 Model formulation for energy condition and neutral section

On each interval, there are three variables related to energy consumptions and regenerated during train operations for each direction. The net energy consumption on each interval (ΔE_i) is contains two parts, the energy provided for train's traction ($\Delta E_{i,t}$) and the energy collected from the regenerative braking ($\Delta E_{i,b}$). In this model, traction energy supplied by the power network is regarded positive and the regenerative energy collected is regarded negative. For each direction, the following two constraints apply.

$$\Delta E_{i,up} = \Delta E_{i,t,up} + \Delta E_{i,b,up} \tag{2}$$

$$\Delta E_{i,down} = \Delta E_{i,t,down} + \Delta E_{i,b,down} \tag{3}$$

A logic variable b is introduced to ensure both traction and braking operation cannot occur simultaneously, by (4) and (5) as follows.

$$0 \leq \Delta E_{i,t} \leq b \times E_{i,t,max} \tag{4}$$

$$0 \leq -\Delta E_{i,b} \leq (1 - b) \times E_{i,b,max} \tag{5}$$

Considering the characteristic of NS, no energy is consumed by the train and the train can only coast without traction and braking force applied. The logic variable z is introduced to represent the locations of NS. On the intervals where NS is located, the corresponding z equals to 0 and z equals to 1 otherwise. ΔE_{max} is a sufficiently large number to represent the maximum energy allowed for an interval.

$$\Delta E_i \leq \Delta E_{max} \times z_i \tag{6}$$

$$\Delta E \geq 0 \times z_i \tag{7}$$

If NS covers more than one interval in this model, the following equations ensure the consecutive intervals occupied by NS to be zero while the others be one.

$$\Sigma_{i=1}^{N} u_i = K \tag{8}$$

$$x_i - y_i = =u_i - u_{i-1} \tag{9}$$

$$x_i = y_{i+K} \tag{10}$$

$$z_i = 1 - u_i \tag{11}$$

K refers to the numbers of intervals occupied by the specific neutral section, and $K\Delta d$ is the length of NS. x_i, y_i are two series of logic variable, with one component of each is 1 and others are zero. The 1's location in x_i denotes the first interval where the NS start and the 1's location in y_i denotes an interval before which the NS end. u_i is a series of logic variable, containing a consecutive series of 1 between the 1's location in x_i and y_i. z_i is the inversion of u_i and represents the NS location.

173

2.3 Model formulation for trains motion and energy conversion

2.3.1 Model for train movement

The train is treated as a particle on each interval. The average speed is calculated by (12).

$$v_i + v_{i+1} = 2v_{i,ave} \tag{12}$$

where v_i and v_{i+1} are the initial and ending instant speeds for each distance interval.

It is assumed that the train is doing uniformly acceleration or deceleration on each interval, and thus its speed and acceleration relationship is shown in Eq. (13).

$$v_{i+1}^2 - v_i^2 = 2a_i \Delta d \tag{13}$$

a_i is positive for acceleration and is negative for deceleration, and it is constrained by the maximum A_{max} and minimum values $-A_{max}$ in (14). The speed is also limited in (15)

$$-A_{max} \leq a_i \leq A_{max} \tag{14}$$

$$0 \leq v_i \leq V_{max} \tag{15}$$

The time spent on each interval is Δt_i, determined in (16). Sum of the time denotes the total time spent on the journey, restricted by maximum time T_{total} allowed in (17).

$$\Delta t_i = \Delta d / v_{i,ave} \tag{16}$$

$$\Sigma_{i=1}^N t_i \leq T_{total} \tag{17}$$

The kinetic energy changed in intervals must not exceed the one determined by the maximum force and maximum power of the electric machines. (18)-(21) gives the constraints.

$$-F_{max}\Delta d \leq f_{i,up}\Delta d + mg\Delta h + 0.5m(v_{i+1,up}^2 - v_{i,up}^2) \leq F_{max}\Delta d \tag{18}$$

$$-P_{max}\Delta t \leq f_{i,up}\Delta d + mg\Delta h + 0.5m(v_{i+1,up}^2 - v_{i,up}^2) \leq P_{max}\Delta t \tag{19}$$

$$-F_{max}\Delta d \leq f_{i,down}\Delta d + mg\Delta h + 0.5m(v_{i,down}^2 - v_{i+1,up}^2) \leq F_{max}\Delta d \tag{20}$$

$$-P_{max}\Delta t \leq f_{i,down}\Delta d + mg\Delta h + 0.5m(v_{i,down}^2 - v_{i+1,up}^2) \leq P_{max}\Delta t \tag{21}$$

F_{max}, P_{max} stand for maximum force and power for traction and $-F_{max}$, $-P_{max}$ stand for the maximum ones for during the braking. $f_{i,up}$, $f_{i,down}$ denotes the drag force defined by the Davis equation (22). Δh is the altitude changed on each interval.

$$f_i = A + Bv_{i,ave} + Cv_{i,ave}^2 \tag{22}$$

This model has nonlinear relationship between speed for each interval for up running and for down running, and their average speed on every intervals, with symbols being v, v^2, v^{-1} and v_{ave}, v_{ave}^2, v_{ave}^{-1}. The SOS2 is employed to realize piecewise linear by (23)-(27). SOS2 is a method proposed by IBM and is used in many researched models (Tan et al., 2018). $V_1, V_2 \ldots V_P$ is a series of speed key point ranged from 0 to V_{max}. α_p, β_p are sets of variables of SOS2.

$$v^2 = \Sigma_{p=1}^P V_p^2 \alpha_p \tag{23}$$

$$v = \Sigma_{p=1}^{P} V_p \alpha_p \tag{24}$$

$$v_{ave} = \Sigma_{p=1}^{P} V_p \beta_p \tag{25}$$

$$\frac{1}{v_{ave}} = \Sigma_{p=1}^{P} \frac{1}{V_p} \beta_p \tag{26}$$

$$v_{ave}^2 = \Sigma_{p=1}^{P} V_p^2 \beta_p \tag{27}$$

2.3.2 *Energy conversion*

The following constraint (28) is based on the Law of Conservation of Energy during energy conversion. Energy is converted in different forms between electrical, kinetic and potential energy with heat generated along the process. η_t refers to the traction energy efficiency and η_b refers to the regenerative braking energy efficiency.

$$\Delta E_{i,t} \eta_t + \Delta E_{i,b}/\eta_b - f_i \Delta d - mg \Delta h - 0.5m(v_{i+1}^2 - v_i^2) \geq 0 \tag{28}$$

Since the energy consumed or regenerated must not exceed the maximum value determined by the maximum allowed traction/braking force and power, constraints (29)-(32) are imposed.

$$0 \leq \Delta E_{i,t} \leq F_{max} \Delta d / \eta_t \tag{29}$$

$$-\eta_b F_{max} \Delta d \leq \Delta E_{i,b} \leq 0 \tag{30}$$

$$\Delta E_{i,t} \leq \frac{P_{max} \Delta d}{v_{i,ave} \eta_t} \tag{31}$$

$$\Delta E_{i,b} \geq -\frac{\eta_b P_{max} \Delta d}{v_{i,ave}} \tag{32}$$

3   RESULT AND DISCUSSION

The optimization result is shown in Figures 2 & 3 based on a typical inter-station high-speed railway journey with a distance of 21 km. The parameters for the model are presented in Table 1. In this case, we assume a 420-m NS must be applied on the journey. A speed limit is supposed, shown in both Figures 2 & 3. The running time for both directions are fixed at 512 s.

With the changing location of NS, its corresponding minimum energy which is the sum of the both directions, is shown in Figure 2. For the circumstances that NS is applied on the 0-2.52 km or on the last 19.74-21 km, the train is not able to operate within the maximum time (512 s). Forced coasting operation during the initial and final distance has resulted in

Table 1.   Parameters of the model.

	m	F_{max}	P_{max}	η_t	η_b	L	L_{NS}	T_{total}	A	B	C
Parameters	t	kN	kW			km	m	s	kN	kNs/m	kNs^2/m^2
Value	480	250	8437	0.9	0.7	21	420	512	10.689	0.28906	0.011282

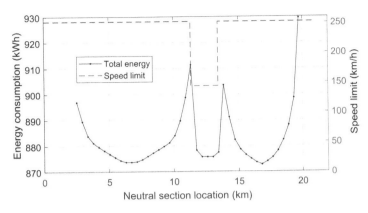

Figure 2. Speed limit and minimum energy consumption for a 512-s journey with various NS locations.

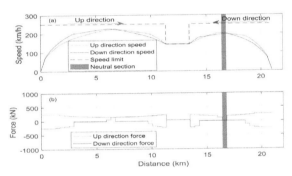

Figure 3. Optimal NS location and its corresponding optimal speed trajectories for up and down directions (a). Tractive effort (b).

infeasible operations for the journey. For all other feasible solutions, the minimum energy consumption is related to the NS location. In particular, for the cases with NS locations on 11 km and 14 km, it is found that the reason for the increase of energy consumption is that the traction operation due to the increase of speed limit is switched into forced coasting resulted by NS. Such a switch causes the train operation to stray away from the original optimal train operations without the impact of NS. Meanwhile, the energy consumption with NS located near the stations is much higher than that with NS located in the middle of the journey. The original braking and traction determined by the optimal train speed trajectory without NS has been replaced by the forced coasting. The great influence on the trajectory due to such a replacement leads to a significant increase of energy consumptions. However, the forced coasting has less impact within some areas, such as the distance range of 6-8 km shown in Figure 2, where more operations of cruising and coasting are likely to be taken and less traction and braking can be replaced, so there is less impact on the energy consumption of the speed trajectory. In practice, to select NS location, it is suggested to avoid the areas where traction and braking are conducted. In normal train operations, the traction and braking areas can be around the speed limit switching point and the area near stations.

The distance of 2.52-19.74 km is a possible range to apply NS. The mean value of the energy consumption with various NS locations in this range is 882.64 kWh. The optimal value is 872.74 kWh, which means 9.90 kWh can be saved for this 21-km inter-station journey. This is a typical journey of a high-speed railway, and 1.12% of energy consumption is saved compared to the mean value in this example. The maximum value of the energy consumption within the range is 929.6 kWh. Comparing to this, the optimal result can save 56.9 kWh, which means 6.12% of electric energy can be saved. The punctuality of the journey is guaranteed as the delayed time due to the NS will be compensated in other areas as constrained by (17).

The optimized train speed trajectory is shown in Figure 3(a) where the optimized NS location is at 16.38-16.80 km. It can be observed that the train is conducting forced coasting during NS. Its corresponding traction and braking forces are shown in Figure 3(b).

4 CONCLUSION

The design of NS location relies on engineers' experience, therefore an optional location can be found in a feasible range, which satisfies the requirement of safety while the energy consumption saving cannot be realized. This paper combines the neutral section location design with the train operation optimization so as to achieve a minimum energy consumption. This paper proposes a Mixed Integer Linear Programming (MILP) model to achieve the optimal location of NS and the optimal train speed trajectory under engineering constraints including traction characteristics and speed limits. A typical 21-km inter-station journey of high-speed rail is studied using this model. It is found that the optimal case can achieve an energy saving of 1.12% compared with the mean value of different feasible cases with various NS locations. This is because the forced coasting resulted from NS contributes less to increase of energy consumption within some areas, where more operations of cruising and coasting are likely to be taken and less traction and braking can be replaced. In practice, to select NS location, it is suggested to avoid the areas where traction and braking are conducted. For the future work, multiple-NS location optimization problem for multi-station operation over a long railway can also be investigated to improve model practicality.

REFERENCES

Albrecht, A. Howlett, P. Pudney, P. Vu, X. & Zhou, P. 2016a. The key principles of optimal train control-Part 1: Formulation of the model, strategies of optimal type, evolutionary lines, location of optimal switching points. *Transportation Research Part B: Methodological* 94: 482–508.

Albrecht, A. Howlett, P. Pudney, P. Vu, X. & Zhou, P. 2016b. The key principles of optimal train control—Part 2: Existence of an optimal strategy, the local energy minimization principle, uniqueness, computational techniques. *Transportation Research Part B: Methodological* 94:509–538.

British Standard Institution. 2017. *Railway applications — Current collection systems — Technical criteria for the interaction between pantograph and overhead line (to achieve free access).*

Han, H. Sun, P. & Wu, J. 2013. Research on optimizing scheme of neutral section based on genetic algorithm. *In Proceedings of the 32nd Chinese Control Conference.* TCCT, CAA 8042–8047.

KieBling, P. Puschmann, R. Schmieder, A. & Schneider, E. 2018. *Contact Lines for Electric Railways Planning, Design, Implementation, Maintenance, Railway Gazette International VO - 174.* Reed Business Information Ltd.

Song, S. Sun, P. & Wang, Q. 2014. Optimization of the layout of neutral sections based on a differential evolution algorithm. *In Computers in Railways XIV Special Contributions* 65–76.

Tan, Z. Lu, S. Bao, K. Zhang, S. Wu, C. Yang, J. & Xue, F. 2018. Adaptive Partial Train Speed Trajectory Optimization. *Energies* 11(12): 3302.

Wu, C. Lu, S. Xue, F. & Jiang, L. 2018. Optimization of Speed Profile and Energy Interaction at Stations for a Train Vehicle with On-board Energy Storage Device. *In 2018 IEEE Intelligent Vehicles Symposium (IV)* 1–6.

Yan, N. Kang, M. & Liao, G. 2013. Heavy-duty freight train longitudinal impulse influence after electric locomotive passing phase separation and optimization strategy. *Technology and Market* 7:5.

Ye, H. & Liu, R. 2017. Nonlinear programming methods based on closed-form expressions for optimal train control. *Transportation Research Part C: Emerging Technologies* 82:102–123.

Sustainable Buildings and Structures: Building a Sustainable Tomorrow – Papadikis et al. (Eds)
© 2020 Taylor & Francis Group, London, ISBN 978-0-367-43019-1

Investigation on mechanical characterizations of metal-coated lattice structure

X. Wang, F. Yuan & M. Chen
Department of Industrial Design, Xi'an Jiaotong-Liverpool University, Suzhou, P.R. China

J. He, P. Wang, Y. Yu & J. Li
Key Laboratory of MEMS of the Ministry of Education, Southeast University, Nanjing, P.R. China

ABSTRACT: Compared with traditional structures, lattice components produced by additive manufacturing have an outstanding mechanical performance with ultra-low density per unit. This lattice structure consisting of periodic unit cells can carry loads in tension or compression. In this study, the numerical simulation is used to explore the effect of the metal coating on the lattice units, and the experiment has been conducted to validate the mechanical properties of metal-coated lattice structure. The results of the simulation indicate that the coating technology can enhance the relative stiffness to density for lattice structures. Nevertheless, the experimental results exhibit the controversial conclusion, which was possibly caused by plating technologies for lattice with thin rods.

1 INTRODUCTION

The development of 3D printing (additive manufacturing (AM) technology) provides more possibilities to produce complex structures. For the perspective of environmental sustainability, AM is also a promising sustainable technology for the future, as it is efficient to reduce not only the consumption of raw materials, but also the energy consumption during transportations and installations. The features of AM include non-pollination(Lozano 2008), energy conservation (Tang et al. 2016), less resource consumption and less emissions(Yang & Li 2018).

Moreover, the AM technology makes the design of structural members more diverse, like the inconsistent cross-sectional area, streamlined curve and non-solid configurations. One typical model was the lattice structure, which was derived from natural cellular structure, such as honeycomb and foam like components. With reference to this natural cellular structure, the artificial lattice components were fabricated to achieve the expectable mechanical properties as ultra-lightweight material. Therefore, the lattice structure has enormous potential for industrial and structural applications (Son et al. 2017, Song et al. 2018, Zheng et al. 2014). The morphological effect of the lattice is widely investigated, such as body-centred cubic (BCC), face-centred cubic (FCC), on the mechanical properties (Junyi & Balint 2016, Xiao et al. 2018).

Additionally, there is one emerging application used in various industrial areas, namely the lattice structure with metallic coating to achieve the light-weight and high-strength criteria (Song et al. 2018). For micro- and nanoscale lattice structures, the multi-scale approach can be applied for the prediction of mechanical properties (Zheng et al. 2016). There is limited work involving the coated lattice structure using both simulation and experimental methods.

In this study, without the consideration of the size effect of the lattice units, the performance of macro- or meso-scale lattice members with/without thin-film coating under tensile forces are studied quantitatively through both of numerical simulation and experiment.

2 METHODOLOGY

There are two steps in this study; numerical simulation was applied to compare the solid structural component (SSC) and lattice structural component (LSC); furthermore, the influence of coating was investigated through experimentation.

2.1 Nickel metal coating

Metallic coating is deposited on the surface of test sample via electroless plating process that has been investigated in our previous work (Li et al. 2019). The process includes four major steps: (1) etching, (2) sensitization/activation, (3) acceleration, and (4) plating. The test samples were first ultrasonically cleaned in isopropanol (C_3H_8O), and then immersed in strong $KMnO_4$ etchant for roughening the sample surface. After rinsing with deionized water (DI) water, the samples are sensitized/activated by dipping into a Pd/Sn colloidal catalyst solution. Then, samples are submerged in acceleration solution for releasing Pd elementary particles. After thoroughly DI water rinsing, the samples are immersed in alkaline nickel bath for metal plating.

2.2 Dimensions of samples

According to ASTM D638-14 (ASTM D638-14 2015) standard test methods for tensile properties of plastics, the numerical models were established in ANSYS. Additionally, at least three samples were prepared for the experiment. Figure 1 and Figure 2 illustrate that LSC by using the lattice unit with 3.5 mm of cubic and 1 mm-diameter rod. The Young's modulus and Poisson's ratio for 3D printed plastic matrix and metallic coating are (2.46GPa, 0.23) and (50GPa, 0.28), respectively. Here, the thickness of film coating is considered from $5\mu m$ to $20\mu m$.

2.3 Numerical simulation

Modelling of metallic film
The stiffness behavior of surface coating was assumed as membrane and bending, namely one-layer element on the top of matrix.

$$E = \begin{bmatrix} kGh & 0 \\ 0 & kGh \end{bmatrix} \tag{1}$$

where, k is the shear-correction factor, G is shear modulus and h is thickness of the shell.

Boundary condition
The model as previously described was established in ANSYS software and the metal coating was set as a surface coating (stiffness behavior was assumed as membrane and bending).

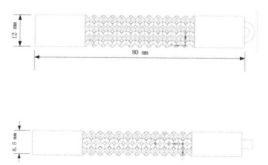

Figure 1. Dimension of testing samples.

Figure 2.   Lattice unit.

Table 1.   Validation of mesh sensitivity.

	Element number	Maximum stress (MPa)	Relative Error
Case-1	136607	153.09	
Case-2	230152	159.49	4%
Case-3	447007	167.26	4.8%

Considering the results of the experiment, 2 mm of constant displacement was applied on the top surface of the model and the surface on the other side was fixed.

Mesh verification
Table 1 shows the sensitivity of the mesh elements for the solid structure, the relative error of case-2 compared with case-1 and the relative error of case-3 compared with case-2 were 4% and 4.8%, respectively. Considering precision and efficiency, the case-2 with 0.2 mm of element size is selected.

2.4   *Experiment processes*

The study aims to investigate the effect of the coating on the lattice structure component (LSC), as well as the effect of different condition of coating treatment (processing time and temperature of corrosive liquid) on coated lattice structural component (CLSC), therefore three groups of CLSC with was designed for cross-reference. According to ASTM D638-14(ASTM D638-14, 2015), the tensile experiments were performed by UTM2503, with a loading rate of 2mm/min. LSC only with etching (E-LSC) was designed to investigate the influence of etching on the matrix.

3   RESULTS AND DISCUSSION

3.1   *Numerical simulation results*

Figure 3 shows the equivalent stress distribution on three models. The position of the maximum stress is close to the fatigue model in Figure 5 and Figure 6. Moreover, under the same

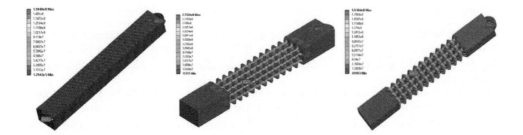

Figure 3.   The equivalent stress distribution on the designed component.

displacement load, CLSC present a comparable maximum stress to SSC, while the maximum stress in LSC increase rapidly due to density reduction.

Furthermore, theoretically the linear or quadratic relationship between the relative compressive or tensile stiffness and the density can be present (Zheng et al. 2014):

$$\frac{E}{E_s} \propto \left(\frac{\rho}{\rho_s}\right)^n, \frac{\rho}{\rho_s} = \frac{V}{V_s} \tag{2}$$

where, E is the Young's modulus or stiffness, ρ is the density and s presents the bulk character of solid material properties.

Therefore, as Figure 4 shows the relative stiffness of CLSC was comparable to that of SSC, while the stiffness of LSC achieved a large increment. Table 2 shows the reduction of tension stiffness and volume of LSC and CLSC compared with those of SSC. The simulation results demonstrate that the use of metal coating on LSC can relieve the decrease of strength caused by the reduction of density of LSC.

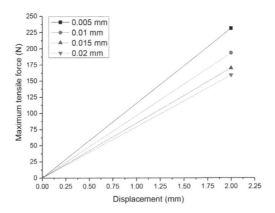

Figure 4.   Critical load factor of the samples.

Table 2.   The equivalent stiffness and volume of models.

	Tension Stiffness (N/mm)	%	Volume (m³)	%	$\frac{E}{E_s}/\frac{\rho}{\rho_s}$
SSC	79.75		8.22×10^{-6}		1
LSC	127.63	60	6.02×10^{-6}		2.19
CLSC	96.52	21.03	6.02×10^{-6}	26.8	0.55

3.2 Experimental results

There is a brittle crack occurring when the LSC samples fail at the maximum load, while CLSC performs a relatively favorable ductility. It is possible to indicate that the coating on LSC can relieve the spalling of the failure of LSC, i.e. the position where the crack occurred by using the coating.

Figure 5(b) shows that the maximum load for LSC is around 251.78 N. Furthermore, by comparing the results from Figure 6, it seems that the coating process has a negative influence on the tensile strength of LSC. However, the possible reason causing such results is probably not because of the coating materials, but the coating process; especially surface etching on matrix, results in the degradation of the matrix. Furthermore, the results of another control group as shown in Figure 5(c), where the samples are only treated by corrosive liquid, demonstrate the reduction of the maximum tensile strength compared with the experimental results of LSC.

Figure 6 & 7 demonstrate that the tensile strength decreases when processing time increases, the maximum load decreases by 26.86% when processing time increases by 10 mins. Based on these results, it is possible to demonstrate that the coating technologies induce damage to the matrix.

The average maximum load and equivalent stiffness are listed in Table 3. As the table shows, the maximum load decreases by 62.27% at most compared with group-4. Nevertheless, the stiffness of CLSC displays a slight improvement. The maximum enhancement of equivalent stiffness is around 28.26%.

According to the results of Zheng et al. (2014) and experimental results of Song et al. (2018) which indicate that the yield strength or elastic modulus was enhanced slightly, the metal-coated lattice in this study achieved a conflicting result. From Figure 5(b) and 5(c), the possible reason is verified according to the comparison between LSC and E-LSC, the remarkable degradation of the strength occurred for the matrix material with the process of etching. This

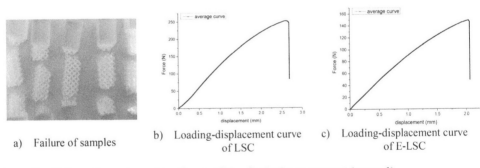

a) Failure of samples   b) Loading-displacement curve of LSC   c) Loading-displacement curve of E-LSC

Figure 5. Failure of samples and loading condition for lattice component (group-1).

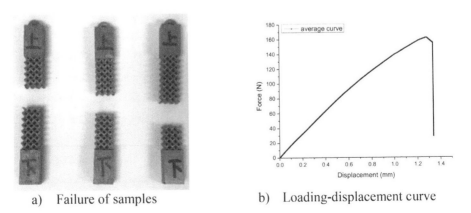

a)   Failure of samples   b)   Loading-displacement curve

Figure 6. Failure of samples and loading condition for coated lattice component (30°C and 10 min).

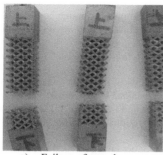

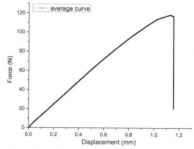

| a) Failure of samples | b) Loading-displacement curve |

Figure 7.   Failure of samples and loading condition for coated lattice component (30°C and 20 min).

Table 3.   The ultimate force and equivalent stiffness for tested samples.

	Group-1		Group-2	
			CLSC	
	LSC	E-LSC with 30°C, 10 min	30°C, 10 min	30°C, 20 min
Average Maximum force (N)	251.78	148.64	161.89	118.4
Equivalent stiffness (N/mm)	103.46	75.27	130.82	108.75

possible mechanism is also the reason leading to the difference between the numerical simulation and the experimental results, where for numerical simulation the interface between coating and matrix was assumed to be perfect contact.

4   CONCLUSION

According to the experimental and simulation results, several conclusions were obtained: the results of numerical simulation indicate that metal coating can enhance the relative stiffness to density of LSC. The experimental results show the decrease of CLS strength due to the metal coating. Additionally, the tensile strength decreases with the increase of etching period, the possible reason being etching damage to the matrix of LSC based on the experimental results. Further investigation, such as SEM, should be conducted.

ACKNOWLEDGEMENTS

The authors thank the funding support NSFC (51805447, 61504024), Jiangsu Department of Science and Technology (BK20170418) and XJTLU Key Program Special Fund (KSF-E-01), Fundamental Research Funds for the Central Universities (2242017K40060, 2242019k1G002), the Innovative and Entrepreneurial Talent Plan of Jiangsu Province, China (Grant No. 1106000206), and the science and technique program foundation for elite returned overseas Chinese scholars, Nanjing, China.

REFERENCES

ASTM D638-14. 2015. *Standard Test Method for Tensile Properties of Plastics.*
Junyi, L. & Balint, D.S. 2016. A parametric study of the mechanical and dispersion properties of cubic lattice structures. *International Journal of Solids & Structures* 91: 55–71.

Li, J. Wang, Y. Xiang, G. Liu, H. & He, J. 2019. Hybrid additive manufacturing method for selective plating of freeform circuitry on 3D printed plastic structure. *Advanced Materials Technologies* 4:1800529.

Lozano, R. 2008. Envisioning sustainability three-dimensionally. *Journal of Cleaner Production* 16: 1838–1846.

Son, K.N. Weibel, J.A. Kumaresan, V. & Garimella, S.V. 2017. Design of multifunctional lattice-frame materials for compact heat exchangers. *International Journal of Heat and Mass Transfer* 115: 619–629.

Song, J. Gao, L. Cao, K. Zhang, H. Xu, S. Jiang, C. Surjadi, J.U. Xu, Y. & Lu, Y. 2018. Metal-coated hybrid meso-lattice composites and their mechanical characterizations. *Composite Structures* 203: 750–763.

Tang, Y. Mak, K. & Zhao, Y.F. 2016. A framework to reduce product environmental impact through design optimization for additive manufacturing. *Journal of Cleaner Production* 137: 1560–1572.

Xiao, Z. Yang, Y. Xiao, R. Bai, Y. Song, C. & Wang, D. 2018. Evaluation of topology-optimized lattice structures manufactured via selective laser melting. *Materials & Design* 143: 27–37.

Yang, Y. & Li, L. 2018. Total volatile organic compound emission evaluation and control for stereolithography additive manufacturing process. *Journal of Cleaner Production* 170: 1268–1278.

Zheng, X. Lee, H. Weisgraber, T.H. Shusteff, M. DeOtte, J. Duoss, E.B. Kuntz, J.D. Biener, M.M. Ge, Q. & Jackson, J.A. 2014. Ultralight, ultrastiff mechanical metamaterials. *Science* 344: 1373–1377.

Zheng, X. Smith, W. Jackson, J. Moran, B. Cui, H. Chen, D. Ye, J. Fang, N. Rodriguez, N. & Weisgraber, T. 2016. Multiscale metallic metamaterials. *Nature materials* 15:1100-1106.

Potential for the design of an energy saving facade system using agglomerated cork: Implications in subtropical climates

F. Afonso
Guangzhou University, Guangzhou, China

ABSTRACT: Agglomerated cork is a known material by its contribution to the sustainment of the environment, not only because it is a wholly natural material, without chemical additives, but also because its industrial process of production results from the lowest quality residues of cork or industrial waste material, unsuitable for other applications. It is a reusable material, which means, the cork facade elements can be converted into a new agglomerated material, demonstrating a huge potential for adaptation to existing buildings following a reversible process. It is durable, lightweight, water resistant, low-cost material, some of the properties that may qualify it as suitable for application in large surfaces of vertical construction façades. The aim of this article is to analyze the mechanical, thermal and acoustic characteristics of cork composites against site-specific climatic conditions of subtropical climates and its suitability as an external coating system for residential buildings with the goal to reduce the energy consumption for cooling the inner environment. In high-density cities like Guangzhou, Shenzhen and Hong Kong the majority of the buildings starting from the 1960s until early 21st century (Brach & Song 2006), did not integrate thermal insulation systems into external walls, producing a high level of heat transfer through the external façade from the outside environment during spring and summer seasons. Due to the extremely fast urban growth of the modern Chinese city, little importance is given to the quality of the external walls in current residential building construction. For at least during six months each year the consumption of energy due to air conditioning in Guangdong province is extremely high. The study concluded that substantial energy could be saved by implementing an external coating upgrade to existing buildings. Additionally, this study details the result obtained through software for energy simulations (Design Builder, ENVI-met) demonstrating the potential of this project to produce homogeneous and comfortable inside temperatures, which cools the indoor ambient temperature in summer time.

1 INTRODUCTION

1.1 *Context*

This research studies the potential of a computational and experimental performance analysis method for retrofitting residential buildings based on cork. This is aimed at improving the energy performance of buildings in subtropical climates. The study focuses on Guangdong Province, China. As a consequence of rapid urbanization, Guangdong mega cities have been transformed into congested and overpopulated areas leading to a number of environmental problems such as urban heat island (UHI) and heat stress (Ward 2016). The UHI effect arises mainly from the heat storage capacity of built areas and surface temperatures (Theeuwes 2014). Hence, cooling the building surfaces' temperature by retrofitting the existing buildings could significantly contribute to UHI mitigation, improve the indoor air temperature and lower cooling energy demand (Zinzi & Agnoli 2012).

In Guangdong's province main cities, the building stock accounts for more than 60% of the carbon footprint, which are responsible for 89% total primary energy consumption in cities like Guangzhou and directly linked with the production of greenhouse gas emissions. The majority of the buildings produce a high level of heat transfer through the external façade from the outside environment during spring and summer. For at least six months each year the consumption of energy due to air conditioning in Guangzhou accounts for more than 30% of the total energy consumption.

Based on this data, the main question of this research is how to reduce the energy consumption of residential buildings by retrofitting the building envelope (exterior walls and roof).

1.2 *Method*

A computational method will be used to examine overheating risk and energy consumption of buildings under such conditions for a more sustainable building design. The study will provide a parametric method to identify and quantify critical design features so that specific guidelines can be developed for retrofitting the building envelope (exterior surfaces and roof). The parametric method will take in consideration a layer of cork as the building envelope. Cork is a raw vegetable material, biodegradable and totally recyclable. It is durable, water resistant, water tight, presents very low thermal conductivity and its chemical structure makes it a CO_2 fixer (Dent & Sherr 2014), some of the properties that may qualify it as adaptable to subtropical climatic conditions. At the same time, the cork oak is a Chinese native species and cork has been harvested and cork agglomerates produced commercially in China for decades, which make the project feasible in this region (Brach & Song 2006).

2 RELATED WORK AND HISTORICAL OVERVIEW

2.1 *Background of agglomerated cork building envelopes*

There are not many approaches to the investigation of cork as a material for building envelopes, besides the direct application of expanded cork agglomerates in buildings, or singular experiments without any scientific research output from architects and the cork industry (as an example: the Portuguese Pavilion in the 2010 World Expo Shanghai, designed by Carlos Veloso, Quinta de Portal Winery, designed by Álvaro Siza, and recently the Herzog & de Meuron and Ai Weiwei's London Serpentine Pavilion). Cork agglomerates and cork composites are being used in construction mainly as insulation materials, as thermal and acoustic insulation, as a core material in concrete structures, or as finishes for interior walls and pavements, but it is underused in façade systems (Castro et al. 2010, Moreira et al. 2010). There is no specific methodology that has hitherto applied in the investigation of cork composites on envelope systems in Asia or worldwide (Silva et al. 2005).

3 RESEARCH METHOD AND TECHNICAL ROUTE

3.1 *Experimental method – the Cork House sustainable envelope system*

Agglomerated cork demonstrates a potential for application in architectural envelope systems. In fact, the experiments presented in this paper are the result of previous work in the field of architecture. In particular, the Cork House, a built project by the author, in the north region of Portugal, was a pioneer experiment applying agglomerated cork block as a full coating system - façade and rooftop - for a residential project. A density of 100kgm^{-3} represents a thermal conductivity of 4.0 x10^{-5}kj. Within the Cork House project the author substituted conventional 60mm polystyrene insulation for an equivalent 150mm thickness of 100 kgm^{-3} density cork blocks, thus generating a façade that performs at the same time as an insulation system (Afonso & Magalhães 2012). The thermal and mechanical characteristics of this specific cork agglomerate (very similar in composition to the ones applied within Nuspa project) make this material suitable as an exterior façade system for temperate and dry climates. Figure 1 shows the facade system developed for the Cork House project, as a single layer of

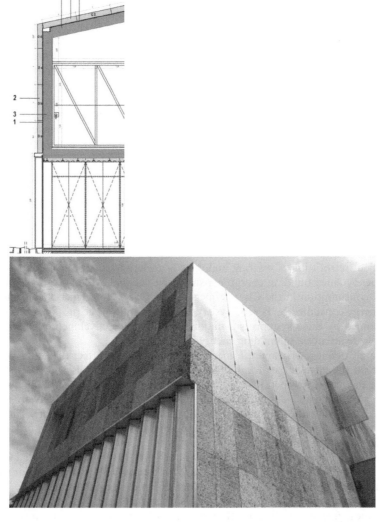

Figure 1. Left: section of the facade system (1-waterproof coating and Adesan-viero glue layer, 2-150mm agglomerated cork panels 120 kg m^{-3} density, 3-concrete structure. Right: detail image of the built house.

agglomerated cork blocks attached directly to the external wall of the building using a specially developed glue based on mineral binders, co-acrylic polymers alkali-resistant (adherence >1,0 N/mm^2).

3.2 *Energy simulations and contributions to sustainability*

The evaluation by life cycle assessment of environmental impacts has been performed after the completion of this project. Results of the impact assessment for the life cycle of a 1m^2 of Cork House envelope:

- Abiotic Depletion Potential: kg Sb-eq. U (unit), 6.25E-01 CtG (cradle to gate)
- Global Warming Potential: kg CO_2 -eq. U, 3.02E+01 CtG
- Ozone Layer Depletion Potential: kg R11-eq. U, 6.09E-06 CtG
- Photochemical Ozone Creation Potential: kg Ethene-eq. U, 4.40E-02 CtG

The unit basis of the calculations correspond to a $1m^2$ of façade composed of 150mm thickness agglomerated cork panels. Environmental impacts evaluated are abiotic depletion, global warming, ozone layer depletion and photochemical ozone creation using the CML 2001 methodology according to the European standard EN 15804 about sustainability of construction works. The cork façade system has a tendency to lower the inner temperature conditions, with a difference of 5°C in comparison with other traditional façade solutions that employ a double wall system with 60mm extruded polystyrene insulation. Being a natural insulation material and an exterior finishing at the same time, agglomerated cork eliminates the facade insulation layer while still complying with the requirements of the European building codes (EN Eurocodes). The benefits are to ensure better indoor conditions and reduce energy consumption, particularly in warm temperate moist climate summer conditions present in Portugal.

4 RESEARCH METHOD OF AN ENVELOPE SYSTEM IN SUBTROPICAL CLIMATE

4.1 *Experimental method*

Energy-efficient retrofitting of existing buildings is an important contribution to the sustainability of society since building energy consumption accounts for more than one fourth of total energy consumption in the Chinese mega-cities. Of the total energy consumption from buildings, energy consumed by HVAC systems accounts for a large proportion, thus reducing energy consumed by these systems is a major issue. Exterior envelope design is a topic that draws worldwide attention in building energy.

This research defines a parametric method to simulate the effect of a thin layer of cork agglomerate on the envelope of the building (exterior surface, walls and roof): this method will improve the indoor temperature and cooling demand of residential buildings under subtropical climates, using a co-simulation approach with customized API software (to be developed) and existing simulations methods (like ENVI-met and Design Builder). The objective is to quantify the real impact of the computational simulation methodology in order to perform a responsible system to improve the indoor conditions.

The isothermal characteristics and the conductivity of the agglomerated cork prototypes were investigated under different conditions: temperature and environmental conditions in air and ambient relative humidity. This task helped to verify the behavior of cork agglomerates under high humidity levels. The fact that cork has strong watertight cell membranes give us confidence that it will not absorb water in subtropical humid climates.

4.2 *Energy simulations for envelope systems in subtropical climatic conditions*

Previous studies were conducted by the author addressing the ability of cork agglomerated envelopes to perform in high-humid subtropical climates. The majority of the applications of cork as façade systems currently are located in Mediterranean and Sub Mediterranean climate regions in Europe. The author executed a selection of scale model tests adjusted to specific environmental conditions in the Hong Kong facade testing laboratory R.E.D. (Research Engineering Development). Isothermal, mechanical resistance and impermeability tests were performed under temperature and environmental conditions in air at ambient relative humidity. Thermal transitions of cork have been tested by calorimetric analysis. Cork contains more than 3.5% adsorbed water at ambient relative humidity, which is likely to produce modifications in its physical properties, but that does not affect its mechanical properties at temperatures between 30–80°C. Within this range of values the thermal conductivity $(Wm^{-1}K^{-1})$ is 0.045. Heat treatment with water was tested, and concluded that water absorption softens the cell walls of cork agglomerates only at very high temperatures, during boiling (Compressive modulus, heat treated at 100°C, 28 days, is 11MPa, in the radial direction) (Afonso et al. 2016). This allowed the verification of the suitability of the introduction of this material under the subtropical climatic conditions (Peel 2007).

In this research paper, a conventional semi-light facade typology was modeled with and without the layer of cork panels, based on MoWiTT type tests designed by the Lawrence Berkeley laboratories at the University of California (Klems 1984). This simulation has been set according to the constraints of each extreme orientation (North, South, East and West) in Hong Kong, China. The conclusions drawn are utilised to calibrate the simulation tool (Design Builder software). The wall has been modeled with the data of a semi-light facade improved with the cork module prototype, and has been simulated over a year in the four previously mentioned extreme orientations in the city of Hong Kong without energy input or cooling elements. Figure 2 shows the final results. The thermal comfort inside the cork model prototype is stable and homogeneous, with registered temperatures between 15 and 25 °C.

This test also allowed to conclude that for facades including the cork panel prototype (with a density of 100kgm^{-3} and 8cm thickness), the thickness of the extruded polystyrene insulation layer would be 2 cm, when a value of thermal conductivity (Wm^{-1}K^{-1}) 0.3 is required. By contrast, for the same facade but without the cork panel addition, the thickness of the extruded polystyrene insulation should be 12.59 cm.

This research shows that if we improve a standard semi-light multilayer facade composed of a extruded polystyrene insulation layer with a thermal conductivity (Wm^{-1}K^{-1}) value 0.031, by adding the cork panel prototype, we will obtain a reduction of 10.59 cm of the necessary thermal insulation for the facade. This data have been obtained considering Hong Kong's climate conditions.

The computer simulations were also performed without energy input or cooling elements considered, and comparing the standard semi-light facade, and the same facade improved with the cork panel subjected to summer test conditions in the city of Hong Kong. It has determined that thermal comfort inside the building, using the cork wall prototype, is stable and homogeneous, which may offer an improvement of 5 °C temperature with respect to not using the cork prototype panel (Figure 3). Both facades have been simulated from the 10th to the 17th of July, in Hong Kong, with a south orientation.

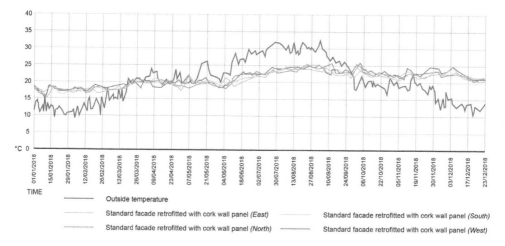

Figure 2.   Thermal behavior inside an adiabatic box.

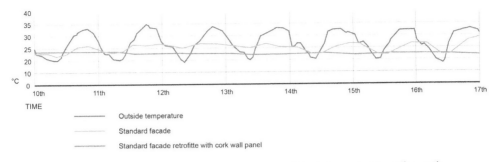

Figure 3.   Comparison of the thermal behavior inside two adiabatic boxes in July 10th to 17th.

5   CONCLUSIONS

The China government and professional associations are trying to solve the problem of energy inefficiency of buildings, but the past initiatives and building envelopes were constructed using archaic techniques poorly adapted to present energy demands.

The output of this research is the development of a customized computational methodology to simulate a new envelope system that is adaptable to the climate of the Guangdong mega cities. The impact will be: maximization of the energy efficiency of the facade surfaces, decrease of the building energy consumption. This could lead to lowering of CO_2 emissions of the buildings and to cost savings. As a future work, this study could evolve into a new research on the reduction in energy use and CO_2 emissions.

This computational method will allow the simulation of a non-intrusive facade system based on a thin layer of cork, for retrofitting of buildings. At the same time, the cork layer allows personalized configurations for different facade typologies.

This study also allows us to conclude:

- The application of cork in building envelopes is of great importance to design innovation, producing a cost effective, environmentally friendly and recyclable technology for reducing the energy demand of the building.
- Envelope systems built on natural composite materials such as agglomerated cork are suitable for application in distinct climatic conditions.

Given the fact that cork is a 100% natural and recyclable material, the released granules can be collected to produce new agglomerates. The waste material resulting from the subtractive fabrication is part of a sustainable process.

We can discuss how to develop further the conceptual ideas of this study in order to provide planers and researchers in the field of architecture and design an objective evaluation tool for the advantages of introducing cork-related products in facade systems as a contribution for the sustainment of the environment.

REFERENCES

Afonso, F. & Magalhães, V. 2012. Cork House. In C. Sant'Ana & M. Leiria (eds), *Arquitectos Anóni mos*: 15-20. Lisbon: Caleidoscópio.

Afonso, F. Kirill, J. Alzira, P. & Aurora, C.M. 2016. Cork matters: towards new applications of cork agglomerates in architecture and product design. *Proceedings of the 50th International Conference of the Architectural Science Association*. Adelaide: The University of Adelaide.

Brach, A. & Song, H. 2006. eFloras: New directions for online floras exemplified by the Flora of China Project. *Taxon* 55(1): 188-192.

Castro, O. Silva. J.M. Devezas, T. Silva, A. & Gil, L. 2010. Cork agglomerates as an ideal core material in lightweight structures. *Materials & Design* 31: 425–432.

Dent, A. & Sherr, L. 2014. *Material innovation in architecture*. London: Thames & Hudson.

Klems, J.H. 1984. Measurements of fenestration net energy performance: considerations leading to development of the mobile window thermal test (MoWiTT) facility. *Journal of Solar Energy Engineering* 8 (3): 165-173.

Moreira, R. de Melo, F.J.Q. & Dias Rodrigues, J.F. 2010. Static and dynamic characterization of composition cork for sandwich beam cores. *Journal of Materials Science* 45: 3350–3366.

Silva, S.P. Sabino, M.A. Fernandes, E.M. Correlo, V.M. Boesel, L.F. & Reis, R.L. 2005. Cork: properties, capabilities and applications. *International Materials Reviews* 50(6): 345–365.

Theeuwes, N.E. Steeneveld, G.J. Ronda, R.J. Heusinkveld, B.G. Van Hove, L.W.A. & Holtslag, A.A.M. 2014. Seasonal dependence of the urban heat island on the street canyon aspect ratio. *Quarterly Journal of the Royal Meteorological Society* 140(684): 2197–2210.

Ward, K. Lauf, S. Keinschmit, B. & Endlicher, W. 2016. Heat waves and urban heat islands in Europe: a review of relevant drivers. *Science of The Total Environment* 569-570: 527–539.

Zinzi, M. & Agnoli, S. 2012. An energy and comfort comparison between passive cooling and mitigation urban heat island techniques for residential buildings in the Mediterranean region. *Energy and Buildings* 55: 66–76.

Sustainable Buildings and Structures: Building a Sustainable Tomorrow – Papadikis et al. (Eds)
© 2020 Taylor & Francis Group, London, ISBN 978-0-367-43019-1

Use of multi-agent system to switch driving strategy in rail transit and procedure simulation

Y.D. Guo, C. Zhang & S.F. Lu
Xi'an Jiaotong-Liverpool University, Suzhou, Jiangsu, China

ABSTRACT: The present paper proposes a multi-agent control system for rail transit. The multi-agent system (MAS) consists of multiple train agents, station agents and a central agent. The implementation of the proposed system is based on distributed optimal control. In the proposed system, each train agent can directly obtain information from the neighboring train agents. Each train agent consists of five subsystems, which are train data set, safety inspection system, timetable inspection system, energy optimization system and trajectory generate system. The train agent can optimize the speed trajectory according to the running state of adjacent trains with the cooperation of the five subsystems. The built-in algorithm of the proposed system can infer the driving strategy (such as time-priority or distance-priority) that should be adopted based on specific situations, which provides a good anti-disturbance ability. Furthermore, the distributed optimization method enables the system to perform a multi-objective optimization in a short time when the system is disturbed.

1 INTRODUCTION

1.1 *Background & relevant research*

The traditional communication-based train control (CBTC) system consists of two subsystems, which are on-board side system and ground side system respectively. The train control could be realized through the assistance of the cooperation between the two sub-systems. When there are multiple trains in the same area, the ground side system needs to interact with all the trains in the relevant area simultaneously. The on-board side system needs to transmit the current position of the train to the ground side system, which needs to determine the movement authorization of each train through calculation and transfer it back to each train.

The present research on the optimization of the traditional rail transit system can be divided into two categories. The first category is timetable optimization, and the other category is train speed optimization. The optimized timetable and speed trajectory for each train could be obtained through these two kinds of optimization. The train driver (or automated train operation system (ATO)) then follows the pre-determined speed trajectory during operation procedure. However, rail transit inevitably faces unexpected interference during the operation process, and some disruptions may change the original running plan of trains. Therefore, the ground side system needs to receive the information of all trains in the corresponding area, analyze the collected information, and calculate the movement authorization for each of them. In addition, the speed trajectory optimization of trains is a typical multi-objective optimization problem, during which safety, punctuality, parking accuracy, energy efficiency and passenger comfort should be considered simultaneously (Wang et al. 2014). At present, the algorithms used to optimize the train speed trajectory can be divided to three types, which are mathematical programming, Pontragin's maximum principle, and heuristic algorithms. Based on the relevant theoretical deduction and simulation, the following significant conclusions is derived: The total energy consumption is related to the speed switching point (Jiaxin & Howlett 1992). The optimum driving strategy of a train consist of maximum acceleration-cruising-coasting-maximum braking (Albrecht et al. 2016, Howlett et al. 2009), and the coasting phase should be started as early as possible (Liu & Golovitcher 2003). In 2019, Yida et al. (2019a &b) attempted to use some multi-agent systems to improve the efficiency and flexibility of train speed trajectory optimization.

Scientists in this fields have made great efforts and made remarkable achievements. However, most of the existing algorithms have high requirements on computational power, which are not able to meet the industrial demand of multi-train optimization when the running plan is disturbed.

1.2 *Motivation and contribution*

The present paper proposes a multi-agent system which could provide intelligence decisions. The MAS system includes a central agent, multiple station agents and multiple train agents. Among them, the central agent and station agents can be understood as the ground side system in the traditional rail transit, which is responsible for sending track data, temporary speed limit and other information to the train. Train agent can be considered as the on-board system, responsible for collaborating with drivers and auxiliary control of train traction and braking. Each train agent can directly communicate with the adjacent agents, and real-time optimization of train speed trajectory can be carried out through the collected data and the collaboration of the internal algorithms.

Compared with the traditional train control system, the system proposed in this paper can switch the driving strategy and provide the speed trajectory optimization with a higher degree of optimization under the premise of some large disturbances.

The rest of the present paper consists of three parts: methodology, case study and conclusion.

2 METHODOLOGY

2.1 *Multi-agent structure*

Figure 1 shows a collaboration example between trains and stations of the proposed multi-agent system (MAS). The figure includes three stations (Station A, Station B, and Station C),

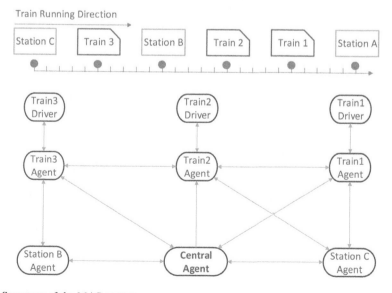

Figure 1.   Structure of the MAS system.

and three trains (Train 1, Train 2 and Train 3). The running direction of the three trains is from Station C to Station A. At moment shown in the figure, Train 1 and Train 2 are running Station A and Station B, whereas Train 3 is running between Station B and Station C. Each train agent in the diagram has direct connections with the driver, central agent, the agent of the next target station, and the agent of the following train. Trains in this mode have two methods of interacting with adjacent trains. The first approach is to transmit information through the direct connected channel. This channel is also the most commonly used channel for train interaction in the framework. The other approach is to employ the central agent as a medium to transfer information between trains. When a train did not receive the expected information through the first approach, the train agent would try to communicate with the central agent to transfer necessary information. The specific switching rule of the two approaches would be introduced in 3.2.

2.2 Algorithm of train agents

This section describes the decision-making process of train agents. The interaction between Train 1 Agent, Train 2 Agent, Train 3 Agent, and Train 2 Driver is shown in Figure 2. In the figure, Train 2 Agent is divided into five modules: Train 2 Data Base, Safety Check System (MBS Based), Time Optimization System, MICP Optimization Systems, and Trajectory Generate System.

An example is adapted here to help the reader better understand the provided sequence diagram. Suppose two trains are running on the same track between stations A and B, Train 1 is the leading train, and Train 2 is the following one. According to the predefined schedule, Train 1 should leave station A at 10:00 and arrive at station B at 12:00. Train 2 should leave station A at 10:30 and arrive at station B at 12:30. In the running process, the two trains need to meet the security requirements at any time based on MBS. Train 1 Agent and Train 2 Agent transfer information in near-real time during the process, and transfer abnormal information to relevant drivers when the predefined plan needs to be changed or aberrant situation occurs.

When Train 1 Agent drives according to the predefined velocity trajectory, Train 2 Agent can also drive according to the expected velocity trajectory. Therefore, Train 1 Agent is responsible for transmitting the current driving plan to Train 2 Agent without waiting for the information returned by Train 2 Agent when executing command 1 in Figure 2.

However, when Train 1 Agent finds Train 1 unable to drive according to the predefined velocity trajectory due to some unexpected reasons, it is necessary for Train1 Agent to firstly calculate the updated speed trajectory of Train1 according to specific conditions, and finally transfer the new trajectory to the database of Train 2 Agent. It should be noted that the trajectory of Train1 has changed in this case, the Train 2 trajectory may need to be changed accordingly to adapt to the new environment. Therefore, Train 1 Agent is expected to receive the information returned by Train 2 Agent. If Train 1 Agent successfully receives the returned information, it proves that the information channel is normal. If no reply is received within the expected time, it indicates that the information channel is abnormal or the relevant optimization demand takes too long. If such case occurs, Train2 will be forced to stop by the Central Agent to ensure safety.

Assume that Train 1 is in normal operation between 10:00 and 11:00, a line failure occurs at 11:00, and Train1 needs to stop for 15 minutes to solve the fault.

Then from 10:00 to 11:00, Train1 agent continuously sends the position of Train 1 to Train 2 Agent and reminds Train 2 Agent that Train 1 is driving continuously according to the predefined speed trajectory (the process No. 1 in Figure 2). Train 1 Agent received the signal from Train 1 Driver at 11:00. According to the signal, Train 1 Agent know that Train 1 needs to stop for 15 minutes to deal with line failure. Thus Train 1 Agent optimizes train speed according to the new situation and transfers the newly generated speed trajectory to Train 2 Database. Train 2 Agent first needs to transfer the new trajectory of Train1 to the Safety Inspection System (the process No. 3 in Figure 2). The system will compare the predefined Train 2 trajectory with the new Train 1 trajectory to check whether the Train 2 trajectory can still meet the safety requirements.

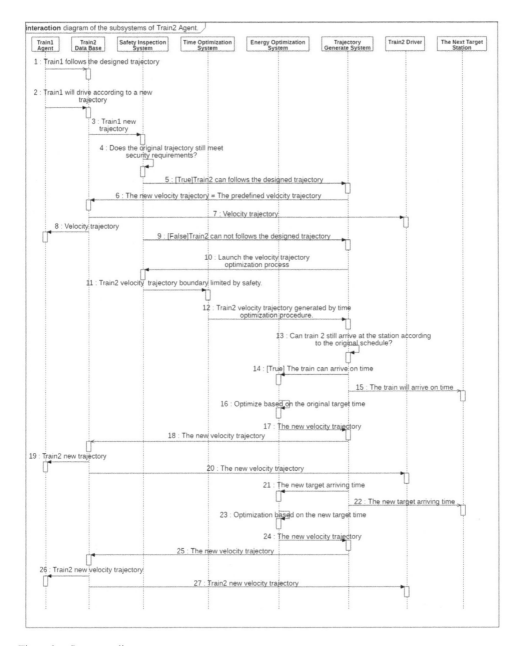

Figure 2. Sequence diagram.

If the safety requirements can be met, Train 2 will adopt the 4-8 process in Figure 2 and still take the predefined speed trajectory. Because Train Agents assumes that the predefined trajectory has been highly optimized offline, and the process of offline optimization already considered the time table and global optimization.

However, if the results of safety inspection show that the predefined speed trajectory of Train 2 no longer meets the safety requirements, the system will transfer the inspection results to the Trajectory Generation System. Then the system launches a new trajectory generation process. The Safety Inspection System then passes the generated trajectory boundary that

195

restricted by security to the Time Optimization System. The Time Optimization System will first calculate the earliest time that the train can reach Station B.

It could be found out that whether the train will be late or not by comparing the planned arrival time with the earliest arrival time. The train would able to ensure punctuality if the earliest arrival time is earlier than the planned arrival time, which can be expressed by Equation (1):

$$T_{earliest} - T_{planned} \leq 0 \tag{1}$$

If the situation here satisfies the formula (1) (for example, Train 2 Agent determines its earliest arrival time would be 12:20, which is 10 minutes earlier than the planned 12:30), then the excess time can be used to optimize energy through the energy optimization system. So the Time Optimization System will pass instructions to the energy optimization system. The energy optimization system will take the arrival time specified in the predetermined schedule as the limiting condition, generate the speed trajectory based on the energy optimization, and transfer the speed trajectory to the database of Train 2, which will be transferred back to the driver of Train 2 and the agent of Train1 respectively.

If the train fails to meet the conditions shown in Equation (1), Train 2 Agent will know that delay is inevitable. In this case, the train will take the shortest time obtained in this step as the limiting condition and transfer it to the energy optimization system for energy optimization calculation, so as to obtain the energy optimization trajectory under the premise of minimizing delay. Finally, the obtained speed trajectory is transferred to the Train 2 database, which will be transferred to the Train 1 Agent and Train 2 Driver.

During the whole process, parameters such as maximum acceleration are limited to ensure passenger comfort.

3 CASE STUDY

The Huagang Station (departure Station) and West Jiangling Road Station (target Station) of Suzhou metro line 4 are selected to demonstrate the system. The distance between the two stations is around 3km. Assume that the specific data of the metro line are as follows: (1) The maximum speed limit between the two stations is 20 m/s. According to the predefined schedule and speed trajectory, the trains running between the two stations were supposed to have an accelerating - cruising at 15m/s -coasting- braking driving strategy. (2) The headway between Train1 and Train2 is 100 seconds. (3) the weights of both trains are 80 ton. (4) The Davis coefficient of the two trains adopts the common values of A group of urban rail trains, A=3.6449, B=0.001710, and C=0.01134. (5) At any time, the two trains need to meet the security condition restricted in the moving block system, and the block is 500m.

Train 1 left Station 1 at t=0 (s), and Train 2 left Station 1 100 seconds later. The two trains first accelerated to the original planned cruising speed and then start to cruise at the speed. The driver of Train 1 found that a temporary stop was needed to deal with an emergency faulty. Based on his own experience, the driver gives the Train 1 Agent an instruction that the estimated stopping time is 90 seconds. After receiving the instruction, Train 1 Agent recalculated the velocity trajectory and transmitted the newly generated trajectory to Train 2 Agent. After that, Train 2 Agent found that in order to meet the safety requirements of MBS, Train 2 must decelerate, and Train 2 will be delayed due to this deceleration. Train 2 Agent first adopted the coasting strategy. When Train 2 slows down to the critical cruising speed, it begins to cruise with a speed-hold mode. The critical velocity not only meets the requirement of Train 2 to avoid additional kinetic energy loss, but also meets the safety specification of MBS. In scenario 1, Train 1 fixed the problem at t=260 (s) and began to accelerate. When the speed of Train 2 reached the speed at which Train 1 was cruising, the two trains began to accelerate simultaneously. At that moment, both trains were in a state of delay, so the agents of both trains adopted the least time driving strategy. The two trains first accelerate to the maximum speed limit (20m/s) and cruise with the speed. Finally, both trains took maximum

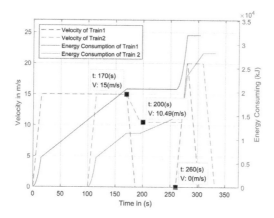

Figure 3. Train velocity and energy consumption for scenario 1.

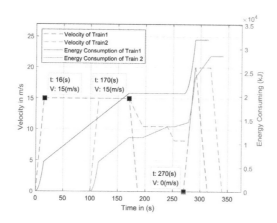

Figure 4. Train velocity and energy consumption for scenario 2.

braking when they approaching the target station. The first 240 seconds of scenario 2 are the same as in scenario 1. However, at the time of t=240s, the driver of Train 1 found that the accident situation requires some extra time to deal with the fault. The new estimated movement time is t=270. Therefore, follows similar process, Train 1 Agent sends the new situation to Train 2 Agent, and the driver of Train 2 starts to coast again according to the new situation, until reaching the new critical speed of 8.2 m/s. At t=270 seconds, Train 1 solved the fault and began to move. The two cars adopted the driving strategy with the shortest time and arrived at the station one by one. The speed trajectory and total energy consumption of the two trains in scenario 1 and 2 are shown in Figures 3 and 4 respectively.

4 CONCLUSION

In this paper, a multi-agent rail transit control system including train agent, station agent and central agent is proposed and its internal algorithm diagram is introduced. Each train agent in the system can directly exchange information with other adjacent train agents in the system and make decisions according to sub-systems. Furthermore, the train agent can obtain information from the driver and provide the driver with driving strategy suggestions. The proposed system has the potential to improve data exchange efficiency and optimization speed. Suzhou metro line 4 is taken as an example to discuss the feasibility of the system. The results show that the system can accomplish multi-objective optimization and switch driving strategies. In the future, the station agent and central agent will be designed, and the collaboration rules between different agents will be established.

ACKNOWLEDGEMENT

We would like to acknowledge the support from the Key Program Special Fund of Xi'an Jiaotong-Liverpool University, with project code KSF-E-04.

REFERENCES

Albrecht, A.R. Howlett, P. Pudney, P. Vu, X. & Zhou, P. 2016. The Key Principles of optimal train control—Part 1: Formulation of the model, strategies of optimal type, evolutionary lines, location of optimal switching points. *Transportation Research Part B: Methodological* 94:482–508.

Howlett, P.G. Pudney, P.J. & Vu, X. 2009. Local energy minimization in optimal train control. *Automatica* 45(11): 2692–2698.

Jiaxin, C. & Howlett, P. 1992. Application of critical velocities to the minimisation of fuel consumption in the control of trains. *Automatica* 28(1):165–169.

Liu, R. & Golovitcher, I.M. 2003. Energy-efficient operation of rail vehicles. *Transportation Research Part A: Policy and Practice* 37(10): 917–932.

Wang, Y. & Bart, D.S. 2014. Optimal trajectory planning for trains under fixed and moving signaling systems using mixed integer linear programming. *Control Engineering Practice* 22(1):44–56.

Yida, G. Cheng, Z. & Shaofeng, L. 2019a. Enhancing sustainability of rail transit system by applying multi-agent system. *The 2019 ASCE International Conference on Computing in Civil Engineering.*

Yida, G. Cheng, Z. & Shaofeng, L. 2019b. The application of a multi-agent system in control systems of rail transit. *The 2019 Transportation Research Board.*

Sustainable Buildings and Structures: Building a Sustainable Tomorrow – Papadikis et al. (Eds)
© 2020 Taylor & Francis Group, London, ISBN 978-0-367-43019-1

Multi-agent systems for energy efficient train and train station interaction modelling

Y. Guo, Q. Wang & C. Zhang
Xi'an Jiaotong-Liverpool University, Suzhou, Jiangsu, China

ABSTRACT: Energy-efficient train control and timetabling are typical research problems in the field of rail transit. The multi-objective optimization of rail transit network under interference is always a challenge due to the excessive demand for computing power, which makes it difficult to realize multi-train optimization when the system is disturbed during practical operation process. In the present paper, a multi-agent system (MAS) is proposed to establish an interactive mechanism to reduce the energy consumption in rail transit field. In this system, every train and station is considered as an agent with certain autonomy. Through cooperation and negotiation, each train can adjust local time schedule according to specific situations. In these processes, several factors (such as passenger flow and energy consumption) are comprehensively considered to generate optimized solutions. The result of the case study shows that the energy efficiency is improved when a station has a high passenger density and longer boarding time is needed.

1 INTRODUCTION

1.1 *Background*

The research on energy conservation in the field of rail transit can be divided into two categories. The first category is the optimization of train speed trajectory, which is known as energy efficient train control (EETC). The process is to optimize the train speed trajectory by taking energy consumption as the objective function with several limiting conditions (distance, time, and speed limit). The other category of optimization is energy efficient train timetabling (EETT) (Scheepmaker et al. 2017). This category is based on EETC theory to reduce energy consumption by optimizing the time schedule. The process of EETT needs to consider factors such as passenger flow, maximum passenger waiting time, relevant theories of EETC, and minimum safety distance between two trains. The current main optimization approaches include approximate and decomposition techniques, mixed integer programming, multi-objective evolutionary algorithm, genetic algorithm (Tan et al. 2018). At present, the research in related fields mainly focuses on the optimization of train timetable based on the historical passenger flow data, safety distance between different stations and relevant EETC theory. Especially, adjusting the headway between trains according to the historical passenger flow can significantly reduce the no-load rate and improve the energy efficiency. However, the number of passengers at a station varies every day, hence the flexibility schedule adjusted based on real-time passenger flow is considered a trend for the next stage of research. Apart from the advantages it may bring, the flexible schedule has the following technical difficulties. First of all, the realization of elastic schedule requires consideration of multiple factors such as safety, energy consumption, and passenger flow. Multi-objective optimization often requires a long computational time, which makes it difficult to generate an optimized schedule in a short time during actual operation process. Furthermore, real-time passenger density data can be collected at a station continuously, but adjustment of timetables and train headway has no immediate impact on a specific station. Furthermore, for the subway, the schedule adjustment needs to comprehensively consider the impact on all relevant stations on the line. Different stations at the same time may have conflicting demands because of different passenger density. Therefore, the schedule adjustment needs to balance the demand of the affected stations. In addition, each station may have opposite demands at some moments, which requires that the impact of all stations should be taken into account when making real-time adjustments to the schedule.

In this paper, a multi-agent system is introduced and the system is expected to overcome some difficulties of the traditional rail transit control system in achieving flexible timetable, and to improve the efficiency of trains. The focus of this paper is on improving energy efficiency by realizing the interaction between the station and the train based on multi-agent technology.

2 LITERATURE REVIEW

2.1 *Passenger density technology*

Before 1950s, the photograph technology is the only way to detect the traffic flow density. Presence-type detectors and processing of signal pulses method is a significant methodology which can calculate percent occupancy in the early 1960s (STUDIES 1963). After 1960s, the measurement and calculation of in-put and out-put method combined with photograph technology have been widely used (May 1990). Input-output algorithm has been applied to detect traffic flow by the calculation of density from speed and flow (Gazis & Szeto 1971). Metro Count device input-output is the similar approach with the input-output method (Mohamed 2013). Table 1 summarized the mentioned technology used in passenger flow research.

2.2 *Algorithms*

In addition to passenger density data, developing proper algorithms are essential for EETT optimization. Station sequential model, train sequential model and space-time sequential model are three basic traffic models used to analyze train dynamic problems (Campion et al. 1985).The schedule of metro transportation contains relatively complex relations which is supposed to use approximate decomposition techniques to figure out an result that closes to optimized schedule (Kroon et al. 2013), station and train model could solve the problem by considering series of stations. In 2005, Kwan & Chang (2005) proposed to minimize the total energy consumption and passengers by a multi-objective evolutionary model. Carey & Crawford (2007) generated a multi-Station and multi-line model with heuristic-based algorithms to

Table 1. Summary and comparison of method and model for passenger flow control in metro system.

Methodology	Traffic Model	Utilization	Research Problem	Publication
Photograph technology	/	Traffic flow density detection	Passenger flow density control	(May 1990)
Input-output count algorithm	/	Traffic flow density detection	Passenger flow density control	(Gazis & Szeto 1971)
Metro Count device input-output algorithm		Passenger flow detection	Passenger flow density control	(Mohamed 2013)
Presence-type detectors and processing of signal pulses	/	Measurement and analysis of density	Compute percent occupancy	(STUDIES 1963)
Video filming	/	Estimation of macroscopic density	Passenger flow density control	(Thabet 2010)
Piecewise linear function	Piecewise linear function	Uninterrupted flow situations	Traffic flow density	(TRB 2010)

Table 2. Summary and comparison of method and model for scheduling problems in metro system.

Methodology	Traffic Model	Utilization	Research Problem	Publication
Approximate and Decomposition techniques	Station And train model	A series of train and stations scheduling	High-dimension problem	(Kroon et al. 2013)
Mixed integer programming (MIP)	Space-time based model	Train scheduling	Train scheduling problem	(Cacchiani et al. 2016)
Integer linear programing	Space-time based model	Train unit assignment scheduling	Train circulation problem	(Cacchiani et al. 2010)
Multi-objective evolutionary algorithm		Optimization of cost and passenger dissatisfaction	Energy problem	(Kwan & Chang 2005)
Multi-station multi-line model and heuristic-based algorithms	Station model	Train scheduling	Busy metro system problem	(Carey & Crawford 2007)
Genetic Algorithm and Simulation	Nonlinear	Train energy consumption	Timetable	(Liang et al. 2014)
Dynamic programming-based algorithm		Irregular train scheduling	Irregular problem	(Yin et al. 2016)
Multi-agent System and Mixed integer linear programing		Multi-objective optimization	Multi-objective problem	(Guo et al. 2019a, 2019b)

optimize the train schedule for busy metro system. In 2010, Cacchiani et al. (2010) raised the train circulation problem could be solved by integer linear programming with the space-time model, it is efficient to generate a schedule for train unit assignment. Furthermore, Cacchiani et al. (2016) applied a mixed integer programming (MIP) for coordinating the departure time for each train at each station, which take the headway as the control variable. Moreover, the most common train schedule is certain time headway for applications, the irregular headway schedule is more reasonable because it is more effective and efficient to improve the traffic environment (Cacchiani et al. 2016). In 2019, the multi-agent algorithm proposed by Guo et al. 2019(a) & (b) established a direct connection between trains, and the station can obtain the estimation of the real-time arrival time of trains through the direct connection between trains and stations. These algorithms are summarized in Table 2.

3 METHODOLOGY

The proposed multi-agent system is shown in Figure 1. The figure includes several Station Agents, Train Agents and a Central Agent. Among them, the Station Agent stores the historical data of passenger density at the station, and can estimate the boarding time of the train by comparing it with the real-time passenger density data. The Station Agent needs to continuously communicate with the incoming train, and pass the estimated boarding time to the coming train. Two trains are presented in Figure 1 to explain the communication mechanism. the trains need to exchange information to meet the safety requirements of the moving block system. In addition, each Train Agent and Station Agent has information interaction channels with the Central Agent. The Central Agent uses these channels to collect necessary information and saves them into database, which can provide data for the further optimization of the system.

Based on the information collected from Station Agents, each Train Agent makes decisions and provide optimized speed trajectory for the corresponding train, thereby reducing energy consumption.

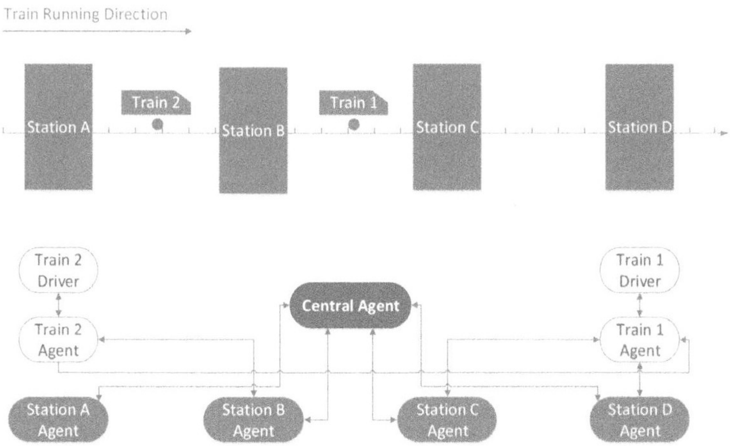

Figure 1.   Adjust the schedule according to historical information and passenger flow forecast.

4   CASE STUDY

A case study of one train with three stations is carried out to investigate the proposed methodology. The train is heading from Station A, passing Station B, and to Station C. The distances between Station A, Station B, and Station C are 3000 m, and 2000 m, respectively. According to the given schedule, the travel time from Station A to B is 180 seconds. The travel time from Station B to Station C is 120 seconds. The train has to wait at the station for 40 seconds for passengers to get on and off.

The quadratic form of Davis equation is used to obtain the resistance of the trains, as shown in equation (1).

$$f_{drag} = A + Bv + Cv^2 \tag{1}$$

where: A=3.6449, B=0.001710, C=0.01134 for typical urban train. Therefore, the total energy consumption would be:

$$E = \int_0^T \left[(ma + f_{drag})^+ \times v(t) \right] dt \tag{2}$$

where:

$$x^+ = \begin{cases} x & if \ x \geq 0 \\ 0 & if \ x < 0 \end{cases} \tag{3}$$

Three scenarios are discussed in this case study. In the first scenario, the train runs according to the original plan and normally leaves Station B after waiting 40 seconds at Station B (for passengers to board). This is the case where the train operates in full accordance with the original plan without interference, and the total energy consumption was 93,401 KJ.

In the second scenario, the architecture of multi-agent system proposed is employed, hence Station Agent B could communicate with the train agent continuously. The agent of Station B estimated that the train will stay at Station B for an additional 20 seconds by combining the historical data and the real-time station entrance data, and then passed the information to the train in advance. The train agent analyzed the condition and found that compare with shortening 20 seconds between Station B and C, shortening 20 seconds between Station A and B would save more energy. Hence the train adjusted its speed trajectory when it left Station A,

202

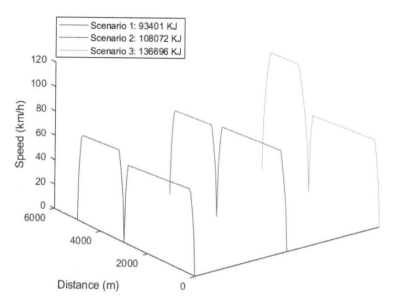

Figure 2. The speed trajectory and total energy consumption of the three scenarios.

and arrived at Station B 20 seconds ahead of schedule. Then the train goes to Station C as planned. The original speed trajectory is adopted between Station B and C and the train arrived at Station C on time. The total energy consumption during the process is 108072 KJ, the train uses 16% more energy than the original plan to catch up the extra 20 seconds.

The third scenario shows a case where the MAS system is not used, in which the train cannot be directly informed of boarding information from the station in advance. Due to the overcrowding of Station B, the train left Station B 20 seconds later due to the delay of boarding time. In order to ensure that the train arrives at Station C on time, the train takes the total energy consumption as the objective equation and re-generated the speed trajectory between Station B and Station C according to the new target duration. The train then followed the new speed trajectory and arrived at Station C on time. In this scenario, the total energy consumption of the train is 1,36,696 KJ. The train used 46% more energy than planned to compensate for the extra 20 seconds boarding time at Station B.

Figure 2 shows the speed track and total energy consumption of the train in the three scenarios discussed. The velocity trajectory of the train between each station is optimized by taking energy as the objective equation. In the three scenarios, the difference in energy consumption between the same two stations is due to different running duration.

5 CONCLUSION

According to the literature review, it is believed that the non-periodic schedule is one of the trends of EETT research. However, the high computational complexity of correlation optimization makes this field of research difficult. The multi-agent system proposed in this paper realizes the local adjustment of schedule by transforming the central optimization into multiple local optimization. Due to the high flexibility of multi-agent system, this system has good extensibility. The result of case study shows that the system can save energy by micro adjustment of the original schedule. In the future, it is planned to combine the station train interactive MAS system proposed in the present paper and train interactive MAS system to realize the interactive model between multiple trains and stations.

ACKNOWLEDGEMENT

We would like to acknowledge the support from the Key Program Special Fund of Xi'an Jiao-tong-Liverpool University, project code KSF-E-04.

REFERENCES

Cacchiani, V. Alberto, C. & Paolo, T. 2010. Solving a real-world train-unit assignment problem. *Mathematical Programming* 124(1–2):207–231.

Cacchiani, V. Fabio, F. & Martin, P.K. 2016. Approaches to a real-world train timetabling problem in a railway node. *Omega* 58:97–110.

Campion, G. V.Van Breusegem, P.P. & Georges, B. 1985. Traffic regulation of an underground railway transportation system by state feedback. *Optimal Control Applications and Methods* 6(4):385–402.

Carey, M. & Ivan, C. 2007. Scheduling trains on a network of busy complex stations. *Transportation Research Part B: Methodological* 41(2):159–178.

Gazis, D. C. & M. W. Szeto. 1971. Estimation of traffic densities at the lincoln tunnel from time series of flow and speed data. *In Proceedings of Stochastic Point Process Conference, Yorktown, NY.*

Guo, Y.D. Cheng, Z. & Shaofeng, L. 2019a. Enhancing sustainability of rail transit system by applying multi-agent system. *In The 2019 ASCE International Conference on Computing in Civil Engineering.*

Guo, Y.D. Cheng, Z. & Shaofeng, L. 2019b. The application of a multi-agent system in control systems of rail transit. *In The 2019 Transportation Research Board.*

Kroon, L.G. Leon, W.P. Peeters, J.C.W. & Rob, A.Z. 2013. Flexible connections in pesp models for cyclic passenger railway timetabling. *Transportation Science* 48(1):136–154.

Kwan. Chungmin & Chang, C. 2005. Application of evolutionary algorithm on a transportation scheduling problem-the mass rapid transit. in 2005. *IEEE Congress on Evolutionary Computation* 2: 987–994.

Liang, Z.C. Qingyuan, W. & Xuan, L. 2014. Energy-efficient handling of electric multiple unit based on maximum principle. *33rd Chinese Control Conference (CCC)* 3415–3422.

May, A.D. 1990. *Traffic Flow Fundamentals.*

Mohamed, S.R.M. 2013. *Generalized Traffic Model for Major Roads in Egypt and Its Utilization in Macroscopic Simulation.*

Scheepmaker, G.M. Rob, M.P.G. & Leo G. K. 2017. Review of Energy-Efficient Train Control and Timetabling. *European Journal of Operational Research* 257(2):355–376.

STUDIES, PRELIMINARY OPERATIONAL. 1963. *Development and Evaluation of Congress Street Expressway Pilot Detection System.*

Thabet, O.A. 2010. *Modeling and Macroscopic Simulation of Traffic Streams on Multi-Lane Highways.*

TRB. 2010. *Highway Capacity Manual.*

Yin, J.T. Tao, T. Lixing, Y. Ziyou, G. & Bin, R. 2016. Energy-efficient metro train rescheduling with uncertain time-variant passenger demands: An approximate dynamic programming approach. *Transportation Research Part B: Methodological* 91:178–210.

Green and low carbon buildings

Sustainable Buildings and Structures: Building a Sustainable Tomorrow – Papadikis et al. (Eds)
© 2020 Taylor & Francis Group, London, ISBN 978-0-367-43019-1

Semantic enrichment of city information models with LiDAR-based rooftop albedo

F. Xue, W. Lu & T. Tan
Department of Real Estate and Construction, The University of Hong Kong, Hong Kong SAR, China

K. Chen
Department of Construction Management, Huazhong University of Science and Technology, China

ABSTRACT: In the era of smart city, semantically rich city information models (CIMs) are demanded as a critical information hub. Roof albedo, a semantic property measures how much solar radiation is reflected, is vital to various urban sustainability topics, including heat island, local climate, green roof, and urban morphology. This paper presents an approach that enriches LiDAR-based albedo to rooftop models for CIM. First, we apply Chen et al. (2018)'s method to the reconstruction of the geometries of rooftop elements. Then, albedos of roofs and rooftop elements are estimated from the mean reflectance in LiDAR data. A pilot study was conducted in an urban area in Hong Kong. The results showed that the building models created by the presented approach were satisfactory in terms of rooftop elements and roof albedos. The results from the present approach can provide sustainability study the details of 3D geometries and albedos in an urban area.

1 INTRODUCTION

Roofs usually constitute 20~25% urban surfaces in typical metropolises (Rose et al. 2003). Rooftop albedo that measures how much solar radiation is reflected (other than absorbed) by roof coatings leads to a negative radiative forcing. At a worldwide level, high-albedo roofs can offset billions of tons of CO_2 emissions as well as save billions of dollars of energy bills every year (Akbari et al. 2009). Therefore, some governments, such as the California Energy Commission (2005), have required all new or retrofitted roofs to be white or reflective.

To monitor rooftop albedos and utilize them in simulations, semantically rich building/city information models (BIMs/CIMs) are demanded as data hubs. CIM is the digital presentation of the physical and functional characteristics of a city area, like BIM for a building (NBIMS 2018, Chen et al. 2018). CIM plays the data hub role in various smart city applications (Xu et al. 2014 Cheng et al. 2016). As important physical and functional components in CIM, the rooftop geometries and albedos are vital to various urban sustainability topics such as heat island, local climate, green roof, and urban morphology (Santamouris 2014, Stewart et al. 2014, Baniassadi et al. 2018).

Since only a few buildings have their digital models of as-built conditions, CIM often involves reconstruction of measurement data from sensors like satellite camera, satellite radar, or airborne light detection and ranging (LiDAR) (Pătrăucean et al. 2015, Xu et al. 2018). The reconstruction methods can be broadly categorized as either data-driven or model-driven, where a data-driven method is to perform modeling based on the preprocessed measurement data, and a model-driven method compares, recognizes, and fits known components to the data (Xue et al. 2018). Recently, researchers have made considerable progress on both types of reconstruction by using advanced methods like *a priori* rules, shape regulations, machine learning, and evolutionary computation (Pătrăucean et al. 2015). One example is Chen (2018), in which rooftop elements are regularized as perpendicular or parallel to the major edges of a building footprint.

Semantic enrichment of albedo to roofs is the subsequent problem. Conventional methods used to employ multiple pyranometers (or albedometers) or scanning radiometers to measure the value of albedo (NASA 2014). Some researchers also validated other types of sensors like near-infrared (NIR) sensors and thermocouples, where a city's LiDAR data is measured by ultraviolet-vis-NIR bands (Levinson et al. 2014). Therefore, the reflectance (or absorption equivalently) can confidently approximate the rooftop albedo and reveal the rooftop materials (e.g., coatings and green roofs) if carefully calibrated (Levinson et al. 2014, Tan et al. 2019).

However, many CIMs or 3D city maps equip neither the rooftop details nor albedo. For example, Google Maps, as well as many other topographic maps in geographic information systems (GIS), have box-shaped 2.5D building models, on which rooftop elements and albedos are omitted. The Google Earth has photo-realistic 3D mesh models of building in most metropolises, but the albedo information is missing too. The information gap between the CIMs and sustainability study's demands thus calls for rooftop albedo research for CIM.

This paper reconstructs and enriches the rooftop models with LiDAR-based albedo estimation for creating CIMs by extending Chen et al. (2018). The results are rooftop elements with geometry and albedo for CIM. Section 2 reviews the related works in literature. Section 3 describes our methods. A pilot case is shown in Section 4 and conclusions appear at the end of the paper.

2 RELATED WORKS

The task of processing 3D LiDAR data for 3D models usually starts with segmenting the points into patches of objects (e.g., buildings, roofs, and rooftop elements). This segmentation is generally called "semantic segmentation," primarily based on the geometry (e.g., normal of a surface, connectivity, planarity) (Cao et al. 2017) or the reflectance value captured by the LiDAR sensors (Sun & Salvaggio 2013). The patches can be associated with other types of data. For instance, a topographic map can offer the surveyed ground truth building footprints for filtering noise and 2.5D extrusion (Ledoux & Meijers 2011). Some architectural knowledge like pre-defined roof styles (e.g., flat, skillion, gable, hip, and gambrel) and parallel and perpendicular features can help in the 3D reconstruction too (Xiong et al. 2014, Sampath & Shan 2009, Chen et al. 2018).

Albedo can also be inferred from LiDAR data. It is because about 43% of the solar radiation output is visible light (400~800 nm) and 49% is infrared (> 800 nm) (Mohanakumar 2008); while LiDAR involves the visible spectrum and sometimes infrared. Levinson et al. (2014) conducted an experiment comparing three albedo estimation methods using a *Perkin–Elmer Lambda 900* UV–vis–NIR spectrometer. Their findings showed that the simple average value of reflection is surprisingly accurate – only with a 0.006 mean error (out of 1.0) and 0.021 root-mean-square error (RMSE) from the ground truth; the error can be narrowed further to RMSE = 0.014 if removing white materials and using the polynomial regression. In this paper, we will use the average albedo across the visible light and infrared spectrums.

3 METHODS

Figure 1 shows the framework of the approach presented in this paper. Technologically, the approach is a pipeline extending Chen et al. (2018) for LiDAR-based rooftop albedo modeling. The inputs include multiple sets of urban data. Apart from the LiDAR data, topographic map (2D or 2.5D) and architectural regularity belong to the inputs too. The outputs after the four-step pipeline are cloud-based 3D map and data exchange of rooftop model in GeoJSON format, which is an open standard format designed for representing simple geographical features on top of JavaScript Object Notation (JSON).

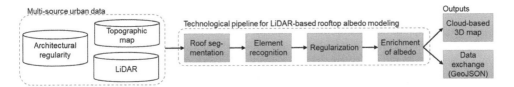

Figure 1. The pipeline of the presented approach in this paper.

The first three steps, from 'roof segmentation' to 'regularization' in Figure 1, duplicate the geometric modeling of rooftop elements in Chen et al. (2018). The last step, semantic enrichment of albedo aims to exploit the "intensity" value in LiDAR data. The intensity records how many laser photons were reflected from the surface of the target location. Although intensity is a standard "scalar" property in the laser (LAS) standard (ASPRS 2011) of LiDAR format, the definition and range of intensity can vary significantly from one LiDAR equipment to another. For example, some LiDAR limits the intensity between 0 and 1, while some have a range between 0 and 128. Therefore, we use the normalized mean intensity (in a range 0 to 1) of the upper surface of a rooftop element as its approximate albedo, i.e.

$$albedo = \sum\nolimits_{i=1 \text{ to } n} intensity_i / intensity_{\text{MAX}}, \tag{1}$$

where n is the number of LiDAR points on a top surface, and $intensity_{\text{MAX}}$ is the highest value of the defined range of intensity in the LiDAR data.

4 A CASE IN HONG KONG

The study area in this paper was a squared area around the University of Hong Kong (HKU) Main Campus, about 0.3 km^2 in the Central Western District, Hong Kong. As shown in Figure 2, the small area included hundreds of high-rise buildings, groundcover vegetation, hills, roads, flyovers, and ferry facilities. The study area consisted of various urban landscapes, including high-density urban blocks.

The LiDAR data, as shown in Figure 2b, was collected by the Civil Engineering and Development Department (CEDD) of the Hong Kong SAR Government (CEDD 2015). The intensity in the LiDAR data was measured by an airborne Optech 3100 LidAR sensor, which was a wavelength range of 400-2000 nm using a portable field spectro-goniometer *ASD FieldSpec*

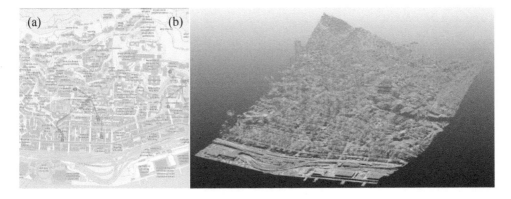

Figure 2. The study area around HKU campus in central western district, Hong Kong, (a) map of the target area, (b) LiDAR 3D data (cooler color indicates higher albedo).

Pro (Ahokas et al. 2006). We use the average reflectance as the estimated albedo, as shown in Eq. (1), according to the validation in Levinson et al. (2014).

The results are the albedo models of 1,087 blocks of buildings and 1,288 rooftop elements on top of them. It should be noted that a complex building's blocks and parts, e.g., podium and blocks, are defined as multiple block models. We also converted the models into GeoJSON formats and visualized on a web 3D library *OSMBuildings* (version 3.0.1, https://github.com/OSM Buildings/OSMBuildings). Figure 3 shows the visualized albedo models of the study area.

Beside of the albedo, each roof model was also associated with the topographical map of Hong Kong (in HKGS1980 coordinate system) and *Open Street Map* (in WGS1984). As a result, more semantic properties can be enriched to the model of the building. Figure 4 shows the Knowles Building, HKU, in which the offices of the authors reside, in the web visualization system. Apart from the geometric dimensions, one can read the more properties from the mouse tooltip (Figure 4a): Name, building IDs in the topographic map and *Open Street Map*, roof albedo (0.351), type of building, storeys (including the level of basements). In comparison to the model in Google Earth, as shown in Figure 4b, all rooftop elements including the parapet walls, elevators' machine rooms, water tank, and cooling towers, except for one circled in red. Since the albedo is not available in Google Earth, and the GeoJSON is an open GIS format, the albedo models presented in this paper can facilitate more in sustainability study.

Figure 3. Screenshot of the LiDAR-based albedo models of roofs and rooftop elements visualization (cooler color indicates higher albedo).

Figure 4. Example of the output model. (a) Roof and rooftop elements of knowles building, HKU (cooler color indicates higher albedo), (b) the referential 3D building model in google earth, where the circled is a missing element in (a).

5 CONCLUSION

Roof albedo that measures how much solar radiation is reflected is vital to various urban sustainability topics, including heat island, local climate, green roof, and urban morphology. However, the detailed models of roof geometries and rooftop albedos are not well prepared to enable such studies. This paper focuses on the LiDAR data, which involves roofs' reflectance on visible light and NIR spectrums. We present an approach that reconstructs rooftop geometry and enriches LiDAR-based albedo for CIM. A pilot study confirmed the methodological feasibility, and the results were encouraging. Future directions include reconstruction of irregularly shaped rooftop elements such as satellite dishes and the data interoperability with energy modeling software.

ACKNOWLEDGEMENTS

The work presented in this paper was supported by the Hong Kong Research Grant Council (No. 17200218) and The University of Hong Kong (Nos. 201711159016, 102009741). The work was completed when the coauthor K.C. was with The University of Hong Kong.

REFERENCES

Ahokas, E. Kaasalainen, S. Hyyppä, J. & Suomalainen, J. 2006. Calibration of the Optech ALTM 3100 laser scanner intensity data using brightness targets. *International Archives of Photogrammetry, Remote Sensing and Spatial Information Sciences* 36(Part 1).

Akbari, H. Menon, S. & Rosenfeld, A. 2009. Global cooling: increasing world-wide urban albedos to offset CO_2. *Climatic change* 94(3-4):275-286.

ASPRS. (2011). *LAS specification version 1.4–R13*. Bethesda, MD, USA: American Society for Photogram-metry and Remote Sensing.

Baniassadi, A. Heusinger, J. & Sailor, D.J. 2018. Building energy savings potential of a hybrid roofing system involving high albedo, moisture retaining foam materials. *Energy and Buildings* 169:283-294.

California Energy Commission. 2005. *California Energy Commission Building Energy Efficiency Standards, For Residential and Nonresidential. Publication No. CEC-400-2006-015*. Sacramento, CA, USA: California Energy Commission.

Cao, R. Zhang, Y. Liu, X. & Zhao, Z. 2017. 3D building roof reconstruction from airborne LiDAR point clouds: A framework based on a spatial database. *International Journal of Geographical Information Science* 31(7):1359-1380.

CEDD. 2015. *The CEDD 2010 LiDAR Survey (private communication)*. Hong Kong: Civil Engineering and Development Department.

Chen, K. Lu, W. Xue, F. Tang, P. & Li, L.H. 2018. Automatic building information model reconstruction in high-density urban areas: Augmenting multi-source data with architectural knowledge. *Automation in Construction* 93:22-34.

Cheng, J.C. Lu, Q. & Deng, Y. 2016. Analytical review and evaluation of civil information modeling. *Automation in Construction* 67:31-47.

Ledoux, H. & Meijers, M. 2011. Topologically consistent 3D city models obtained by extrusion. *International Journal of Geographical Information Science* 24(5):557-574.

Levinson, R. Chen, S. Berdahl, P. Rosado, P. & Medina, L.A. 2014. Reflectometer measurement of roofing aggregate albedo. *Solar Energy* 100:159-171.

Mohanakumar, K. 2008. *Stratosphere troposphere interactions: An Introduction*. Springer.

NASA. 2014. *Measuring Earth's Albedo*. NASA, USA.

NBIMS. 2018. *United States National Building Information Modeling Standard*. Retrieved July 25, 2019, from Whole Building Design Guide: http://www.wbdg.org/building-information-modeling-bim.

Pătrăucean, V. Armeni, I. Nahangi, M. Yeung, J. Brilakis, I. & Haas, C. 2015. State of research in automatic as-built modelling. *Advanced Engineering Informatics* 29(2):162-171.

Rose, L.S. Akbari, H. & Taha, H. 2003. *Characterizing the fabric of the urban environment: a case study of Greater Houston, Texas*. Berkeley, CA, USA: Lawrence Berkeley National Laboratory.

Sampath, A. & Shan, J. 2009. Segmentation and reconstruction of polyhedral building roofs from aerial LiDAR point clouds. *IEEE Transactions on Geoscience and Remote Sensing* 48(3):1554-1567.

Santamouris, M. 2014. Cooling the cities–a review of reflective and green roof mitigation technologies to fight heat island and improve comfort in urban environments. *Solar energy* 103:682-703.

Stewart, I.D. Oke, T.R. & Krayenhoff, E.S. 2014. Evaluation of the 'local climate zone'scheme using temperature observations and model simulations. *International journal of climatology* 24(4):1062-1080.

Sun, S. & Salvaggio, C. 2013. Aerial 3D building detection and modeling from airborne LiDAR point clouds. *IEEE Journal of Selected Topics in Applied Earth Observations and Remote Sensing* 6(3):1440-1449.

Tan, T. Chen, K. Lu, W. & Xue, F. 2019. Semantic enrichment for rooftop modeling using aerial LiDAR reflectance. *Proceedings of the 2019 IEEE International Conference on Signal Processing, Communications and Computing, in press*. IEEE.

Xiong, B. Elberink, S.O. & Vosselman, G. 2014. A graph edit dictionary for correcting errors in roof topology graphs reconstructed from point clouds. *ISPRS Journal of Photogrammetry and Remote Sensing* 93: 227-242.

Xu, X. Ding, L. Luo, H. & Ma, L. 2014. From building information modeling to city information modeling. *Journal of Information Technology in Construction* 19:292-307.

Xue, F. Lu, W. & Chen, K. 2018. Automatic Generation of Semantically Rich As-Built Building Information Models Using 2D Images: A Derivative-Free Optimization Approach. *Computer-Aided Civil and Infrastructure Engineering* 33(11):926-942.

Sustainable Buildings and Structures: Building a Sustainable Tomorrow – Papadikis et al. (Eds)
© 2020 Taylor & Francis Group, London, ISBN 978-0-367-43019-1

Experimental study on thermal comfort improvement of building envelope with PCM energy storage

H. Ye, Y. Wang & F. Qian
College of Architecture and Urban Planning, Tongji University, Shanghai, China

ABSTRACT: Two lightweight reduced scale model rooms of sandwiched color steel plate were built as comparison experimental chambers. PCM (phase change temperature: 17 °C) encapsulated by aluminum foil bag were attached to the inner wall of the south side of the PCM chamber, and the application effect of PCM in Shanghai winter was tested. The results show that the PCM has a certain attenuation effect on the fluctuation of the air temperature and the average temperature of the inner surface. The degree of its effect depends on the weather. PCM reduce the average heat flow through windows by 5.3% and the average heat flow through walls by 4.35%. With the use of PCM, the temperature fluctuation range decreased, and the percentage of thermal comfort dissatisfaction decreased by about 10%. For grade I thermal comfort time percentage increased by 3.3 on average, and for grade II thermal comfort time percentage increased by 19.0 on average.

1 INTRODUCTION

As a kind of energy storage material, PCMs (Phase Change Material) have significant advantages in reducing building energy consumption and improving energy efficiency. Applying this type of energy storage materials to the building energy conservation is of great significance for the establishment of a resource-saving society (Yang et al. 2015), and has broad application prospects and market demands in terms of solar energy utilization, waste heat recovery, and power load regulation (Wang & Wang 2014). According to the characteristics of the PCMs, it is obvious that the larger daily temperature range is the more favorable for heat storage and heat release of PCMs. For regions with hot summer and cold winter, it is necessary to go into the environmental conditions of using specific PCM and the energy-saving effect.

The application of PCMs in buildings is mainly to combine PCMs with building envelope to form Phase Change Energy Storage Building Envelope (PCESBE) for adjusting building indoor temperature. PCESBE can increase the thermal storage capacity of building envelope greatly, so that the fluctuation range of indoor temperature is weakened and the action time is delayed, thereby improving the temperature self-adjusting ability of the building and improving the indoor thermal environment to achieve energy-saving and thermal comfort.

For the development of PCMs for building envelope, it is important to select suitable PCMs. It is generally considered to have the following characteristics (Khudhair & Farid 2004, Sharma et al. 2009, Sari & Karaipekli 2007). (1) The latent heat of phase change must be high, so it can store or release more heat during the phase change. (2) The phase change process has good reversibility, small expansion and contraction, less super-cooling or overheating. (3) A suitable phase change temperature, which can meet the specific temperature to be controlled. (4) A higher thermal conductivity, density and specific heat capacity. (5) The PCMs is non-toxic, non-corrosive, low cost and convenient to manufacture. At present, many research institutes in China have developed a variety of products, and there are many choices for phase change temperature. The PCMs has developed to a certain extent. It is believed that in the near future the performance of PCMs will be better to meet the needs of practical using.

Due to the dense population, the trend of development of high-rise that requires building envelope to be light is obvious in the hot summer and cold winter areas. However, thermal capacity of ordinary light materials is low and it leads to strong fluctuation of indoor temperature. It not only causes the indoor thermal environment to be uncomfortable, but also increases the air conditioning load, resulting in an increase in building energy consumption. The application of PCMs in building envelope is an effective way to improve indoor environmental comfort, reduce energy consumption and reduce negative impacts on the environment. However, for regions with hot summer and cold winter, the daily range is small (generally between 6 and 8 °C), and the effect of PCMs needs to be verified.

Li et al. (2009) glued the PCM with EPS insulation material got lightweight phase change walls. In Chongqing area, the indoor thermal environment comparison experiment was carried out in an experimental chamber with PCM and an experimental chamber without PCM. It showed that PCM could enhance the thermal inertia of building envelope significantly and improve the thermal comfort of the room. Combined with night ventilation technology, it can effectively dissipate the heat accumulated during the day to the outside. Compared with ordinary rooms, Indoor temperature was reduced by up to 11 °C, and the energy saving effect was remarkable.

Zhu et al. (2016) studied the thermal performance of building envelope with PCM under typical climatic conditions in Shanghai. They used two kinds of PCMs. One's phase change temperatures was 29 °C used in ceiling, floor and west wall. The other's was 18 °C used in east and north wall. Compared with the ordinary room, the experimental results showed that the PCMs can reduce the indoor temperature variation by 4.3 °C in summer and 14.2 °C in winter, both the indoor thermal environment in winter and summer were improved.

Zhang et al. (2016) studied the winter application effect of composite phase change concrete blocks based on the outdoor integrated temperature in Nanjing. For air-conditioned rooms, the composite phase change concrete block (with built-in phase change temperature 19 °C PCM gypsum board) has a 43% reduction in heat flow compared to the inner wall surface of the hollow concrete block, and the total heat dissipation of the room is reduced by 45%. For non-air-conditioned rooms, the composite phase change concrete block (with a built-in phase change temperature of 15 °C PCM gypsum board) has an average indoor air temperature increase of 2.45 °C and a temperature fluctuation reduction of 82%. Their other experiment of phase change energy storage gypsum board in Nanjing showed that the energy saving effect is the best in summer when the phase change temperature of the PCM wall is 26 °C in summer, which can reduce the heat gain of the chamber by 37.73%. The energy saving effect of the room is the best in winter if the phase change temperature of the wall is 7 °C, which can reduce the heat gain of the room by 28.71% (Zhang et al. 2015).

In recent years, the research and application of phase change energy storage in buildings has been further diversified (Hua et al. 2018, Guo et al. 2017, Hua 2017, Xue 2017, Liu et al. 2017). For example, PCMs was used with the integration of building roofs, floors, walls and ventilation heating systems (Pasupathy et al. 2008, Valizadeh et al. 2019, Saxena et al. 2019, Klimes et al. 2019, Zhao et al. 2019, Ryms et al. 2019). Many researches showed that the suitable PCMs and structural forms in the building envelope could achieve energy-saving effect in winter and summer. However, suitable phase change temperature and reasonable structural forms require deep research in combination with meteorological parameters and building materials. Our motivation is to discover the law by a serial of tests. This paper attempts to conduct comparative experiments from the perspective of thermal comfort improvement.

2 EXPERIMENTAL DESIGN

2.1 *Experimental module*

The research method was comparative experiment that the two experimental chambers were made of same materials and dimensions, as shown in Figure 1. Considering the need for PCMs with different package size to be installed on each surface, the experimental chamber

Figure 1.   Testing chambers on building roof.

adopted a square body with an outer length, width and height of 1300 mm. The frame was made of aluminum alloy and the wall material was sandwich panel used for the industrial clean room. The inner and outer sides of the panel are 0.5 mm thick color steel plate, and the middle was 50 mm thick polyurethane insulation layer. In order to simulate normal room, a double-glazed observation window of 500 mm (W) × 450 mm (H) on the south wall to allow solar radiation heat to enter the experimental chamber with area ratio of 0.18. A 400 mm (W) ×700 mm (H) inspection door was opened on the north wall of the experimental chamber.

2.2   PCM

The PCM used in this experiment was the product sold on the domestic market, and the origin is Beijing Guangyu. The main ingredients are several alcohol compounds. The packaging material is aluminum foil bag. Each bag is 300 mm long 150 mm wide and has a mass of 0.5 kg/bag. The PCM model is MG17, its phase change temperature is 16~18 °C, the latent heat of phase change is 160 kJ/kg, and the material density is 835 kg/m^3. Considering that the PCM need to change phase without an artificial heat source, chose a phase change temperature that is lower than the normal indoor comfortable temperature in winter. The PCM is pasted on the inner wall of the south side by tape. See Figure 2.

Figure 2.   Bagged PCM attached on inside wall.

2.3 Test equipment

This experiment was mainly to test the energy storage and temperature adjusting performance of PCM under winter conditions. Some problems had been found through the preliminary experiments in the previous stage, and then the test plan was adjusted. Due to the small size of the experimental chamber, the temperature rise rapidly after receiving solar radiation. The temperature field in the chamber was non-uniform, resulting in a large temperature gradient. Therefore, experiment used a multi-point sampling method. During the experiment, the air temperature, globe temperature inside chamber and heat flow was measured. The instruments were placed in each experimental chamber. A weather station was placed near the experimental chamber to measure the outdoor air temperature and solar irradiance.

2.4 Overall introduction about the experiment

The experiment was completed in Tongji University, Shanghai. The experimental chambers were placed on the roof of Wenyuan Building (3F, height about 12 m), and there were no high-rise buildings covering it. The distance between two chambers is 2 meters. The experimental data records began on January 28, 2019 and lasted until March 23, 2019, for a total of 54 days. The sampling time of the test data was 5 min. Because the weather during the 2019 experiment was out of ordinary, most of them were rainy weather, so the following experimental data analysis was performed for typical weather.

3 ANALYSIS OF TEST DATA

3.1 Comparison of heat flow changes in the experimental chamber

In order to clarify the heat gain and loss path in the experimental chamber, the heat flux meter was used to monitor the heat flow in the middle of the window and the west wall surface. The comparison of window heat flow between PCM chamber and control chamber is shown in Figure 3. The comparison of west wall heat flow is shown in Figure 4. Overall, the heat flow of the window mainly flowed into the experimental chamber, and the heat flow of walls mainly flowed out. The main way to obtain heat in the experimental chamber is that the solar radiation entered window directly, and the obtained heat was lost through the wall to reach heat balance. The heat flux value during the experiment was averaged to get Table 1.

In the above Table 1, the heat flow was positive for flowing into the experimental chamber, and the negative heat flow for heat dissipation. The area of the window which heat is flowing into was small, and the area of walls that emits heat was much larger. Further analysis (Figures 3 and 4) showed that the heat flow through the window was large when the solar radiation was strong. It lost small heat when there was no solar radiation. In particular, in the

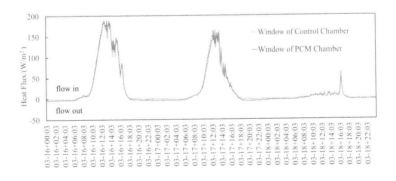

Figure 3. Comparison of window heat flux between control chamber and PCM chamber.

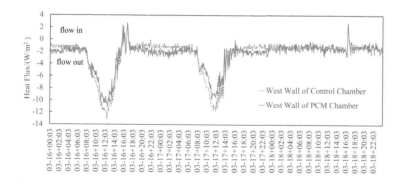

Figure 4. Comparison of west wall heat flux between control chamber and PCM chamber.

Table 1. The average heat flux of testing chamber, W/m².

	The control chamber	The PCM chamber	PCM performance
The south widow	+15.0	+14.2	-5.3%
The west wall	-2.07	-1.98	-4.35%

PCM chamber, the PCM released heat to maintain a higher temperature than the control chamber, so heat dissipation was large at night. The average heat flow of heat gain through window was slightly smaller than the control chamber, which was reduced by 5.3%.

The heat flow through the wall was relatively simple because it almost dissipated heat all the time. At noon when the solar radiation was strong, the heat release was also the largest at the highest temperature in the chamber. The control chamber had a higher heat flow due to the higher air temperature in the chamber. The rest of the time, outward heat flow was larger in PCM chamber just like the heat flow of the window. The surface of west wall was affected by the solar radiation at around 4 pm, and the heat flow had a transient inward flow. Overall, the average heat flow of the PCM chamber wall was less than the control chamber slightly, which was reduced by 4.35%.

3.2 Comparison of PMV and PPD in the experimental cabin

Since the PCM chamber and the control chamber had a large difference between the air temperature and the average temperature of the inner surface, the thermal comfort conditions in the two experimental chambers were compared. Analysis was performed by comparison of PMV (Predicted Mean Vote) and PPD (Predicted Percentage of Dissatisfaction) values. For the convenience of calculation, the human activity, clothing thermal resistance and indoor wind speed were uniformly valued according to the general conditions of winter, that is, the metabolic rate was 1.0 met, the clothing thermal resistance was 1.8 clo, and the indoor wind speed was 0.05 m/s. The temperature and humidity were measured by the multi-point average in the experimental chamber, and the mean radiant temperature T_{mrt} was calculated according to the measured globe temperature T_g according to the formula (1):

$$T_{mrt} = T_g + 2.44 \cdot \sqrt{v} \cdot (T_g - T_a) \tag{1}$$

In the formula (1), v was the indoor wind speed, and T_a was the air temperature. PMV, PPD and the percentage of time to meet thermal comfort in this article were calculated with the Ladybug and Honeybee plug-ins on the Grasshopper platform. Because it was calculated according to the formula, the PMV value might exceed the range of -3 to 3. Taking February 6 as an example, the results were shown in Figures 5 and 6.

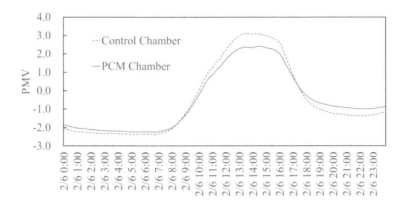

Figure 5. Comparison of PMV between control chamber and PCM chamber

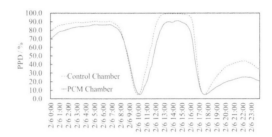

Figure 6. Comparison of PPD between control chamber and PCM chamber.

The fluctuation of PMV was similar to the fluctuation of temperature in experimental chamber. It would be cold and uncomfortable during the low temperature period at night, and be thermal discomfort during the high temperature period at noon. Due to the light fluctuations of temperature in the PCM chamber, the fluctuation range of the PMV was also relatively gentle. Reflecting on the PPD, the percentage of unsatisfactory in PCM chamber at the same time was about 10% lower than the control chamber, as shown in Figure 6.

The Chinese "code for design of heating, ventilation and air conditioning in civil buildings" stipulates that indoor thermal comfort levels be classified into Class I and Class II. Class I corresponds to $-0.5 \leq PMV \leq 0.5$, and Class II corresponds to $-1 \leq PMV \leq 1$. Comparing the temperature of two test chambers in several days, based on the above-mentioned activity level, clothing thermal resistance and indoor wind speed to convert into the percentage of time for different levels of thermal comfort. The results are shown in Table 2. It can be seen that whether it is for Class I or Class II, the percentage of thermal comfort time in the PCM chamber was improved, and the average percentage of Class I thermal comfort time increased by about 3.3, and Class II increased by about 19.0. As for the most obvious improvement on March 18, it can be seen from Figure 7 that the PMV in PCM chamber at night was almost maintained above -1.0, due to the heat storage of the PCM. It can be seen that the use of PCM can reduce the running time of the heating system to a certain extent, achieve the energy-saving effect of heating.

Table 2. Comparison of thermal comfort time percentages of different grades, %.

| Data | Weather | -0.5≤PMV≤0.5 | | -1≤PMV≤1 | |
		The control chamber	The PCM chamber	The control chamber	The PCM chamber
Feb. 5	lightrain	11.8	14.24	26.39	28.47
Feb. 6	cloudy	6.25	8.33	14.93	37.5
Feb. 7	moderate rain	0	0	12.15	24.65
March 16	fine	9.72	15.28	27.43	34.38
March 17	fine	6.94	9.72	16.32	37.5
March 18	lightrain	16.32	23.26	47.22	95.83
average		8.505	11.805	24.073	43.055

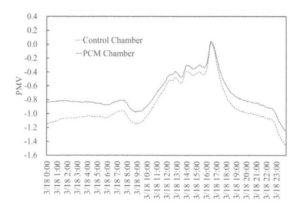

Figure 7. Comparison 2 of PMV between control chamber and PCM chamber.

4 CONCLUSIONS

Based on the comparative experiment, the following conclusions can be drawn for using PCM in the lightweight building envelope in Shanghai:

1. The PCM can reduce the high temperature in the day and boost the low temperature at night, and it affects the heat flow through windows and walls. The average heat flow through window decreases by 5.3%, and the average heat flow through walls decreases by 4.35%. Therefore, it is necessary to strengthen the insulation of transparent envelope of PCM heat storage buildings at night.
2. Due to gentle temperature fluctuation of PCM chamber, percentage of thermal comfort time increases to a certain extent. Reflected on PPD, the percentage of dissatisfied thermal comfort was reduced by 10%. Thermal comfort time in average increase of 3.3 percentage for Class I, and 19.0 percentage to Class II. Using PCM can reduce the running time of the heating system to achieve energy saving effect.

ACKNOWLEDGEMENT

This work is supported by 'National Key R&D Program of China' (Grant No. 2017YFC0702308).

REFERENCES

Guo, J.L. Xu, H. & Liu, G. 2017. Integrated design and application of phase change material and building envelope energy storage. *Building Energy Conservation* 45(10): 38-42.

Hua, X.M. 2017. Research on heat transfer performance and energy analysis of combined phase change in building envelope in four distinctive season district. Donghua University.

Hua, X.M. Diao, Y.F. & Ji, L. 2018. Research on temperature field of phase change material embedded in envelope. *Building Energy & Environment* 37(03): 5-9.

Khudhair, A.M, & Farid, M.M. 2004. A review on energy conservation in building applications with thermal storage by latent heat using phase change materials. *Energy Conversion and Management* 45(2): 263-275.

Klimes, L. Charvat, P. & Ostry, M. 2019. Thermally activated wall panels with microencapsulated PCM: comparison of 1D and 3D models. *Journal of Building Performance Simulation* 12(4): 404-419.

Li, B.Z. Zhuang, C.L. & Deng A.Z. 2009. Improvement of indoor thermal environment in light weight building combining phase change material wall and night ventilation. *Journal of Civil, Architectural & Environmental Engineering* 03:109-113.

Liu, C.Y. Li, D. & Shi, W. 2017. Analysis of optical and thermal properties of semi-transparent PCM-Glazed Unit. *Acta Energiae Solaris Sinica* 38(02):416-422.

Pasupathy, A. Athanasius, L. Velraj, R. & Seeniraj, R.V. 2008. Experimental investigation and numerical simulation analysis on the thermal performance of a building roof incorporating phase change material (PCM) for thermal management. *Applied Thermal Engineering* 28(5-6): 556-565.

Ryms, M. & Klugmann-Radziemska, E. 2019. Possibilities and benefits of a new method of modifying conventional building materials with phase-change materials (PCMs). *Construction and Building Materials* 211: 1013-1024.

Sari, A. & Karaipekli, A. 2007. Thermal conductivity and latent heat thermal energy storage characteristics of paraffin/expanded graphite composite as phase change material. *Applied Thermal Engineering* 27(8-9): 1271-1277.

Saxena, R. Rakshit, D. & Kaushik, S.C. 2019. Phase change material (PCM) incorporated bricks for energy conservation in composite climate: A sustainable building solution. *Solar Energy* 183: 276-284.

Sharma, A. Tyagi, V.V. Chen, & C.R. Buddhi, D. 2009. Review on thermal energy storage with phase change materials and applications. *Renewable & Sustainable Energy Reviews* 13(2): 318-345.

Valizadeh, S. Ehsani, M. & Angji, M.T. 2019. Development and thermal performance of wood-HPDE-PCM Nano-capsule floor for passive cooling in building. *Energy Sources Part A-Recovery Utilization and Environmental Effects* 41(17): 2114-2127.

Wang, Y.X. & Wang, Z.Q. 2014. Research progress of organic phase change materials and its composite technology. *Material Review* S2: 213-215.

Xue, Y. 2017. Using optimized phase-change material to improve building enclosure structure quality in Northern Cold Regions of China. *Architecture Technology* 48(04): 360-363.

Yang, R. Wan, Y. & Yang, Y.Z. 2015. Review on energy conservation with organic phase change materials in building applications. *Material Review* S1:136-140.

Zhang, W.W. Cheng, J.J. & Zhang, Y. 2016. Research on application effect of Composite phase change concrete block in Nanjing. *China Concrete and Cement Products* 04:66-71.

Zhang, W.W. Fang, X.W. & Zhou, Q.Y. 2015. The Impact Analysis in Building Palisade Structure with Phase Change Energy Storage Plasterboard. *Science Technology and Engineering* 01:112-115.

Zhao, J. Yuan, Y. Haghighat, F. Lu, J. & Feng, G. 2019. Investigation of energy performance and operational schemes of a Tibet-focused PCM-integrated solar heating system employing a dynamic energy simulation model. *Energy* 172:141-154.

Zhu, X.Y. Meng, E.L. & Cao, Y. 2016. Study of the thermal performance of a new type of PCM wall. *Journal of Suzhou University of Science and Technology (Engineering and Technology)* 04: 17-21.

Sustainable Buildings and Structures: Building a Sustainable Tomorrow – Papadikis et al. (Eds)
© 2020 Taylor & Francis Group, London, ISBN 978-0-367-43019-1

Effect of fiber type on mechanical properties of magnesium phosphate cements

T. Geng, Z.Q. Jiang, J. Li & L.W. Zhang
School of Civil Engineering, Guangzhou University, Guangzhou Guangdong, China

ABSTRACT: Magnesium phosphate cement (MPC) is a new type of cement with fast curing time, high early strength, strong adhesion, and good durability, which is used commonly to rapidly repair roads and bridges. However, MPC has high dry shrinkage, poor ductility and poor practicality. Adding fine fibers is an effective method to improve the mechanical properties and ductility of cement materials. Thus, five different fiber types with the same length (10mm) and proportion (1.2%) were added to MPC to overcome the problems mentioned above in this study. The fiber that exhibit the best mechanical properties after addition of magnesium phosphate cement and its concentration were determined by measuring the flexural and compressive strength of fiber reinforced MPC. The compressive strength and flexural strength increased by 30% and 85% respectively. These results can be used to guide the reinforcement during the construction of bridges and roads.

1 INTRODUCTION

From the 1920s to the 1990s, China experienced a massive bridge and road-building program. At present, most of these bridges and roads are degraded due to huge traffic flow and harsh environment. Thus, how to effectively repair these deficient bridges and roads has become an urgent problem and with the rapid development of economy, there will be a large number of road damage under great pressure, and magnesium phosphate cement with better mechanical properties will be needed to repair it in order to continue to meet the needs of traffic. Magnesium Phosphate Cement (MPC) is a new type of cementitious material (Li et al. 2008) characterized by fast setting, uniform texture, high strength, good bonding performance, and good stability. MPC presents strong potential for the quick repair of roads and bridges. Fibers (Fang et al. 2017) are often used in order to make MPC work better. However, most research focuses on the influence of fiber length and volume on the mechanical properties of MPC (Feng et al. 2014). Given this, the effects of different fibers on MPC were investigated in this study. A total of five fiber types are employed here including carbon fiber (CF), glass fiber (GF), steel fiber (SF), basalt fiber (BF), and polypropylene fiber (PF). The five kinds of fibers in this paper are abbreviated as CF GF SF BF PF. The effect of fiber type, fiber length, and fiber volume was investigated and the optimum fiber was determined by comparing improvement effects of different fibers on MPC properties, in order to better serve roads, bridges and other projects under long-term heavy load.

2 EXPERIMENTAL MATERIALS, INSTRUMENTS, METHODS

2.1 *Experimental basic mix ratio and material parameters*

According to the relevant literature and the authors experience, the basic mix ratio is as follows: magnesium-phosphorus molar ratio (m/p) = 5, boron-magnesium mass ratio (b/m) = 10%, water-cement ratio = 0.15 (Feng et al. 2014). The mass proportion of fly ash is 15%, and more details are shown in Tables 1-3.

Table 1. Chemical composition of fly ash.

Ingredient	SiO	AlO	FeO	CaO	MgO	TiO	SO
Content (%)	39.5	27.5	12.8	15.3	1.5	2.8	0.6

Table 2. Chemical composition of magnesium oxide materials.

Oxide	MgO	Al_2O_3	Fe_2O_3	CaO	SiO_2
Content (%)	91.00	1.90	1.60	1. 45	3.70

Table 3. Related parameters of various types of fibers.

	Length	Density
Fiber type	mm	g/cm^3
Carbon fiber	10	1.80
Steel fiber	10	7.8
Basalt fiber	10	2.699
Polypropylene fibers	10	0.91
Glass fiber	10	1.38

2.2 Experimental plan

The volume concentration of each fiber was limited into 1.2%, and fiber length was limited into 10 mm. A total of 18 MPC specimens were tested, which have a size of 40 mm in width, 40 mm in height, and 160 mm in length (Sun et al. 2016). The specimens were divided into six groups according to CF, SF (Hu & Sheikh 2018), BF (Ahmad & Chen 2019), PF, GF, and non-fiber (NF). Standard flexural and compressive tests were carried out after natural curing of 7 days. Table 4 offered details of the specimens.

Table 4. Test piece number.

Fiber type	Numbering	Borax content %	Fiber length mm	Fiber content %
Carbon Fiber	M-CF-1	10	10	1.2
	M-CF-2	10	10	1.2
	M-CF-3	10	10	1.2
Steel Fiber	M-SF-1	10	10	1.2
	M-SF-2	10	10	1.2
	M-SF-3	10	10	1.2
Polypropylene Fibers	M-PF-1	10	10	1.2
	M-PF-2	10	10	1.2
	M-PF-3	10	10	1.2
Basalt Fiber	M-BF-1	10	10	1.2
	M-BF-2	10	10	1.2
	M-BF-3	10	10	1.2
Glass Fiber	M-GF-1	10	10	1.2
	M-GF-2	10	10	1.2
	M-GF-3	10	10	1.2
Non Fiber	M-1	10	10	1.2
	M-2	10	10	1.2
	M-3	10	10	1.2

* Borax (Sodium Tetraborate decahydrate)

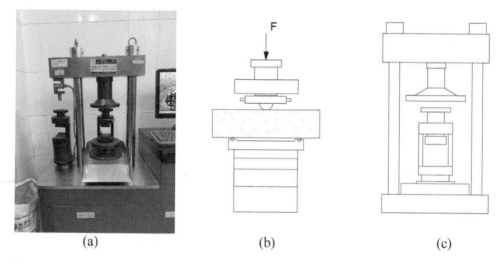

(a)   (b)   (c)

Figure 1. Laboratory instruments.

2.3 Experimental instruments

In this experiment, the YAW-300B electro-hydraulic testing machine is used, as shown in
Figure 1(a), whereas the load rate is set to 50 N/s, (b) used to determine the flexural strength
of MPC and (c) used to determine the compressive strength of MPC.

2.4 Specimen making

In the laboratory, potassium dihydrogen phosphate, borax and fly ash are taken, mixed
evenly in the mixer, and water is slowly added to the mixture for wet mixing. Water and
powder are fully combined to form slurry with good fluidity, and then the slurry is poured
into magnesium oxide. The fibers were then added and stirred slowly. When the fibers are
mixed evenly with the slurry, the mixing is stopped and the slurry poured into the test mould
immediately, and then the test piece is placed on the vibrating table for vibration. After standing, the formwork is removed, and the specimen is cured, and then the bending resistance and
compression test are carried out.

MPC is a gel-like substance with cement characteristics formed by acidic magnesium oxide
powder and alkaline potassium dihydrogen phosphate powder in acid-base neutralization
reaction (Li et al. 2018) in water, releasing a large amount of heat. The reaction is rapid, and
the slurry is condensed at a very fast rate. Therefore, it is necessary to add Borax to retard the
reaction for obtaining sufficient setting time (Sun et al. 2014). Additionally, fly ash is also
needed to ensure the complete reaction (Zhao & Li 2008)

3 EXPERIMENTAL RESULTS AND ANALYSIS

3.1 Failure mode

Figure 2 shows the flexural failure mode of specimens without fiber. The fracture surface of
the specimens was flat, and almost parallel to the cross section. The specimens were pulled
apart directly. In the compression test, the specimens were crushed into pyramid-shape frag-
ments which can also be observed in plain concrete, as shown in Figure 3.

For CF specimens in the flexural test, the fracture surface was irregular and appeared to
contain some micro voids. The reason for this is that the rough surface of CF deteriorated the
fluidity of MPC slurry. This deteriorated slurry led to a heterogeneous MPC specimens.

Figure 2.　Flexural failure (NF).　　　　　　　　　Figure 3.　Compressive failure (NF).

Therefore, cracks occurred and developed along weak points, resulting in an irregular fracture surface. CF on the fracture surface was broken but not pulled out. This indicates that CF has a strong bonding strength with MPC, as shown in Figure 4. In the compression test, as shown in Figure 5, although CF specimens separated into pieces, these pieces were still connected to each other. The specimens only presented some cracks and were deformed. This may be caused by the splicing effect of CF.

Specimens with BF have a similar flexural failure mode with NF specimens. The fracture surface was relatively flat and parallel to the cross section. Additionally, unlike CF specimens, a few fibers were pulled out but not totally (Figure 6). In the compression test, the BF specimens experience little deformation, and only a few cracks appeared in their edges (Figure 7)

The GF specimens have a similar flexural fracture surface to BF specimens. However, for GF specimens, most fibers were broken and only a few fibers were pulled out. In the compression test, GF specimens bulged to both sides, and their broken pieces also did not detach.

For PF specimens, they remained intact after the flexural test although cracks penetrated the entire section. This is due to the PF holding broken members and preventing the members from detaching (Figure 8). After the members were separated, it was observed that a large amount of PF was pulled out instead of being pulled apart (Figure 9). This implies that PF specimens have great ductility. In the compression test, only a few cracks were observed at the edge of PF specimens.

For SF specimens in flexural test, there was an obvious transverse crack. However, the specimens were not separated into two pieces. SF were not pulled out and remained good. Moreover, the broken specimens still can resist higher loads. In the compression test, no significant change could be observed in specimens (Figure 10).

Figure 4.　CF flexural failure.　　　　　　　　Figure 5.　CF compressive failure.

Figure 6.　Basalt fiber test piece.　　　　　　　Figure 7.　Glass fiber test piece.

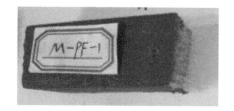

Figure 8.   PF flexural failure mode.   Figure 9.   Separated PF specimens.

Figure 10.   SF specimens flexural failure mode.

3.2   Compressive strength

The average compressive strength of six groups of specimens is shown in Figure 11. (The mixture of fiber and magnesium phosphate cement is more uniform, and their flexure strength and compressive strength are not more than 5%)

It can be seen from Figure 11 that the CF, GF and PF had a minor impact on MPC compressive strength and only increased it by about 1%, 3%, and 5%, respectively. The best effect on the improvement of the compressive strength is SF, which increased the compressive strength by 30%, followed by BF, which increased the compressive strength by 14%.

3.3   Flexural strength

The average flexural strength of the six groups of specimens are shown in Figure 12 (the mixture of fiber and magnesium phosphate cement is more uniform, and their Flexure strength and compressive strength are not more than 5%).

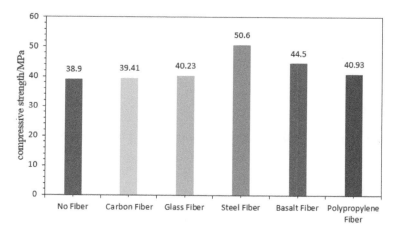

Figure 11.   Compressive strength.

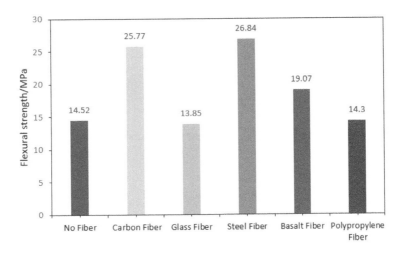

Figure 12. Flexure strength.

It can be seen from Figure 12, that SF have the most significant improvement on the flexural strength of MPC, which increased the flexural strength by 85%. The followed fiber was CF, increasing the flexural strength by 77%. BF provided lower improvement than the two former fibers, which improved the flexural strength by 31%. GF and PF have almost no improvement on the flexural strength.

4 CONCLUSION

In this paper, the standard compressive and flexural tests were carried out to identify the effects of five fiber types on MPC properties. The best fiber was determined and conclusions were as follows:

1. SF can improve MPC's compressive and flexural strength most effectively, and improve its strength.
2. CF only improved the flexural strength effectively, whilst it degraded MPC slurry's fluidity that led to a worse appearance and durability of MPC.
3. BF did not present an obvious improvement on MPC's flexural strength and compressive strength but was conducive to casting.
4. Both of PF and GF did not improve MPC's flexural strength and compressive strength significantly. The pouring behavior into the mold is easy and the forming surface is better, but the compatibility between them and magnesium phosphate cement is poor. Slurry fluidity is poor, casting is difficult, the forming surface is a little flat, and there are fewer holes.

ACKNOWLEDGEMENTS

This work was supported by the National Natural Science Foundation of China (Project No. 51608137).

REFERENCES

Ahmad, M.R. & Chen, B. 2019. A comprehensive study of basalt fiber reinforced magnesium phosphate cement incorporating ultrafine fly ash. *Journal of Composites Part B: Engineering* 168:204-217.

Fang, Y. & Chen, B. 2017. The enhancement and mechanism of glass fiber on mechanical properties of magnesium phosphate cement mortar. *Materials Reports* 31(24):6-9.

Feng, Hu. & Zhao, X.C. 2014. Basic mechanical behavior of micro steel fiber reinforced magnesia phosphate cement mortar. *New Building materials* 45(9):16-20.

Hu F.M. & Neaz Sheikh, M. 2018. Interface bond performance of steel fiber embedded in magnesium phosphate cementitious composite. *Construction and Building Materials* 185:648-660.

Li, H.B. & Yang, T.J. 2018. Main factors affecting the performance of magnesium phosphate cement and its development prospects. *Construction Science and Technology* 19:55-60.

Li, Sa. & Yi, Feng. 2008. Research progress of fiber reinforced magnesium phosphate cement. *Shanxi*

Sun, C.J. & Lin, X.J. 2016. Influence of the ratio of KH_2PO_4 to MgO, ratio of water to bind, content change of borax on water resistance of magnesium-potassium phosphate cement. *Journal of Fuzhou University (Natural Science Edition)* 44(06):856-862.

Sun, M.S. & Guan, Y. 2014. Effect of low temperature sintered magnesia adding B element on properties of magnesium phosphate cement. *New Building Materials* 45(05):88-91.

Zhao, J.T. & Li, X.G. 2008. Effect of fly ash on magnesium phosphate cement. *Bulletin of the Chinese Ceramic Society* 37(02):695-700.

Sustainable Buildings and Structures: Building a Sustainable Tomorrow – Papadikis et al. (Eds)
© 2020 Taylor & Francis Group, London, ISBN 978-0-367-43019-1

BIM model for evaluating green building through its life cycle

X. Liu, L. Lin & Y. Gao
School of Civil Engineering, Suzhou University of Science and Technology, Suzhou, China

ABSTRACT: Environmental problems, energy problems and so on are increasingly obvious, and green buildings are greatly promoted. This paper introduces the green building evaluation system at home and abroad. On the basis of discussing the feasibility analysis of BIM technology applied to the green building evaluation system under the whole life cycle, the information contained in the BIM model is matched with the green building evaluation system, and the framework model of combination of the two in the whole life cycle is constructed, and the way and process of green building evaluation based on BIM technology is presented.

1 INTRODUCTION

In recent years, whether it is the policy direction of Beijing-Tianjin-Hebei integration, or the hot topics such as urban haze, APEC blue and so on, the social concern has been directed to the topic of environmental protection. As the pillar industry of the country, the construction industry not only consumes a lot of energy, but also emits pollutants to the outside world. Under the general policy of environmental protection, the development of green building has become an inevitable trend. In order to standardize and guide and promote the development of green building industry, the evaluation system of green building plays an important role. BIM technology has been put forward since 2002, after more than ten years of rapid development, its application in the field of architecture has developed from the initial design stage of collision inspection to the whole life cycle of construction projects (Wu 2013).

The evaluation standard of green building in China (GB/T 503782019) refers to the application of BIM technology as the evaluation score. At present, the combination of the two has become the research direction of many scholars. Zhao (2019) puts forward the corresponding relationship between green building evaluation index and building information model; Zhao Qin et al. (2017) proposed a process model for indoor environment; Wu et al. (2017) proposed a model for Automatic Evaluation of the Land-saving Design of the Green Building, Most of them are evaluated at a certain stage or aspect of the building. The purpose of this paper is to discuss the relationship between BIM green building evaluation and green building evaluation in the whole life cycle, so as to better realize the evaluation of green building and promote the development of green building.

2 RESEARCH STATUS

2.1 *Green building evaluation system*

Green building evaluation system plays an important role in the development of green building, and can standardize and guide the green building industry. Aware of the important role of green building evaluation standards, countries all over the world have formulated evaluation systems to adapt to the national conditions of various countries, such as LEED, in the United States, BREEAM, in the United Kingdom and CASBEE in Japan. In 2006 and 2014, China promulgated the 'Green Building Evaluation Standard'. On the basis of it, the newly promulgated "Green Building Evaluation Standard" system in 2019 is more perfect, from the previous

Table 1. Comparison of green building evaluation systems in countries.

Name	CASBEE	BREEAM	LEED	Green building evaluation standard
Scope of assessment	Whole life process of buildings	Whole life process of buildings	Whole life of the building	Whole life of the building
Evaluation indicators	Category Q indicators: Indoor environmental quality The quality of service Outdoor environmental quality LR class indicators: energy Resources and materials External environment of building site	Management Health and comfort energy transport water The raw materials Land use pollution Regional ecological	Choose sustainable construction sites Water saving Energy and atmospheric environment Materials and resources The indoor environment Consistent with LEED innovation	Save land and outdoor environment Energy conservation and energy utilization Water conservation and water resources utilization Material saving and material
Rating	Good Very good Good Slightly less Poor	Excellent Good Good Qualified	LEED platinum certification LEED gold certification LEED silver certification LEED certification	Level 1 The secondary Level 3

six dimensions to seven dimensions, each kind of index has control items and scoring items, which can realize the comprehensive evaluation of green buildings. The comparison of green building evaluation systems in countries as shown in Table 1.

2.2 BIM

Building Information Modeling-BIM is the digital representation of physical and functional characteristics of the construction project. It's the shared information resource. Provide the information basis for the multi-stage design, construction and operation of the project. Multistage stakeholders can achieve collaborative cooperation through BIM information insertion, extraction, update and modification (National Institute of Building Science).

The application of BIM technology in the construction field has been developed from the initial design phase to the full life cycle of the construction project, including construction and operation management. The United States Building SMART Consortium summarizes the twenty-five applications of BIM technology, including: current modeling, cost budget, phase planning, planning text preparation, field analysis, design demonstration, design modeling, energy analysis, structural analysis, sunshine analysis, equipment analysis, specification verification, 3D coordination, Site use planning, construction system design, digital processing, three-dimensional control and planning, record model, maintenance plan, building system analysis, asset management, space management tracking, disaster plan, etc.

At present, BIM software is mainly foreign development software, which can be divided into BIM modeling software and BIM management software. The mainstream software for establishing visualization model is ArchiCAD, Revit, Bentley, etc (Azhar et al. 2008); among them, the Revit software developed by Autodesk has good market application and can be said to be the leader of the industry. The concept of "virtual architecture", which was developed by Graphisoft as early as 1980s, coincides with the idea of BIM technology.

The application of building information model changes the way of cooperation among the participants, so that all parties can improve production efficiency (Cao et al. 2015). BIM technology is causing a new change in the construction industry.

3 GREEN BUILDING EVALUATION SYSTEM

In this paper, the whole life cycle refers to the initial design stage, the middle construction stage and the later operation stage of the construction project. The basis of the application of BIM in the green building evaluation system is that BIM can provide a lot of information for evaluation. BIM model contains a lot of data information, including component size, building materials, cost and so on. As long as the model reaches a certain degree of detail and credibility, it can provide the basis for the evaluation (Azhar et al. 2011). In addition, most of the BIM software supports the interactive operation of information, supports the IFC (Industry Foundation Class) standard and the Green Building XML (GBXML) protocol, so that there is a good interface between the building information model and the large number of third-party analysis application software, and the data in the building information model can be transferred to the analysis software, so that the efficiency and the accuracy of the evaluation are improved (Fu 2017).

3.1 *Construction of BIM energy consumption simulation model at the early stage of design can estimate the green building design level*

In the early stage of design, BIM can simulate the energy consumption and realize the pre-evaluation. BIM has a powerful energy consumption simulation project based on its rich information, and can realize the energy consumption simulation of the building in the design stage and even the conceptual design stage (Schlueter and Thesseling 2009), so as to realize the pre-evaluation. If the building meets the standards of green building design, designers can compare and select the optimal scheme on this basis. If not, the designer can change the design and promote the optimization and improvement of the design scheme. Combining BIM with green building evaluation system can make architect and engineer know what grade the building reaches in the early stage, and ensure the realization of green building.

3.2 *Real-time monitoring of BIM virtual construction during construction process and timely correction of deviation*

In the course of construction, the BIM can be constructed and monitored to ensure the completion of the various indexes. The BIM has a time dimension to achieve a virtual construction of 4D. BIM technology can monitor the completion of various indexes in the design in real time. In the process of green building construction, the visual model and data are mutually reinforcing, and the BIM is used to conduct real-time evaluation with reference to the green building evaluation system, and the problems occurring or the deviation from the design are corrected in time, so that the realization of the corresponding level of the green building is strongly guaranteed.

3.3 *After completion and operation, BIM model continuously updates information to facilitate the evaluation process*

In the course of construction, the BIM can be constructed and monitored to ensure the completion of the various indexes. The BIM has a time dimension to achieve a virtual construction of 4D. BIM technology can monitor the completion of various indexes in the design in real time. In the process of green building construction, the visual model and data are mutually reinforcing, and the BIM is used to conduct real-time evaluation with reference to the green building evaluation system, and the problems occurring or the deviation from the design are corrected in time, so that the realization of the corresponding level of the green building is strongly guaranteed.

4 THE CORRESPONDENCE BETWEEN GREEN BUILDING EVALUATION SYSTEM AND BIM INFORMATION IN THE FULL LIFE CYCLE

Because in different life cycle stages of green building, the specific scoring parameters in the evaluation system of green building (land saving and outdoor environment, energy saving and energy utilization, water saving and water resources utilization, material saving and material resource utilization, indoor environmental quality, construction management, operation management) are not completely different, such as water saving and water resources utilization. The water-saving irrigation method can be measured and optimized in the design stage, while the grade of sanitary appliances should be measured according to the specific facilities provided by the construction party in the construction stage. Therefore, in the process of comprehensive evaluation of green buildings based on the whole life cycle, it is necessary to analyze different evaluation indexes according to different life cycle stages of green buildings. In addition, in the process of information acquisition and information correspondence, some evaluation indexes can be obtained by simple operation from the data of BIM model, such as volume rate, green ground rate and so on, while others need to go through parameter input, simulation analysis and other processes before they can be obtained, such as wind speed in pedestrian area, indoor noise level and so on.

Table 2. Correspondence between green building evaluation index and BIM model information.

Life cycle	Green building evaluation index	Specific scoring items	Corresponding BIM information
Design	ground and outdoor environment	Floor area ratio, green land ratio, ratio of underground building area to above ground building area, wind speed in pedestrian area, visible light reflection of glass curtain wall, etc	Project information, site area, statistical details, etc
	Energy conservation and energy utilization	The thermal performance of the envelope, the ratio of domestic hot water provided by renewable energy, and the energy efficiency ratio of cooling and heat source units are adopted in the lighting system	Equipment parameters, components materials, etc
	Water conservation and water resource utilization	Rainwater utilization rate, non-traditional water utilization rate, Water-saving irrigation methods, water-saving cooling technology application, etc	Equipment parameters, statistical details, etc
	Material saving and material resource utilization	Integration design ratio of civil engineering and decoration engineering Reusable partition utilization rate, high strength building structural material ratio, etc	Component material, statistical details, etc
	Indoor environment quality	Sound insulation performance, lighting coefficient, air distribution parameters	Equipment parameters, statistical details, etc
Construction	Water conservation and water resource utilization	Grade of sanitary apparatus, pressure of water supply point, etc	Equipment parameters, statistical details, etc
	Material saving and material resource utilization	Proportion of prefabricated components, producing area of building materials, etc	Component material, statistical details, etc
	Construction management	Noise reduction measures, solid waste emissions, construction water consumption, etc	Statistics, etc.
Operation	Water conservation and water resource utilization	Average daily water amount, over pressure outflow phenomenon, etc	Statistics, etc.
	Indoor environment quality	Indoor noise level, pollutant concentration, carbon dioxide concentration, etc	Statistics, etc.
	Operation management	Ventilation and air conditioning system parameters	Replacement cycle, equipment parameters, usage, etc

According to the stages divided by the life cycle of green building, the evaluation indexes, specific scoring items and BIM information in the green building evaluation system (GB/T 50378-2019) in each stage are combined, as shown in Table 2.

5 BIM FRAMEWORK MODEL CONSTRUCTION FOR GREEN BUILDING EVALUATION SYSTEM IN THE WHOLE LIFE CYCLE

The BIM framework model for the green building evaluation system in the full life cycle corresponds to different project stages, extracts relevant information from the BIM model, corresponds to the indicators of the green building evaluation system, and analyzes and evaluates the project. Further research, revision and decision making are carried out according to the results obtained. The application of BIM in the green building evaluation system is shown in Figure 1.

In the design stage, the BIM model is constructed and the BIM database is formed. The database contains accurate and rich information about the building, including equipment parameters, engineering information, component materials, statistical details and so on. On the basis of the complete establishment of BIM model, the information is imported into Ecotect, Green Building Studio, EnergyPlus and other analysis software for calculation and analysis, and it is evaluated in combination with 2019 Green Building Evaluation Standard. The results are fed back to the professional designers so that they can adjust the design according to the preset green building grade and select the optimal scheme. When the design is formed, new data are generated and the BIM model is updated to guide the follow-up work.

In the construction phase, the virtual construction function of the BIM is applied. With the development of construction work, the BIM data is also updated. And the data information in the updated model can be compared with the green building evaluation standard at any time, and the information comprises the information of each design parameter and the construction stage. Through the real-time evaluation and monitoring in the construction process, the generation of the deviation can be effectively avoided, and the realization of the green building target can be better guaranteed.

During the operational phase, the data in 1 year of operation is entered, and the BIM model is updated again. The BIM model at this time contains detailed information covering the entire life cycle, which is more informative and accurate. The whole building can be evaluated conveniently and comprehensively by extracting, analyzing, and cooperating with the field survey.

The BIM framework model for the green building evaluation system in the full life cycle is shown in Figure 2.

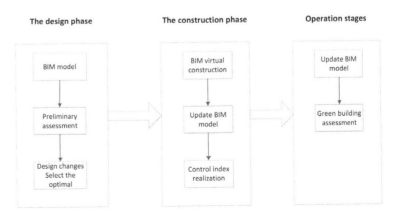

Figure 1.   Application of BIM in green building evaluation system with full life cycle.

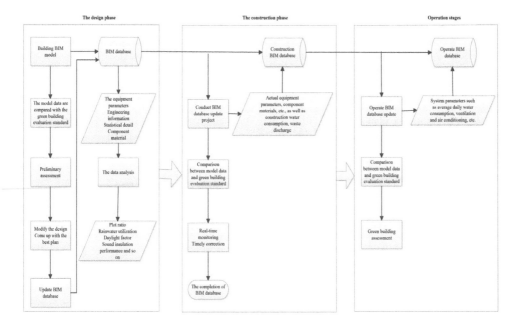

The design phase · The construction phase · Operation stages

Figure 2. Under the whole life cycle in view of the green building evaluation system framework of BIM model.

6 CONCLUSION

Green building evaluation system plays an important role in promoting the development of green building industry, and the combination of BIM technology and green building evaluation system provides technical support for it. By constructing the three-stage BIM application model of design, construction and operation, the paper comprehensively covers the whole life cycle of buildings, and preliminarily points out the application approach of BIM in the green building evaluation system. In the design stage, BIM is applied to achieve pre-evaluation; In the construction stage, BIM is applied to carry out virtual construction, monitor the construction process and correct deviation timely; After entering the operation stage, green building grade assessment will be conducted conveniently according to the updated BIM model.BIM contains a large amount of data and information. Through data acquisition, data analysis and index comparison, the evaluation of green buildings at all stages can be achieved. At present, the green building industry and the BIM technology in our country is still in development, with the development of the industry, technology progress, to vigorously develop BIM software, direct implementation in BIM software according to the standard of green building assessment score function, which makes the combination of BIM and green building evaluation system better, to promote the development of the green building industry.

REFERENCES

Azhar, S. Carlton, W.A. & Olsen, D. 2011. Building information modeling for sustainable design and LEED (R) rating analysis. *Automation in Construction* 20(2): 217-224.
Azhar, S. Nadeem, A. & Mok, J.Y.N. 2008. Building information modeling (BIM): A new paradigm for visual interactive modeling and simulation for construction projects. *Proceedings of First International Conference on Construction in Developing Countries* 435-446.
Cao, D. Wang, G. 2015. Practices and effectiveness of building information modelling in construction projects in China. *Automation in Construction* (49): 113-122.

Fu, Y, Zhang, S. & Shi, Q. 2017. Optimum design of water systems for a high-rise residential building cluster combined with technologies of BIM and green buildings. *China Water & Wastewater* 33(06):83-87.

National Institute of building science, United States National Building Information Modeling Standard, Versional-Part 1.

Schlueter, A. & Thesseling, F. 2009. Building information model based energy/exergy performance assessment in early design stages. *Automation in Construction* 18(2):153-163.

Wu, J. Li, X. Cao, L. & Zhang, S. 2017. BIM-based Automatic Evaluation of the Land-saving Design of the Green Building. *Building Science* 33(04):123-128.

Wu, D. 2013. Building knowledge modeling: Integrating knowledge In BIM. Proceedings of the 30th CIB W78 International Conference.

Zhao, Q. Tian, Q. Liu, Y. & Hei, X. 2017. BIM technology application of construction management under a new assessment standard for green building. *Journal of Xi'an University of Technology* 33(02):211-219.

Zhao, W. 2019. Dynamic integration mechanism of building information modeling technology and low carbon green building evaluation index system. *Science Technology and Engineering* 9(3):196-201.

Sustainable Buildings and Structures: Building a Sustainable Tomorrow – Papadikis et al. (Eds)
© 2020 Taylor & Francis Group, London, ISBN 978-0-367-43019-1

BEAM upgrade and cost premium for green building development

H. Wang & W. Tang
School of Public Administration, Zhejiang University of Finance & Economics, Hangzhou, China

X. Zhang
Department of Public Policy, City University of Hong Kong, Hong Kong

ABSTRACT: The world has witnessed the tremendous momentum of green building development. One of the trends is that the definitions of green building are evolving as long with the upgrading of building environment assessment models (BEAMs). Because of the increasing "green" requirements, the cost adhered to the green materials, technologies and management practices is increasing, particularly in the exploration phase. Although the cost premium for the delivery of green buildings is prohibitive in the green building movement, it is expected to be reduced over time due to scale effect and technology maturation, particularly at the exploitation phase. This study provides a critical review of 55 empirical studies published up to the end of 2018. The empirical studies involved have developed two main types of estimation methods: the *ex-ante* approach and the *ex-post* approach. It is found that the actual cost premium is much lower than generally believed, and the cost premium obtained by using the *ex-post* approach is on average higher than that using the *ex-ante* approach. It is also found that the pull power of exploitation is weaker than that of exploration, as the cost premium is increasing at an annual rate of approximately 10%, probably due to increased green requirements.

1 INTRODUCTION

While the world has witnessed the tremendous momentum of green building development, it is noticed that the incremental upfront costs of the design and construction of green buildings comparing to their conventional counterparts, defined as the cost premium, is reported to be a major concern hindering realization of their potential benefits (e.g. Dwaikat & Ali 2016, Olubunmi et al. 2016). Several trade and academic publications report inconclusive results of the amount of cost premium involved (Kats et al. 2003, 2006, Dwaikat & Ali 2016, Matthiessen & Morris 2004, 2007, Steven Winter Associates 2004). Varying estimation methods are used globally, along with differing project features, project teams' capacity for green practices and support of incentive policies (Kats, 2003, 2006, Kats et al. 2003, 2010, Kats & Perlman 2005, Matthiessen & Morris 2004, 2007). Meanwhile, economies of scale are expected to reduce the cost of purchasing green materials/technologies (Houghton et al., 2009), as are government incentive policies to promote green building development (Shen et al. 2015, Dwaikat & Ali 2016, Olubunmi et al. 2016). Cost premium is among the most critical factors in considering delivering a green building and will gain more research attention (Zuo & Zhao 2014).

In this research, existing literature relating to the cost premium of green buildings is reviewed, with particular attention being paid to empirical studies. A series of further analyses are conducted to examine potential critical factors. The discussion of the results is then provided, followed by some concluding remarks concerning the cost premium of green buildings and implications for future research work.

Building green may require additional financial inputs for energy-saving facilities/equipment, eco-friendly materials, modeling integration, and management and consultancy fees (Circo 2007). These additional financial inputs are conceptualized as the cost premium, also known as the 'first cost premium' or 'upfront cost premium'. While most of the trading reports of the costs for green buildings were released without any definitions of cost premium (e.g. Kats et al. 2003, SWA 2004, WGBC 2013), some have tried to identify it from its constitutions: direct construction costs and indirect costs (e.g. Matthiessen & Morris 2004, 2007, Ugur & Leblebici 2017); 'soft costs' such as architectural, design and consultant's fees; and 'hard costs' including the building structure itself but not site acquisition (NEMC 2003, Fullbrook et al. 2005, WGBC 2013). Houghton et al. (2009) question the widely accepted consensus of "base cost", as the cost of delivering conventional code-compliant buildings, and define the cost premium as 'the additional construction costs associated with design and construction elements included in a project primarily due to their environmental performance'.

Estimation of the cost premium is a very challenging undertaking. Questionnaire surveys could be used to elicit perceptions of the cost premium for green buildings from various professionals. There are also estimation methods based on real life cost data. These comprise two main approaches – ex-post and ex-ante. Ex-post methods evaluate the final upfront cost of a green building and compare it with the "base cost", which could be the budget, the modeled cost or the actual cost of similar conventional projects. Using the original budget or the average local costs for delivering conventional buildings as the "base cost" is called the 'budgeting method' by Matthiessen & Morris (2004) and is used for most of the projects (Dwaikat & Ali 2016). Though the data needed is easily accessible, the reported project budget could be very unreliable since it is highly dependent on financial availability and the objective judgement of design experts. The modeled cost is theoretically more reliable, yet still depends on the construction of design models for both green and non-green versions, which could place an additional financial burden on project teams.

An alternative of re-designing the non-green version is presented in several publications, in that the components of green features are replaced with the conventional design so that the "base cost" equals to the actual green building cost minus the additional cost of the green features. However, comparison of the actual costs of both green buildings and non-green buildings in this way could be limited by difficulties in acquiring the actual costs of appropriate conventional projects. The method can also lead to a similar result as in the budgeting method, since the budgeting decisions might be made based on the actual performance of similar projects.

Based on the three ways of measuring the base cost, the ex-post approach can be further divided into three groups: the budgeting method, the modeled-cost method and the actual-cost comparison method. The three types of evaluation methods reviewed by Dwaikat & Ali (2016) conforms to this taxonomy. In contrast, the ex-ante approach, as the name implies, evaluates the cost premium pertaining to the specific requirements of green buildings before the actual design and construction costs are incurred. This approach can be conducted through the evaluation of itemized additional costs for each green feature and then assembling all of these to form the final cost premium. It has a strict demand for the standards of green features and therefore has the weakness of focusing only on the design details with the risk of losing the big picture of a green building. For example, the individual green materials/technologies (e.g. exterior insulation, double envelope system, water conservation, LED technology, and green roofs), to varying degrees will usually incur additional design and construction costs (UGC 2009, Driscu 2014). However, it has been argued that hiring an experienced project team to develop a more comprehensive design model might provide multiple green features without increasing the financial burden (Kats & Perlman 2005, Houghton et al. 2009). More empirical evidence is needed to determine whether the ex-ante approach results in a larger cost premium evaluation than the ex-post approach.

Cost premium trends over time can also be found by meta-analysis of the large pool of empirical evidence in existing publications. On one hand, it is argued that there is an overall trend towards the reduction in the cost premium of green buildings, since green building standards will be more demanding, green materials and technologies will be more mature and the industry will be more skilled at delivering green buildings (Matthiessen & Morris 2004, 2007, Kats & Perlman 2005, Houghton et al. 2009, WGBC 2013). On the other hand, in a survey by DD & A (2015), the builders responded that the cost premium for greening new homes was increasing in the 2010s. DD & A (2015) argue that the increasing demand for high quality green buildings might be responsible for this phenomenon, together with other possible factors, such as the higher price of green products/services or reduced incentive policies. The rivalry between the demand side for green buildings and the supplier side for providing them still lack of sufficient exploration.

Furthermore, the differing requirements of the various BEAM models will have a significant impact. Most of the publications concerning the cost premium of green buildings focus on the LEED system, with little attention being paid to other BEAM models in the global building market (WGBC 2013). In another view is that the maturity of the green building market is closely related to the establishment and implementation of BEAM model (s). It is hypothesized, therefore, that the cost premium of LEED labeled green buildings will be lower on average than that of other BEAM models.

3 RESEARCH METHODOLOGY

Firstly, relevant publications on the subject of cost premium for green buildings were selected. A preliminary computer search using suitable keywords ('cost premium', 'additional cost', or 'incremental cost' for cost premium, while 'green building' or 'sustainable building' for green building) in the title and abstract was carried out to select relevant publications in various databases, such as ABI/Inform, ISI Web of Knowledge, EBSCO, Emerald and Google Scholar. A manual scanning of the abstracts was then conducted. Articles lacking content on the subject of green buildings cost premiums were removed from the list and the full content of the remaining articles were then retrieved. Review articles were scrutinized in order to refine the list of papers. Using the 'snowballing' (Wohlin 2014) or "citation pearl growing" method (Dolan et al. 2004), the references cited at the end of each review article were compared with our list to minimize the chance of missing important articles. The articles focusing on only one or multiple green products or technologies were also removed, since they could not truly reflect the cost information for whole green buildings. Afterwards, outlier detection using the 3σ rule was conducted, in order to identify the specific sample points that deviate too much from the others, i.e., points lying beyond the interval of mean ± 3 standard deviations (Rousseeuw & Leroy 2005). This resulted in the identification of just one outlier, which is the one with smallest result of -27% (i.e. Hanson, 2010). Finally, a list of relevant articles, including 10 survey reports and another 45 empirical studies published between 2000 and 2018, was established through these rigorous processes.

Secondly, the selected studies were coded using content analysis. Content analysis is a systematic, replicate technique for compressing many words of text into fewer content categories based on explicit rules of coding (Stemler 2001). In this study, Endnote ® and Microsoft ® Excel were used as the working platform, with the basic information of these studies, including Authors, Publication Years, Titles, and Article Types, manually coded and compiled into an Excel file. Based on the research questions to be answered, other items were recorded, including Estimation Methods, BEAM models, Green level, Time, Project Types, Countries or Regions, Sample Size (of green buildings), The overall and individual results for the cost premium were also extracted from the selected articles and collated into the Excel file. To reduce any potential bias, the coding process was manually conducted independently by the authors. Agreement of the coded contents was made by interactions between the coders. Eventually, all the extracted information was stored in the Excel file as the database for further analysis.

Thirdly, a series of statistical analyses were conducted on the data in the Excel file. Meta-analysis refers to a kind of analytical techniques developed to draw conclusions from existing publications (Stanley 2001). It is essential to estimate the effect size of each publication result by sample size and/or accuracy, so that all the results can be weighted to form the overall effect size (Elle 2010). According to the principles of meta-analysis, in the analysis of this study, due to the lack of accuracy information, the sample size was adopted for this purpose. The articles were then clustered into different groups of Article Type, Estimation Method, Project Type, BEAM model, Green Level, Time Effect, Country or Region to measure their impact on the estimation of cost premium, with the same analysis conducted on each group followed by a set of unpaired two sample t-tests with unequal variances to compare the differences between the means of groups.

4 RESULTS AND FINDINGS

4.1 Survey reports

The profiles and results of the 10 survey reports are presented in Table 1. The surveys asked respondents to assign the additional cost (in percentage) of green buildings comparing with conventional buildings. No individual results are available for the first four articles. Using equal weights, the total overall result is 11.20%, i.e., the respondents believed that the delivery of green buildings to be 11.20% more than the conventional buildings. This is echoed in the argument of WGBC (2013), in that the expected cost premium involved is a major obstacle for the green building movement it is too high for practitioners.

4.2 Project case based

There are 24 academic publications and 21 business publications. Most academic articles were published after 2010. This probably reflects the fact that the industry was leading the research into the cost premium amount in the early period, with the academe quickly catching up. Most of the green buildings investigated are in the U.S. and all of the studies published before 2005 focus on the US LEED awarded projects, with the studies in other countries taking place subsequently.

Table 1. Summary of questionnaire survey results.

Citation	Article Type	Year	Sample size	Overall result	Detailed results (Unit: %)				
					<0%	0%	0-5%	5-10%	>10%
Berman, A. (2001)	Report	2001	249	10-15%	-				
Turner Green Buildings (2005)	Report	2005	665	13% (18%) for experienced (inexperienced) executives					
WBCSD (2008)	Report	2008	103 countries	17%	(<10% in developed countries)				
Coetzee and Brent (2015)	Journal	2015	193	20.36%	-				
Hwang et al. (2017)	Journal	2017	242+121	5-10%	-				
Breslau (2007)	Report	2007	414	more than 5%	1	8	38	30	22
BD+C (2007)	Report	2007	631	6-10%	1	13	24	27	35
Ahn and Pearce (2007)	Journal	2007	87	-	1	15	23	34	27
Park et al. (2008)	Report	2008	16	6-10%	0	0	37	38	25
DD&A (2015)	Report	2011		8%	3	4	34	43	15
		2014	-	9%	5	4	30	38	22
		2015		12%	4	3	15	42	35

Note: The intervals of detailed results are open at the left boundary and closed at the right boundary; the total overall and detailed results are calculated as the average of correspondent available data, with - denoting missing.

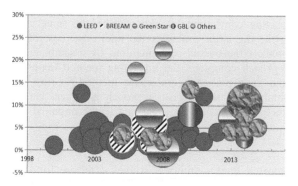

Figure 1.   Cost premium for green buildings between 2000 & 2018.

In Figure 1, the horizontal axis is the publication date and the vertical axis is the overall cost premium result. The circle colors represent the different BEAM models, while the diameters of circles denote the green level of the corresponding BEAM models, with unspecified green levels being assumed to be minimal. Roughly speaking, as expected, the cost premium for green buildings in LEED appears to be a little lower than that of the other BEAM models and more stable over time, while it shows an increasing trend in other BEAM models. Also, it appears that the cost premium for higher green levels does not appear to be higher than for lower levels.

Following the analysis processes in Elle (2010), an empirical study was adopted as the unit for calculating the effect size, and the weighted mean effect size and homogeneity obtained as

$$\text{Weighted Mean Effect Size}: \quad \overline{ES} = \frac{\sum w_i ES_i}{\sum w_i}$$

$$\text{Homogeneity}: \quad Q = \sum w_i \left(ES_i - \overline{ES}\right)^2$$

where w_i is the number of green building projects in the i-th study, and ES_i is the effect size of the i-th study, i.e. the reported average cost premium. The differences were tested for significance by a 2-tailed t-test. Under the assumption of the t-distribution of the effect size, $\frac{\overline{ES}}{\sqrt{Q/n}} \sim t(n-1)$. Given Prob (A) is the probability of event A, $\text{Prob}\left(\frac{\overline{ES}}{\sqrt{Q/n}} > 0\right) \geq 95\% \text{ or } 90\%$ indicates a significantly positive result at the confidence level of 95% or 90%.

Following the above process, the mean effect size is 3.05%. Under the assumption that t (45-1), p value is 0.076, which means the positive result of cost premium for green building is not statistically significant at 95% confidence level, but statistically significant at 90% confidence level.

5   IMPACT FACTORS ON COST PREMIUM EVALUATION

Estimation of the cost premium could be affected by a variety of impact factors, such as the method used, project type, BEAM model, green level and time effect. Further analyses were made on the 45 empirical studies to explicate these effects. Case study is the most popular method in reporting cost premium for green building (e.g. Zhu et al. 201, Sun et al. 2019).

5.1   Estimation method

As aforementioned, there are ex-post and ex-ante approaches to estimation. The former emphasizes the comparison of actual green buildings with the evaluated 'base cost' for designing and constructing conventional buildings with the same characteristics except for the green features, while the latter evaluates the additional cost of delivering decomposed green features and predicts the overall cost premium by assembling the itemized additional costs. The ex-post approach comprises the budgeting method (ex-post 1), the modeled-cost method (ex-post 2) and the actual cost comparison method (ex-post 3). The overall result of the cost premium obtained by the ex-post approach is larger than the ex-ante approach. However, this difference is not significant according to the unpaired two-sample t-test.

5.2 *Project type*

The project types reported in the empirical studies comprise four groups of office buildings, residential buildings, schools and others. The type of school is comprises academic buildings, research laboratories and libraries. 'Others' includes banks, facility buildings, council houses, heritage buildings, commercial centers, healthcare buildings, etc. The decomposed results of the cost premium for the four project types are shown in Annex. Most projects are schools, with the lowest cost premium. Office and residential buildings have the largest cost premiums of around 5%. These are all significant to some extent, with school building type significant at the 95% confidence level and office and residential buildings significant at the 90% confidence level.

5.3 *BEAM models and green levels*

The empirical studies cover 13 kinds of BEAM models, with LEED being the most common. By applying the meta-analysis process again, the overall cost premium evaluated in the LEED labeled green buildings is 2.64%, which is much lower than the 3.64% weighted average of the whole population. The results for Green Star, GBL, BREEAM, and other models are also presented in Annex, among which only the BREEAM result is equivalent to LEED, with other BEAM model results appearing to be larger than LEED. The unpaired two-sample t-test was conducted, and the difference between cost premium for LEED system and others is confirmed to be significant. When statistical analysis is used, the territory scope in early year studies usually is a country where several BEAM models might be several, e.g. LEED and BREAM (e.g. Sun & Shao 2010). The territory scope in the studies of the last three years is narrowed to a smaller administrative region (e.g. Ge et al. 2018), and the BEAM model adopted is single, thanks to the fast growing sample size.

5.4 *Time effect*

As shown in Annex, the decomposed cost premium results increase in the three periods from 2.20% to 5.28%. Discounted by 10 years, the annual rate of increase is around 9.5%. One possible reason provided by DD & A (2015) is that the increasing demand for green features may be responsible for the increasing trend in cost premium. This may well be the case, as a dominant portion of the LEED labeled green buildings completed before 2005 was awarded as Certified or Silver while, from 2006 to 2015, Gold labeled green buildings outnumbered the Certified ones, with the Silver labeled being the most dominant.

6 DISCUSSIONS AND CONCLUSIONS

Concerns over the high upfront cost for green features is frequently mentioned as one of the main obstacles holding back green building initiatives and a great deal of research has focused on their benefit-cost analysis. Following early work by Kate et al. (2003) and Matthiessen & Morris (2004), a number of empirical studies have been published, the results of which indicate that the cost premium appears to be very diverse, positive, not significant, or even negative. The reasons given for the inconclusive results include the different estimation methods used, project types, BEAM models, green levels and time effects.

Totally, 55 empirical studies on the topic of cost premium for green buildings were reviewed, including 10 survey reports and 45 articles. The survey results indicate that most professionals in the building sector consider a considerable positive cost premium is involved, of over 10% on average, for the extra cost of providing green features. Some respondents also believe that this cost premium is increasing. The 45 empirical studies were analyzed by using the sample size as the effect size for each study and the weighted mean effect size was calculated and tested for significance. This echoes some of the previous literature in showing the actual cost premium to be much lower than that perceived by survey respondents. Furthermore, the impact factors on the evaluation of cost premium, including estimation methods, project types, BEAM models, green levels and time effects, were examined by clustering and comparison. The principle of the meta-analysis method was adopted to calculate the weighted mean effect size in specific clusters of impact factors. In comparing the results by the unpaired two-sample t-test with unequal variance, it was found that, in terms of estimation methods, ex-ante and ex-post approaches provide different, but not statistically significant, results. Office and school buildings, except the residential buildings, are confirmed to have a significantly positive cost premium, with only small the differences between the three types of green buildings. The BEAM models have a significant impact on the evaluation of cost premium, with sound green standards, high market maturity and experience accumulation, resulting in the average cost premium for green buildings with LEED or BREEAM certification being significantly lower. As for the impact of green levels, the cost premium for low level green buildings is significantly positive and more stable over time. However, the cost premiums for high level green buildings vary a great deal, although they are not convincingly greater than low level green buildings on average. The high variance of cost premium will exist till the advancements on green design, green technologies, green materials, and green processes are stabilized and standardized. Surprisingly yet reasonably, the overall cost premium for green buildings in the period from 2011 to 2015 is significantly greater than 10 years ago, with an increasing annual rate of around 10%. It is envisaged that future research work will have to consider new horizons of the green building concept and to integrate them in order to reduce the cost premium for green buildings.

ACKNOWLEDGMENT

Thanks for the support from two students, Ms Jingyi Wang and Ms Chenyue He, in terms of data collection and primary analysis.

REFERENCES

Ahn, Y. & Pearce, A. 2007. Green construction: Contractor experiences, expectations, and perceptions. *Journal of Green Building* 2(3):106-122.

BD+C. 2007. *Green buildings research white paper: Building Design + Construction.*

Berman, A. 2001. *Green Buildings: Sustainable Profits from Sustainable Development. Unpublished report, Tilden Consulting.*

Breslau, B. 2007. *Sustainability Perceptions and Trends in the Corporate Real Estate Industry: CoreNet Global and Jones Lang LaSalle.*

Circo, C. J. 2007. *Using mandates and incentives to promote sustainable construction and green building projects in the private sector: A call for more state land use policy initiatives. Penn* St. L. *Rev., 112: 731.*

Coetzee, D. & Brent, A. 2015. Perceptions of professional practitioners and property developers relating to the costs of green buildings in South Africa: technical paper. *Journal of the South African Institution of Civil Engineering* 57(4):12-19.

DD & A. 2015. *Green and Healthier Homes: Engaging Consumers of all Ages in Sustainable Living. Smart Market Report of Dodge Data & Analytics.*

Dolan, P. Shaw, R. Tsuchiya, A. & Williams, A. 2004. QALY Maximization and People's Preferences: A Methodological Review of the Literature. *Health Economics* 14(2):197–208.

Driscu, M. 2014. 'Renaissance' complex-behind the green building idea-design and technology. *SGEM2014 Conference Proceedings* 2:3-10.

Dwaikat, L. & Ali, K. 2016. Green buildings cost premium: A review of empirical evidence. *Energy and Buildings* 110:396–403.

Fullbrook, D. Jackson, Q. & Finlay, G. 2005. *Value case for sustainable building in New Zealand (ME 705). Wellington: Ministry for the Environment.*

Ge, J. Weng, J.T. Zhao, K. Gui, X.C. Li, P. & Lin, M.M. 2018. The development of green building in China and an analysis of the corresponding incremental cost: A case study of Zhejiang Province. *Lowland Technology International* 20(3): 321-330.

Hanson, M.E. 2010. Green schools on ordinary budgets. *School Administrator* 67(7): 32-35.

Houghton, A. Vittori, G. & Guenther, R. 2009. Demystifying first-cost green building premiums in healthcare. *HERD: Health Environments Research & Design Journal* 2(4):10-45.

Howard, J. L. 2003. The Federal Commitment to Green Building: Experiences and Expectations. *Federal Executive, Office of The Federal Environmental Executive, Washington.*

Kahn, M.E., & Kok, N. 2014. The capitalization of green labels in the California housing market. *Regional Science and Urban Economics* 47:25-34.

Kats, G. 2006. *Greening America's schools: Costs and benefits. A Capital E Report.*

Kats, G. & Perlman, J. 2005. *National review of green schools: costs, benefits, and implications for Massachusetts.*

Kats, G. Alevantis, L. Berman, A. Mills, E. & Perlman, J. 2003. *The costs and financial benefits of green buildings: a report to California's sustainable building task force: Capital E.*

Kats, G. Braman, J. & James, M. 2010. *Greening our built world: costs, benefits, and strategies: Island Press.*

Matthiessen, L.F. & Morris, P. 2004. *Costing green: A comprehensive cost database and budgeting methodology. Davis Langdon.*

Matthiessen, L.F. & Morris, P. 2007. *Cost of green revisited: Re-examining the feasibility and cost impact of sustainable design in the light of increased market adoption: Langdon, Davis.*

Meron, N. & Meir, I.A. 2017. Building green schools in Israel. Costs, economic benefits and teacher satisfaction. *Energy and Buildings* 154(1):12-18.

NEMC. 2003. Analyzing the cost of obtaining LEED certification. *Report prepared by Northbridge Environmental Management Consultants for the American Chemistry Council.*

Olubunmi, O.A. Xia, P.B. & Skitmore, M. 2016. Green building incentives: A review. *Renewable and Sustainable Energy Reviews* 59:1611-1621.

Park, C. Nagarajan, S. & Lockwood, C. 2008. *The dollars and sense of green retrofits: Deloitte Development LLC.*

Rousseeuw, P. J. & Leroy, A. M. 2005. *Robust regression and outlier detection (Vol. 589): John Wiley & Sons.*

Sacks, A. Nisbet, A. Ross, J. & Harinarain, N. 2012. Life cycle cost analysis: a case study of Lincoln on the Lake. *Journal of Engineering, Design and Technology* 10(2): 228-254.

Sun, D & Shao, W. 2010. Statistic and study on incremental cost of green building in china. *Eco-city and Green Building* 4: 43-49.

Sun, C.Y. Chen, Y.G. Wang, R.J. Lo, S.C. Yau, J.T. & Wu, Y.W. 2019. Construction Cost of Green Building Certified Residence: A Case Study in Taiwan. *Sustainability* 11:2195.

Turner Green Buildings. 2005. *Survey of Green Building Plus Green Building in K-12 and Higher Education. Market Barometer, Turner Green Buildings, New York.*

UGC. 2009. *The cost of green in NYC*: Urban Green Council (UGC), sponsored by Davis Langdon and New York State Energy Research & Development Authority.

Ugur, L. & Leblebici, N. 2017. An examination of the LEED green building certification system in terms of construction costs. *Renewable and Sustainable Energy Reviews* 73:1364-0321.

WBCSD. 2008. Energy Efficiency in Buildings: Facts and Trends. *World Business Council for Sustainable Development.*

Wohlin, C. 2014. Guidelines for snowballing in systematic literature studies and a replication in software engineering. *In Proceedings of the 18th International Conference on Evaluation and Assessment in Software Engineering (p. 38). ACM.*

Xenergy Inc., & SERA Architects. 2000. Case Study: Green City Buildings-Applying the LEED Rating System: Prepared for Portland Energy Office Portland, Oregon.

Zhu, Y. Lin, B. & Yuan, B. 2010. Low-cost green building practice in China: Library of Shandong Transportation College. *Frontiers of Energy and Power Engineering in China* 4(1):100-105.

Zuo, J. & Zhao, Z.Y. 2014. Green building research–current status and future agenda: A review. *Renewable and Sustainable Energy Reviews* 30:271-281.

Sustainable Buildings and Structures: Building a Sustainable Tomorrow – Papadikis et al. (Eds)
© 2020 Taylor & Francis Group, London, ISBN 978-0-367-43019-1

The performance of the backfills of the borehole heat exchanger of the ground source heat pump system in cooling dominated region of China

W. Ma, J. Hao, C. Zhang & F. Guo
Department of Civil Engineering, Xi'an Jiaotong-Liverpool University, Suzhou, China

H. Wen
Department of Electrical and Electronic Engineering, Xi'an Jiaotong-Liverpool University, Suzhou, China

ABSTRACT: The ground temperature increase is the crucial problem that influences the long-term performance of Ground Source Heat Pump (GSHP) system in cooling dominated region of China. In this report, the numerical simulation model based on the heat rejection model of an in-situ Thermal Response Test (TRT) is created using TRNSYS simulation tool. Seven different thermal conductivity backfills are simulated for 1, 5, 10, 15 and 20 years. The result illustrates that the higher conductivity backfills improves the heat exchange performance between BHE and surrounding ground but it also increases the ground temperature in cooling dominated regions for long terms operation. In the case of Chuzhou city of this study, when the conductivity of backfill is higher than 1.9 W/(m·K) which is closest to the thermal conductivity of the surround ground, there are little growth in ground temperature as well as the heat exchange performance.

1 INTRODUCTION

Ground Source Heat Pump (GSHP) system is recognized as an effective geothermal system using the ground as a heat source (Wang et al. 2019). The vertical closed-loop GSHP is one of the mostly adopted types of the GSHP system for buildings in urban areas where is lacking of land spaces. Its configuration comprises the vertical borehole heat exchanger (BHE) installed inside a borehole and a pump that circulates a solution of water or anti-freeze mixture fluid through the buried pipes (Cui et al. 2018). Thus, the heat is transferred from the ground to the heat carrier fluid. The ground temperature at a suitable depth is independent from the air temperature but strongly depending on the geographical characters. Therefore, GSHP system performance is determined by the thermal interaction between the ground and the BHEs. The design of the BHE is crucial to the performance of GSHP system. The performance of BHE is influenced by the ground thermal characteristics, the backfills and the types of BHE (Desmedt et al. 2011). In the cooling dominated regions, the ground temperature increases is a major problem which influences the long-term performance of vertical closed loop GSHP system (Ma et al. 2019). The optimal design of BHE could effectively control the ground temperature increase problem of GSHP system. The conductivity of backfills is one of major factors for the design of BHE (Erol & Francois 2014).

Backfill materials play a significant role in the heat transfer between the ground and the heat carrier fluid in the pipes for BHEs. Previous studies have shown that backfills with relatively higher thermal conductivity improves the heat exchange rate of the BHE and is expected to be a major approach to shortening the required length of BHEs for GSHP applications in order to save initial investment cost (Jun et al. 2009, Borinaga-Treviño et al. 2013). According to ASHRAE Hand book -HVAC Systems and Equipment, the commonly used backfill materials have the thermal conductivity from 0.7 W/(m·K) to 2.4 W/(m·K), depending on the geographical characteristics of the ground (ASHRAE 2012). The influence of different thermal conductivities of backfills on the ground thermal imbalance are rarely investigated based on the local ground geothermal characteristics in cooling dominated regions.

As a conventional method of measuring the ground thermal characteristics, the thermal response test (TRT) is widely conducted in many countries. The TRT is based on constant heat injection or rejection into an in-situ testing BHE and measures the temperature response during a certain period (Lee et al. 2011). However, for certain factors which are difficult to achieve in a single field test including the different backfills, especially for testing a long-term performance. Numerical simulation method is wildly used nowadays to investigate the effectiveness of different design parameter of GSHP system (Zhang et al. 2018). In this study, Transient system simulation (TRNSYS) software is used to create the model for BHE (Klein et al. 2010). The data recorded of an in-site TRT conducted in Chuzhou city is used to validate the TRNSYS simulation model and used as a case study. The performance of backfills with seven different thermal conductivities for 1, 5, 10, 15 and 20 years based on the TRT of the heat rejection test will be analyzed. The research result could provide valuable reference for selecting backfills for BHE of GSHP system in Chuzhou city. And the methodology of this research could be used for investigate design parameters of BHE for GSHP system.

2 METHOD

2.1 TRT

TRT is an in-situ measurement method to measure the effectiveness of the ground thermal properties (Gehlin 2002). In an international conference in Stockholm, TRT were first proposed by Mogensen which the heat at a constant power was extracted from or injected in a borehole while the inlet and the outlet borehole temperature were measured (Mogenson 1983). An in-situ TRT is typically implemented on a vertical borehole with approximately the same diameter and depth as the heat exchangers planned for the site. A vertical ground-loop heat exchanger has a U-tube inserted into a borehole. Figure 1 illustrated the diagram of TRT. Backfill is placed in the borehole to fill the space that is not occupied by the U-tube. The equipment for an in-situ test is illustrated in Figure 1, where an electric heater serves as a controlled heat source with a constant heat.

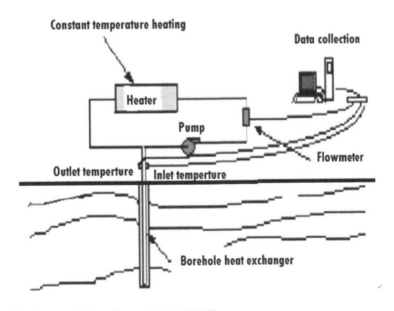

Figure 1.   The diagram of thermal response test (TRT).

Water is pumped through the U-tube by a pump and the Borehole heat exchanger exchanges heat with the ground. The inlet, outlet temperature and flow rate during a test period is recorded by the data collection device.

2.2 *The simulation model created with TRNSYS*

The simulation model was created by TRNSYS based on the TRT of the heat rejection test as shown in Figure 2. The system consists of a vertical U-tube ground heat exchanger (Type 557a), a pipe system, a circulation pump (Type 114) which circulates water flow throughout the pipe, a heater (Type 6) which supplies the constant heat to the water inside the pipe, and a data collection device (Type 65) which continuous collecting data of the inlet and outlet temperature, the external weather data (Type 15) is connected to Type 557a and Type 6 offering the weather of the specific location, and a Equa is created to calculate the difference of inlet and outlet temperature of BHE.

2.3 *The simulation model validation*

The model was validated with the data of the in-situ TRT conducted in Chuzhou city, Anhui, China. The in-situ TRT is conducted comply with the "Ground source heat pump systems engineering technical specifications (2009) (GB 50366-2005)" (MOHURD 2009). The parameters of the TRNSYS model components are based on the data of the TRT and illustrated in Table 1.

The model simulation time is 48 hours comply with the in-situ TRT testing time. The starting temperature of the surrounding ground is 16.45 °C. The constant heat of the heater set temperature is 35 °C. The outlet temperature of TRNSYS simulation result and the in-situ TRT is shown in Figure 3.

As can be seen from Figure 3, the outlet temperature of in-situ TRT is continuously growing but slightly unstable in the first beginning, and gradually keeps stable around 30 °C after

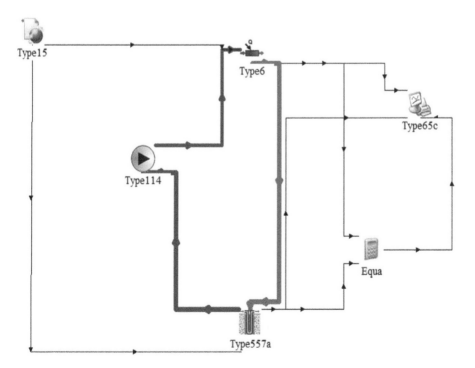

Figure 2. Simulation model created by TRNSYS.

Table 1. Main parameters of the borehole heat exchanger.

Number of boreholes	Borehole depth (m)	Storage heat capacity $[[kJ/(m^3 \cdot K)]]$
1	100	2200
Initial thermal temperature (°C)	Borehole spacing (m)	Number of U-tubes per borehole
16.4	5	2
Backfill conductivity [W/(m·K)]	Borehole Radius(m)	Ground thermal conductivity [W/(m·K)]
1.88	0.13	1.88

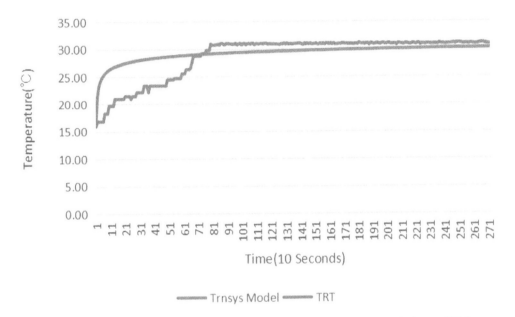

Figure 3. The outlet temperature comparison between TRNSYS simulation and the in-situ TRT.

810 seconds. In another hand, the outlet temperature of TRNSYS is growing very fast in the beginning and gradually keeps stable to around 30 °C. The different between the result of TRT and the simulation model is about 2.6% which is in an allowable rate expect the unstable period in the beginning. Therefore, the simulation model created by TRNSYS demonstrated its effectiveness. The model could fulfill the further simulation investigation and analysis of influence factors for the BHE design.

3 RESULTS

In order to investigate the performance of the BHE and the ground temperature increase of using backfills with different conductivities, simulation analyses are carried out by considering seven backfills with conductivities of 1,1.3, 1.6, 1.9, 2.1, 2.4, 2.7 W/(m·K) which is indicated in Ground Source Heat Pump Systems Engineering Technical Specifications 2009 (GB 50366-2005). The TRNSYS model with these seven different backfills in 1, 5, 10, 15 and 20 years operation are performed the simulation. The simulation result of ground temperature is illustrated in Figure 4.

As illustrated in Figure 4, the ground temperature is higher when the thermal conductivities of the backfills is greater and the ground temperature gradually increases when the operation year is getting longer. However, the ground temperature difference between backfill conductivity 1 and 1.3 W/(m•K) is 0.3 °C, but ground temperature difference between backfill

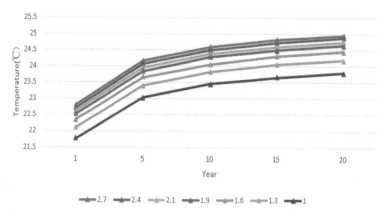

Figure 4. The ground temperature comparison between backfills.

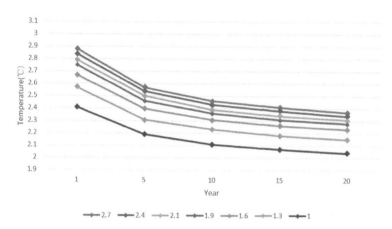

Figure 5. Outlet and inlet temperature difference between backfills.

conductivity 2.4 and 2.7 W/(m•K) is 0.09 °C. Therefore, in this case, when the thermal conductivity of backfill is higher than 1.9 W/(m•K) which is closest to the thermal conductivity of the surrounding ground 1.88 W/(m•K), the ground temperature growth is getting slower.

Figure 5 shows the outlet and inlet temperature difference of BHE using backfills with different conductivity. As can been seen, the higher conductivity of backfill improves the performance of heat exchange of BHE with the surrounding ground. However, when the conductivity is higher than 1.9 W/(m•K) which is the closest to the thermal conductivity of the surrounding ground 1.88 W/(m•K), there is little advantage of using higher conductivity backfill, especially for a long-term operation.

4 CONCLUSION

In this report, the numerical simulation model based on the heat rejection model of TRT is created using TRNSYS simulation tool. Backfills with seven different thermal conductivities are simulated for 1, 5, 10, 15 & 20 years. The model is validated with the data of the in-situ TRT conducted in Chuzhou city, Anhui. The performance of backfills thermal conductivities 1, 1.3, 1.6, 1.9, 2.1, 2.4, 2.7 W/(m·K) for 1, 5, 10, 15 & 20 years based on the TRT of the heat rejection test are simulated. The result illustrates that the higher conductivity backfills improves the heat exchange performance between BHE and surrounding ground but it also increases the ground temperature. In the case of Chuzhou city, when the conductivity of backfill is higher than 1.9 W/(m·K) which is closest to the thermal conductivity of the surround ground, there are little growth in ground temperature as well as the heat exchange performance. The research result could provide valuable reference for selecting backfills for BHE of GSHP system in Chuzhou city. In addition, the methodology of this research could be used by researchers for investigating design parameters of BHE for GSHP system.

REFERENCES

American Society of Heating, Refrigerating and Air-Conditioning Engineers (ASHRAE). 2012. *Geothermal Energy. ASHRAE Handbook HVAC Systems and Equipment. American Society of Heating, Refrigerating and Air-Conditioning Engineers.*

Borinaga-Treviño, R. Pascual-Muñoz, P. Castro-Fresno, D. & Blanco-Fernandez, E. 2013. Borehole thermal response and thermal resistance of four different grouting materials measured with a TRT. *Applied Thermal Engineering* 53(1): 13-20

Cui, Y, Zhu, J, Twaha, S. & Riffat, S. 2018. A comprehensive review on 2D and 3D models of vertical ground heat exchangers. *Renewable and Sustainable Energy Reviews* 94:84-114.

Desmedt, J. Bael, J.V. Hoes, H. & Robeyn, N. 2011. Experimental performance of borehole heat exchangers and grouting materials for ground source heat pumps. *International Journal of Energy Research* 36:1238-1246.

Erol, S, & Francois, B. 2014. Efficiency of various grouting materials for borehole heat exchangers. *Applied Thermal Engineering* 70:788-799.

Gehlin, S. 2002. Thermal response test. *Lulea University of Technology. Doctoral Thesis* 1202-1544.

Jun, L, Xu, Z, Jun, G. & Jie, Y. 2009. Evaluation of heat exchange rate of GHE in geothermal heat pump systems. *Renewable Energy* 34(12): 2898-2904.

Klein, S.A. Beckman, W.A. Mitchell, J.W. Duffie, J.A. Duffie, N.A. & Freeman, T.L. 2010. *Transient System Simulation Tool, Version 17, User Manual. Solar Energy Laboratory University of Wisconsin-Madison. Madison. USA.*

Lee, C. Park, C. Nguyen, T. Sohn, B. Choi, J.M. & Choi, H. 2011. Performance evaluation of closed-loop vertical ground heat exchangers by conducting in-situ thermal response tests. *Renewable Energy* 42: 77-83.

Ma, W, Kim, M.K. & Hao, J. 2019. Numerical simulation modeling of a GSHP and WSHP system for an office building in the hot summer and cold winter region of China: A case study in Suzhou. *Sustainability* 11(12):3282.

Ministry of Housing and Urban-Rural Development of China. 2009. *Technical code for ground source heat pump system (GB50366-2005). Ministry of Housing and Urban-Rural Development of China.*

Mogenson, P. 1983. Fluid to duct wall heat transfer in duct heat storages. *Proceedings of the International Conference on Subsurface Heat Storage in Theory and Practice. Swedish council for building research.*

Wang, G. Wang, W. Lou, J. & Zhang, Y. 2019. Assessment of three types of shallow geothermal resource and ground-source heat-pump applications in provincial capitals in the Yangtze River Basin, China. *Renewable and Sustainable Energy Reviews* 111:392-421

Zhang, C. Wang, Y. Liu, Y. Kong, X. & Wang, Q. 2018. Computational methods for ground thermal response of multiple borehole heat exchangers: A review. *Renewable Energy* 127:461-473.

Intelligent design of climate responsive skin in modern buildings

J.Y. Ying & Z. Zhuang
College of Architecture and Urban Planning, Tongji University, Shanghai, China

ABSTRACT: The climate responsive skin can adapt to changing climates by using intelligent or non-intelligent, active or passive building envelop. This paper reviews state-of-the-art of various kinds of intelligent climate responsive skin designs strategies in modern buildings, and classifies them according to environment performance and operation mode. It is expected to provide the reference for the design and application of climate responsive skin in the future.

1 INTRODUCTION

The harmonious relationship between architecture and climate is one of the most important themes of architectural design. As the only connection between the building and the outdoor environment, the skin takes a paramount role of regulating the internal environment of the building. Recent years, the development of computer simulation, mechanical engineering, materials science and other disciplines makes the climate responsive skin (CRS) possible. In response to the ever-changing climate, CRS is designed to take the potential range of change into account at the early design. By using different solutions, CRS can solve the conflicting requirements between climate and architecture, and has a great potential in the context of "Green Building" in the world.

For the climate adaptive buildings, the skin has attracted many attentions. Since 1981, the concept of building skin was firstly proposed (Micheal 2003). Studies on CRS have achieved a great progress: Delft University of Technology proposed a skin design method of "environment parameterization" in the study of climate-adaptive building interface (Jason 2015); Nimish & Valentina (2009) used a parametric design method to design the multi-layer composite interactive skin and verified its feasibility; Loonen (2013) classified the existing climate adaptive skins and analyzed the influencing factors of climate adaptability; Miao & Feng (2016) proved the energy-saving advantage of the dynamic skin by simulating the building physical environment; Shi & Hu (2017) classified the skin design strategy based on different environmental factors in international solar decathlon competition. However, these studies of CRS are still not enough, especially, due to the lack of data statistics and analysis about their applications in modern buildings. Therefore, this paper collects and summarizes 203 cases to analyze the application status of different epidermis in modern architecture, understand market bias and provide reference for future skin design.

2 CONCEPT AND DEFINITION OF CRS

The difference between CRS and ordinary building envelop is mainly reflected in the design orientation, materials or components. The definition of CRS is a skin that can reversibly change some of its functions, features or behavior over time in response to changing performance requirements and variable boundary conditions, and does this with the aim of improving overall building performance (Loonen 2013). CRS usually improve the building performance and reduce energy consumption by taking intelligent or non-intelligent, active or passive design strategies, which are based on the targeted analysis and simulation of building indoor environment.

The intelligent CRS combines with computer technology, responds to the climate faster, smarter and more controllable. It perceives the environment through sensors and automatically changes the skin shape through computer simulation software to achieve optimization goals. While providing a more comfortable indoor environment, it has more energy-saving potential and is in line with the era of sustainable development.

3 DESIGN STRATEGY

The design strategy of CRS can be divided into two aspects, as shown in Figure 1. The one is Environment Performance Oriented. According to the main responding climatic factors: temperature, humidity, wind, natural illuminance, rainwater and so on, CRS can be divided into four categories: thermal adaptation, optical adaptation, wind adaptation, and wet adaptation. The other is called Operation Mode. According to its operation mode, it can be divided into: active dynamic skin, passive dynamic skin and non-intelligent dynamic skin.

3.1 Environment performance-oriented design

3.1.1 Thermal adaptation

Building energy consumption accounts for about 30% of the total social energy consumption. The building skin plays a big role to reduce the overall building energy use. There are mainly two ways to achieve thermal adaption: regulating solar radiation and changing the thermal performance of the skin. Compared with using high thermal performance skin materials, regulating solar radiation in CRS design is operational, which favors the architects. For example, in the Kiefer Technology Exhibition Centre in Austria, the smart skin system can adjust the opening and closing modes according to the climate change and space requirement. Meanwhile, the changing facade makes the building very sculptural (Figure 2).

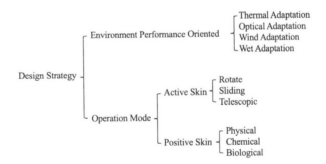

Figure 1. CRS design strategy.

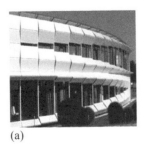 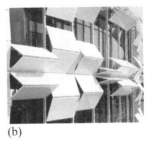

(a)   (b)

Figure 2. Kiefer technic showroom (Zhou 2014).

3.1.2 *Optical adaptation*

Light is the eternal topic of architecture. Only with light could architecture have a sense of time and space. Different light environments give people different psychological feelings. From the perspective of built environmental control, the meaning of optical adaptation is not only to reduce lighting energy consumption, but also to create a comfortable and healthy light environment for people. For example, the world-famous Architecture-Arab World Institute uses CRS whose unit is like an eye to control light, creating a religious environment (Figure 3a). The Korea Pavilion designed by SOMA uses the change of louver form to control the amount of light for adjusting indoor light environment (Figure 3b).

3.1.3 *Wind and wet adaptation*

The wind adaption of CRS embodies two: aesthetic and wind energy utilization. For example, the Crit Building in the USA uses too many aluminum metal sheets on its façade (Figure 4a). Due to the ultra-thin metal sheet material, the weight is extremely light. When the wind blows, the skin can immediately present a ripple flow in the façade, which can attract attention of passers-by.

On the other hand, there are not many researches and practices on wet environment adaptive skins. Professor Steffen Reichert of Stuttgart University pays attention to the morphological changes of pinecone in different humidity environments. The building skin can automatically open and close according to different humidity (Figure 4b).

3.2 *Operation mode design*

3.2.1 *Active dynamic skin*

The active dynamic skin is a comprehensive system consisting of adjustable skin components, climate environment sensors, system control equipment among others, with a process of intelligent analysis of climate adaptation and procedures for how to change the skin components. According to the change mode of the skin components, the active dynamic skin can be divided into three types: rotation, telescopic and sliding. The rotating skin is usually driven or rotated by an internal fixed rod; the telescopic skin change form by controlling lever expand or stack

(a)

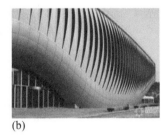
(b)

Figure 3.   Arab World Institute (a) (Miao 2016) and Korea pavilion at expo 2012(b) (Zhou 2014).

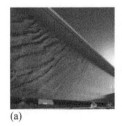
(a)

(b)

Figure 4.   Brisbane airport parking structure (a) (Miao 2016) and Hygroskin-meteorosensitive pavilion (b) (Zhou 2014).

(a)   (b)

Figure 5.   Ai Baha tower (a) (Yang 2015) and heat sensitive metal pavilion (b) (https://www.archdaily.com).

up levers; The sliding skin can move the unit plate on the sliding bar to open or close building façade. For example, Ai Baha Tower is equipped with a CRS that is created from a traditional Middle Eastern pattern wood grille, combined with dynamic sunshade simulation technology according to the sun's running track. So, it can take multiple forms in a day (Figure 5a).

3.2.2 *Passive dynamic skin*
The passive dynamic skin mainly depends on the properties of the skin material itself: physical, chemical and biological. Although the process of passive dynamic skin is instantaneous and variable, it does not consume any conventional energy. As the external environment changes, the epidermis or epidermal material changes spontaneously without any mechanical system (Figure 5b).

4   APPLICATION IN MODERN BUILDINGS

It is not negligible that architecture is an applied discipline in real life. A more market-receiving building skin can make more contributions to energy-saving. At present, the number of practical applications of intelligent CRS in modern buildings is very small. Therefore, 18 actual application cases of smart CRS were found so far through from literature review and the world's authoritative architectural case website: Archdaliy (https://www.archdaily.com) and Pinterest (https://www.pinterest.com), and listed in Table 1. Some of the CRS designs in academic research are not included in the scope of this section because they are not applied to the actual building.

According to statistics, 83.3% of the CRS considers the thermal adaptation, which is the most considered among the four environmental factors. This is because the building's HVAC energy consumption is the biggest part of the total energy consumption. The skin acts as a hot aisle between indoor and outdoor space, and its thermal performance will directly affect the indoor thermal environment. The light adaptation and the wind adaptation are relatively in the second position, accounting for 44.4% and 38.9% respectively, but these two factors are rarely considered separately. There is one building taking light adaptation as the only optimization target, the same as wind adaption. Good indoor light environment gives people a comfortable living and working environment, which is good for their health. The wind environment not only brings fresh air, but also affects the thermal environment. The wet environment was considered the least, and only one building tried it. The lack of relevant skin technology makes it not easy to control of the thermal environment.

Regarding the climate response pattern shown in Figure 6a, 15 CRS used active adaptive design, accounting for 83.3% and the rotating skin is the most application (reaching 47% as shown in Figure 6b). Compared with the passive adaptive design, the active adaptive design has lower requirements on the skin material itself, more controllability and higher application convenience. There are not many materials available for passive adaptation of the skin in the world. Although it has the characteristics of lower energy consumption, the cost of the preliminary research is high. Moreover, the price of the material itself is expensive, and the development cost in the actual application of the building is bigger than the active one.

Table 1. The intelligence design cases of CRS in modern buildings.

Case	Environment performance	Operation mode	Mode type	Designer	Appearance
Kiefer Technic Showroom (Zhou 2014)	Thermal & optical	Active	Sliding	Ernst Giselbrecht	
Al Bahar Towers (Yang 2015)	Thermal & optical	Active	Telescopic	Aedas	
Arab World Institute (Miao 2016)	Thermal & optical	Active	Telescopic	Jean Nouvel	
German Parliament Building Dome (Zhou 2014)	Thermal	Active	Rotate	Foster	
University library (Zhou 2014)	Thermal & optical & wind	Active	Sliding	Foster	
San Francisco Federal Building (Zhou 2014)	Thermal & optical & wind	Active	Rotate	Thom Maine	
Korea Pavilion at Expo 2012 (Zhou 2014)	Optical	Active	Telescopic	SOMA	
Devonshire in Newcastle University (Wang 2011)	Thermal & optical	Active	Rotate	Devereux	
Suffolk Sliding House (Yang 2015)	Thermal	Active	Sliding	DRMM	
CJ Research Center (Zhou 2014)	Thermal	Active	Telescopic	Yazdani Studio	
Karnak Cultural Center (Miao 2016)	Thermal & wind	Active	Rotate	Renzo Piano	
Media-ICT building (Hadeer 2019)	Thermal	Active	Telescopic	Enric Ruiz Geli	
Terrence Donnelly Centre for Cellular and Biomolecular Research (Hadeer 2019)	Thermal & optical & wind	Active	Rotate	Behnisch Architekten	
Debis Headquaters (Wang 2011)	Thermal & wind	Active	Rotate	Renzo Piano	

(*Continued*)

Table 1. (*Continued*)

Case	Environment performance	Operation mode	Mode type	Designer	Appearance
GSW Headquaters (Wang 2011)	Thermal & wind	Active	Rotate	Sauebruch Hutton Architects	
Brisbane Airport Parking Structure (Miao 2016)	Wind	Passive	Physical	Ned Kahn & UAP	
Hygroskin-Meteorosensitive Pavilion (Zhou 2014)	Wet	Passive	Physical	Pro. Mengens	
Metal that Breathes (http:/ www.archdaily.com)	Thermal	Passive	Physical	Doris Kim Sung	

    ■ active ■ passive   ■ Sliding ■ Rotate ■ Telescopic

(a)   (b)

Figure 6. Proportion of different designs (a) and operation modes (b).

5 CONCLUSIONS

The CRS is a relatively new field in architecture. Compared with the existing building skin, it increases the building climate adaptability and improves the building Performance. Through the above data analysis, the following conclusions can be obtained: 1) For climate adaptability, the most effective skin adaptation direction is consistent with the proportion of building energy consumption-thermal adaption, but the light environment and wind environment cannot be ignored. 2) For practical market applications, active climate-adapted skins are more suitable for architectural design before reducing the material cost and increasing controllability of passive climate-adapted skins. 3) Different modes of change have less impact on the application of the epidermis in the market, mainly due to differences in the expression of architectural design. However, due to the relatively complicated technology, the practical application of CRS has not been widely promoted. However, it is undeniable that CRS is the future development direction of building skin.

ACKNOWLEDGEMENT

This work is supported by the National Natural Science Foundation of China Research Award (No.51608370) and the Fundamental Research Funds for the Central Universities.

REFERENCES

Hadeer, S.M.S. 2019. Adaptive building envelopes of multistory buildings as an example of high performance building skins. *Alexandra Engineering Journal* 11:2-8.

Jason, O.V. 2015. Enviromental parametrics: prototyping climate camouflage and adaptive building envelopes. *Time + Architecture* 2: 42-47.

Loonen, R.C. Trčka, M & Cóstola, D. 2013. Climate adaptive building shells: State-of-the-art and future challenges. *Renewable and Sustainable Energy Reviews* 25:483-493.

Miao, Z.T. & Feng, G. & Guo, L.J. 2016. Design of variable Architectural surface in response to external environmental changes. *Ecocity and Green Building* 4: 48-55

Micheal, W. 2003. *Intelligent skins. Dalian: Dalian Technology University Press.*

Nimish, B. 2009. Adaptive building skin systems: A morphogenomic framework for developing real-time interactive building skin systems. *IEEE Computer Society* 119-125

Nimish, B. & Valentina, S. 2009. Performative building skin systems: A morphogenomic approach towards developing real-time adaptive building skin systems. *International Journal of Architectural Computing* 7: 643-675.

Shi, F. Hu, C. & Zheng, W.W. 2017. An analysis of design strategy of operable building skins with dynamic adjustment on environment parameters: The example of the solar decathlon. *New architecture* 2:54-59.

Wang, J.L. 2011. *Biomimicry. Kinetics. Sustainability : Study on kinetic building envelope based on biological acclimatization.* Tianjin: Tianjin University.

Yang, F. 2015. *Based on the climate Adaptability of movable building skin design research.* Shenyang: Shenyang Architecture University.

Zhou, Y.Y. 2014. *Research on theory and several key techniques for kinetic adaptive skin.* Zhejiang: Zhejiang University.

Analysis of key factors causing delay in green buildings: An empirical study in China

M. Sajjad & Z. Pan
Jiangsu University of Science and Technology, Zhenjiang, P.R. China

ABSTRACT: Green buildings, also known as smart or sustainable construction, are one of the emerging topics for the people involved in the construction industry and they are eagerly awaiting further expansion of the industry. Despite the highest focus on environmental sustainability, very few green buildings are analyzed through research and case studies, especially from the perspective of their time-management performance. In this regard, a survey was conducted to collect data to identify and analyze key factors causing schedule delays in green buildings in China. The data was collected via questionnaires and personal interviews. Furthermore, the top five factors causing delays in green buildings in the Chinese context were the level of experience of key parties, poor site management and supervision, unfavorable weather conditions on-site, lack of advanced mechanical equipment and project cost respectively. The major solutions to counter these issues and minimize the risk of schedule delay in green buildings are, (1) the selection of project teams with relevant work experience, (2) technical approaches such as BIM, 4D schedule simulation and Last Planner System should be adopted for better schedule performance and monitoring, (3) weather reports should be regularly updated with the use of GPS and GIS, (4) advanced mechanical equipment should be selected in the pre-planning process to increase productivity and save time, and (5) the project cost should be seriously monitored and controlled via management approaches such as Earned Value Management, Forecasting and Performance Reviews.

1 INTRODUCTION

According to the World Green Building Council (WGBC), a 'green building' is a building that, in its design, construction or operation, reduces or eliminates negative impacts, and can create positive impacts, on our climate and natural environment. Green buildings preserve precious natural resources and improve our quality of life. In CBRE Global Research (June 2015), the development of green buildings in China began relatively late compared to developed countries such as the United States and the United Kingdom. However, the sheer size of the Chinese construction industry (ranked No. 1 in the world), means that green building development has significant room for growth in the coming years.

The two important aspects of construction management are time and cost management, which need to focus on the best utilization of available resources. The pace of progress means we do not have the advantage of time. Delays in construction projects is a global issue, not only in China but other countries around the world faced this issue and there is still an urging need to find a more reliable solution to minimize the risk of schedule delays. China has the largest population in the world, which increases the role and responsibility of the people involved in the construction industry to build a better living environment. The key parties involved in any construction project are clients, contractors, consultants, design engineers, and researchers and they are supposed to put their efforts in this regard.

Delivering any project within the specified period is very important for any construction company. The main motive for choosing this topic is to improve the time-management performance of green buildings by identifying, classifying and analyzing the factors responsible to cause delays and to provide solutions that can increase productivity and ensure timely completion of the work in this specific area.

2 LITERATURE REVIEW

Hwang & Leong (2013) studied the comparison of schedule delay and causal factors between traditional and green construction projects in Singapore. They found that 15.91% of the traditional projects were delayed while 32.29% of the green construction projects were completed behind schedule. The top five factors causing a delay in green projects were the speed of decision making by the client, the speed of decision making involving all project teams, coordination between key parties, and the level of experience of consultants and difficulties in financing project by contractors.

Sambasivan & Soon (2007) identified and evaluated the most significant causes of project delay in the Malaysian construction industry, which are improper planning, poor site management, the inadequate experience of the contractor, inadequate finance of the client and payments for completed work, problems related to subcontractors, material shortage and labor supply.

Al-Kharashi & Skitmore (2009) identified the leading causes of construction project delay in Saudi Arabia by conducting a questionnaire survey administered to contractors, consultants, and clients. They concluded that the lack of financial resources and delay in progress payments by the client has a significant impact on delays in projects.

Haseeb et al. (2011) researched the causes of delay in large construction projects in Pakistan, where the following factors are reported to be the most influential: natural disasters, financial and payment problems, improper planning, poor site management, insufficient experience, shortage of materials and equipment.

3 METHODOLOGY

After a comprehensive literature review, a questionnaire was developed to collect data regarding the proposed study. Looking into various aspects of the smart construction, 34 key factors were selected having more influence to cause a delay in green buildings. These factors were selected in accordance with the similar research conducted in other regions of the world. There were two parts of the questionnaire, Part I and Part II. Part I was about the respondent's profile. Part II was about the factors causing delays in green buildings in China. There were four major respondents, namely, clients, contractors, consultants, and researchers. The data was collected through personal interviews, emails and phone calls. Two hundred questionnaires were distributed in government, private, semi-government organizations, construction companies, and research institute out of which one hundred and ten respondents returned their questionnaires. The response rate was 55%. The participants were briefed about the ranking and grading system of the research. The questionnaire was translated into Chinese language for the local people to collect data that are more precise and to make sure the participants have a full understating of the questions asked.

Relative Importance Index (RII) is used to analyze and rank the factors according to their severity to cause delays in green buildings. Furthermore, the mean, mode and standard deviation (S.D) are also calculated and presented. Based on these analyses and results, a meaningful conclusion is presented and solutions are proposed to counter these issues and to improve schedule management in green buildings. The respondents were asked to mark the factors on the given scale as they think it can cause delays in green buildings and how much influence a particular factor has. In Table 1, 1-5 indicates varying degrees of influence as: not important, slightly, moderately, highly and extremely important respectively, to cause delays.

Table 1. Ranking and assessment criteria.

RII range	Delay impact
>0 - 0.2	Not important
>0.2 - 0.4	Slightly important
>0.4 - 0.6	Moderately important
>0.6 - 0.8	Highly important
>0.8 - 1.0	Extremely important

4 DATA ANALYSIS AND DISCUSSIONS

As part of the aim of this research, factors affecting schedule delays in green buildings in Chinese construction industry are identified and categorized under eight major groups, namely: project-related factors (F1), client-related factors (F2), design-team related factors (F3), consultant related factors (F4), contractor related factors (F5), work-force related factors (F6), equipment and material related factors (F7), and external factors (F8). The evaluation and ranking of these factors are based on the analysis of the data via statistical techniques and approaches. Based on the ranking of the factors, it was possible to identify the causes having the highest influence on time-management performance in green buildings. To this end, the RII of the 34 factors are determined and presented (refer to Table 2) and to validate the results, they are compared to the findings of the relevant researches conducted in other countries, especially Singapore and Malaysia.

RII is used herein to determine the respondent's perceptions regarding the degree of importance of the identified factors. The value of RII ranges from 0-1.

The RII is computed as:

$$RII = \text{summation of } W/A * N,$$

Where, W is the weight given to each factor by the respondents ranges from 1 to 5; A is the highest weight = 5; N is the total number of respondents.

The mean, mode and standard deviation (S.D) of each factor are also calculated using SPSS to deeply analyze the impacts and importance of each factor.

Mean (Average): The sum of all the data entries divided by the number of entries.

Mode: The data entry that occurs with the greatest frequency.

Standard Deviation (S.D): it is the measure of variability and consistency of the sample.

The data analysis is done using Microsoft Excel and SPSS. The results are calculated and presented up to two decimal points for a clear understanding and precise rank order. Whereas, if two factors have the same RII and different rank order, they are calculated up to 10 decimal points to find the minor differences and then ranked accordingly. The results are presented in Table 2.

The above results indicate that there are several significant factors causing delays in green buildings in China. The top five influential factors agreed by the respondents are level of experience of key parties, poor site management and supervision, unfavorable weather condition on-site, lack of advanced mechanical equipment and project cost, respectively.

Respondents' ranked 'level of experience of key parties' as the most influential cause with a relative importance index of 0.92, responsible for delays among the thirty-four key factors. This is not surprising because whether it is a traditional construction or an energy-efficient green building, it is a fact that relevant work experience and awareness of the difficulties and scope of the similar work can help to provide the best outcome and better performance. Smart construction involves the latest engineering approaches in terms of designing software, the use of new materials, the use of advanced mechanical equipment and the use of digital data. The relevant work experience of people involved in can affect time performance.

Table 2. Ranking of key factors causing delays green buildings in China.

Factors	Category	RII	Mean	Mode	S.D	Rank
Level of experience of key parties	F1	0.92	4.62	5	0.52	1
Poor site management and supervision	F5	0.92	4.62	5	0.49	2
Unfavourable weather conditions on-site	F8	0.92	4.58	5	0.73	3
Lack of advanced mechanical equipment	F7	0.90	4.52	5	0.60	4
Project cost	F1	0.89	4.47	5	0.57	5
Delay in progress payment by the client	F2	0.87	4.35	4	0.63	6
Construction method implemented by the contactor	F5	0.87	4.35	4	0.64	7
Imported materials	F7	0.86	4.31	5	0.71	8
Mistakes and delay in producing design documents	F3	0.86	4.29	4	0.70	9
Unskilled labour	F6	0.86	4.28	5	0.72	10
Low labour productivity	F6	0.83	4.17	4	0.65	11
Rework due to defects during construction	F5	0.83	4.13	4	0.71	12
Price fluctuation	F8	0.83	4.24	5	0.80	13
Delay in performing inspection and testing	F4	0.81	4.05	4	0.68	14
Unskilled operator	F7	0.81	4.04	4	0.72	15
Shortage of labor	F6	0.80	4.00	4	0.68	16
Availability of material	F7	0.78	3.91	4	0.80	17
Speed of decision making involving all parties	F1	0.77	3.86	3	0.80	18
Unrealistic project scheduling	F5	0.77	3.84	4	0.66	19
Changes in government regulations and laws	F8	0.75	3.76	4	0.74	20
The complexity of project design	F1	0.73	3.64	4	0.54	21
Time to approve design documents by consultants	F4	0.73	3.64	4	0.69	22
Unforeseen ground conditions	F8	0.71	3.56	4	0.80	23
Delay in approving changes in the scope of work	F4	0.71	3.55	3	0.64	24
Difficulties in financing project	F2	0.69	3.46	3	0.75	25
Lack of communication between key parties	F2	0.69	3.43	3	0.78	26
Changes in materials during construction	F7	0.67	3.37	3	0.52	27
Equipment breakdown	F7	0.61	3.06	3	0.61	28
Delay in the final inspection by the third party	F8	0.61	3.06	3	0.75	29
Misreading of client's requirements by design team	F3	0.58	2.89	3	0.65	30
Client initiated variation of work	F2	0.53	2.65	3	0.60	31
Accidents during construction	F8	0.51	2.60	3	0.55	32
The dispute between key parties	F1	0.50	2.52	2	0.80	33
Project delivery methods	F1	0.47	2.33	2	0.83	34

The second most significant factor causing delays is 'poor site management and supervision by the contractor'. No doubt, the contractor is the one who executes the real work and hence any deficiency in work performance will cause time over-run. Time management, proper supervision and timely execution of activities are keys to the successful completion of the work.

The unfavorable weather conditions on-site and lack of advanced mechanical equipment were ranked as the third and fourth respectively. The severity of the weather conditions on a construction site can cause a significant level of delays. As of China's geographical location, there are different weather conditions in different regions of the country and it can affect schedules up to a great extent. On-time execution of activities as planned can be achieved via the use of the latest construction technology and technical approaches. The use of advanced mechanical equipment will reduce the risk of time over-run.

Project cost ranked fifth in the list. Any development project begins with having enough resources to finance the project until successful completion. China's construction industry is involved in mega construction projects, any delay or difficulty in providing funds will delay the activities. Owners and contractors have to consider the overall project cost and available resources to finalize the project design, scope, and size. Finance is a very important aspect of any industry and has great significance.

Table 3. Top five factors causing delays in green buildings from the perspective of the clients.

Factors	RII	Mean	Mode	S.D	Rank	Overall Rank
Poor site management and supervision	0.99	4.94	5	0.24	1	2
Unfavorable weather condition on site	0.99	4.94	5	0.24	2	3
Unskilled labor	0.95	4.76	5	0.56	3	10
Level of experience of key parties	0.93	4.67	5	0.48	4	1
Price fluctuation	0.92	4.58	5	0.66	5	13

Table 4. Top five factors causing delays in green buildings from the perspective of the contractors.

Factor	RII	Mean	Mode	S.D	Rank	Overall Rank
Imported materials	0.95	4.77	5	0.43	1	8
Unfavorable weather condition on site	0.93	4.63	5	0.49	2	3
Poor site management and supervision	0.92	4.60	5	0.50	3	2
Level of experience of key parties	0.91	4.53	5	0.51	4	1
Lack of high technology mechanical equipment	0.91	4.53	5	0.51	5	4

Table 5. Top five factors causing delays in green buildings from the perspective of the consultants.

Factors	RII	Mean	Mode	S.D	Rank	Overall Rank
Lack of high technology mechanical equipment	0.96	4.80	5	0.41	1	4
Level of experience of key parties	0.95	4.76	5	0.44	2	1
Unfavorable weather condition on-site	0.95	4.76	5	0.44	3	3
Poor site management and supervision	0.94	4.68	5	0.48	4	2
Imported materials	0.92	4.60	5	0.50	5	8

Table 6. Top five factors causing delays in green buildings from the perspective of the researchers.

Factors	RII	Mean	Mode	S.D	Rank	Overall Rank
Unrealistic project scheduling	0.95	4.73	5	0.55	1	19
Changes in materials during construction	0.94	4.68	5	0.48	2	27
Construction methods implemented by contactor	0.92	4.59	5	0.50	3	7
Lack of high technology mechanical equipment	0.90	4.50	5	0.60	4	4
Poor site management and supervision	0.90	4.50	5	0.67	5	2

Based on the above results, it is concluded that the respondent type plays an important role in the determination of delay factors and their impacts. Each respondent type has rated the importance of a particular factor differently according to their experience and perspective. In addition, some factors are commonly rated extremely important (RII>0.8) such as level of experience of key parties and poor site management and supervision.

4.2 *Solutions for improvement of schedule performance in green buildings*

1. Ensure that the contractors, sub-contractors, consultants and design teams have relevant work experience before awarding the tender. This can be done by checking the company's profile that either they have successfully completed similar projects or not.

2. Ensure that the project schedule and resources are seriously monitored and managed. Elements such as schedule, cost, and tasks to be completed should be managed with the use of technical approaches such as Last Planner System (LPS), Building Information Modeling (BIM), and 4D Schedule Simulation. Set regular monitoring intervals, develop a standard format and process for recording work accomplishments. Standard formats and processes improve the accuracy of the information.

3. Update the weather reports regularly and plan the project activities accordingly so that can help to increase productivity. Technological advances, such as Geographic information systems (GIS), Global Positioning System (GPS) and other applications are making weather information more accessible and immediately alerting those in harm's way. The management teams should ensure the use of weather forecasting data while making schedules.

4. In smart construction, speedy work and timely completion of work with quality control are very vital. In order to achieve this, the mechanization of work has to be done, where construction machinery and equipment play a pivotal role. Ensure the use of the latest mechanical equipment that can increase productivity and minimize the risk of time over-run. Proper equipment should be selected in the pre-construction planning process. The important aspects in the selection of the equipment are (1) it should give the best services at low cost, (2) should be easily repairable with low shutdown period, (3) should be capable of doing more than one function, (4) should be of moderate size, and (5) should suit the majority of the requirements of the job.

5. The client and contractor should analyze the overall project cost and consider price fluctuation before the commencement of the work. Project cost monitoring and controlling techniques include (1) Earned Value Management (EVM), a mathematical method which can measure the actual performance of a project in terms of schedule and cost. It provides formulae to forecast the future performance of a project. The forecast is based on the current actual performance; (2) Performance Reviews in projects are required to check the health of a project. This usually involves cost and schedule as the main parameters to assess; (3) To-Complete Performance Index (TCPI) can be used to determine the project performance required to complete the project as planned.

6. Progress payments should be paid on time to facilitate the contractors' ability to finance the work.

7. Check for mistakes and discrepancies in design documents to avoid re-work in drawings as well as actual work.

8. There should be an optimum number of labors for individual activities and motivate the workers with incentives and bonuses.

9. Consultants should not delay the checking, reviewing and approving of design documents leading to delay in construction activities.

10. Select appropriate and skilled workforce to ensure proper execution of work activities.

5 CONCLUSION AND RECOMMENDATIONS

The objectives of this research were to analyze the key factors causing project schedule delays in green buildings and to suggest possible solutions to minimize the risk of delays in sustainable construction. The analysis and results of the collected data and information established that the top five key factors causing delay in green building projects are (1) level of experience of key parties, (2) poor site management and supervision, (3) unfavorable weather condition on-site, (4) lack of advanced mechanical equipment and (5) project cost respectively.

The statistical report in this study enlightens the industry to consider the role of different people involved and the overall period required by the green buildings. Also, factors with high influence to cause delays in green buildings bring forth a focal point for the management teams to enhance its performance.

Finally, it is recommended to analyze different categories of the projects, residential, commercial, and industrial with large sample sizes to design a time management model for sustainable construction to minimize the risk of time over-run.

REFERENCES

Abd El-Razek, M. Bassioni, H. & Mobarak, A. 2008. Causes of Delay in Building Construction Projects in Egypt. *Journal of Construction Engineering Management* 134(11):831–841.

Ahmad, T. Thaheem, M. J. & Anwar, A. (2016). Developing a green-building design approach by selective use of systems and techniques. *Architectural Engineering and Design Management* 12(1):29-50.

Alaghbari, W. Kadir, M. R. A. & Salim, A. Ernawati. 2007. The significant factors causing delay of building construction projects in Malaysia. *Engineering Construction and Architectural Management* 14(2):192-206.

Al-Kharashi, A. & Skitmore, M. 2009. Causes of delays in Saudi Arabian public sector construction projects. *Construction Management and Economics* 27(1):3-23.

Al-Momani, A.H. 2000. Construction delay: a quantitative analysis. *International Journal of Project Management* 18(1): 51-59.

Assaf, S. A. & Al-Hejji, S. 2006. Causes of delay in large construction projects. *International Journal of Project Management* 24:349–357.

Austin, S. A. Baldwin, A. N. & Steele, J. L. 2002. Improving building design through integrated planning and control. *Engineering, Construction and Architectural Management* 9(3):249-258.

Chan, D. W. M. & Kumaraswamy, M. M. 1997. A comparative study of causes of time overruns in Hong Kong construction projects. *International Journal of Project Management* 15(1):55-63.

Hwang, B.G. & Lim, E.S. 2013. Critical success factors for key players and objectives: case study of Singapore. *Journal of Construction Engineering and Management* 193(2):204-215.

Hwang, B. G. & Ng, W. J. 2013. Project management knowledge and skills for green construction: overcoming challenges. *International Journal of Project Management* 31(2): 272-284.

Hwang, B. G.; Zhao, X. & Ng, S. Y. 2013. Identifying the critical factors affecting schedule performance of public housing projects. *Habitat International* 38:214-221.

Hwang, B. G. Zhao, X. & Tan, L. L. G. 2015. Green building projects: Schedule performance, influential factors, and solutions. *Engineering, Construction and Architectural Management* 22(3):327-346.

Hwang, B.G. & Leong, L.P. 2013. Comparison of schedule delay and causal factors between traditional and green construction projects. *Technological and Economic Development of Economy* 19(2):310-330.

Iyer, K.C. & Jha, K.N. 2005. Factors affecting cost performance: Evidence from Indian construction projects. *International Journal of Project Management* 23:283–295.

Kats, G. Alevantis, L. Berman, A. Mills, E. Berkeley, L. & Perlman, J. 2003. The costs and financial benefits of green buildings. *Report to California's Sustainable Building Task Force.*

Li, Y.Y. Chen, P.H. Chew, D. Teo, C.C. & Ding, R.G. 2011. Critical project management factors of AEC firms for delivering green building projects in Singapore. *Journal of Construction Engineering and Management* 137(12):1153–1163.

Wu, W. & Issa, R.R. 2015. BIM execution planning in green building projects: Leed as a use case. *Journal of Management in Engineering*, 31(1):A4014007.

Sustainable Buildings and Structures: Building a Sustainable Tomorrow – Papadikis et al. (Eds)
© 2020 Taylor & Francis Group, London, ISBN 978-0-367-43019-1

Critical success factors for sustainable mass housing projects in developing countries: A case study of Nigeria

M.M. Mukhtar
Department of Quantity Surveying, Abubakar Tafawa Balewa University, Bauchi, Nigeria

M.M. Mustapha
Department of Civil Engineering Ahmadu Bello University, Zaria, Nigeria

ABSTRACT: Many developing countries face challenges of rapid population growth and urbanization increase. This problem leads to slums formation and informal settlements, where houses are built without basic infrastructure facilities and social amenities. Sustainable housing which incorporates environmental, social, cultural and economic fabric of communities will alleviate poverty and improve quality of life of the people. The aim of this study is to identify critical success factors for sustainable mass housing projects in Nigeria. A questionnaire survey was carried out to elicit the perceptions of construction professionals on essential factors influencing development of sustainable mass housing in Nigeria, The data obtained were analyzed with the help of descriptive statistics and factor analysis. Six factors grouping emerged from factor analysis which incorporate the four dimensions of sustainability, namely, environmental dimension, economic dimension, social dimension and cultural dimension. The presence of these factors in the implementation of sustainable mass housing projects is essential.

1 INTRODUCTION

1.1 Background

Housing plays a significant role in improving the quality of life and alleviating poverty most especially in developing countries (Arku 2016). An adequate housing is durable, affordable, safe, secure, habitable and well connected to basic facilities. However, provision of adequate housing is challenging in many developing countries. For instance, it has been estimated that about 1 billion people across the globe, most of whom from developing countries live in slums and poor quality housing and unhealthy environment (UN-Habitat 2010, Mukhtar et al. 2017). In Nigeria alone there is more than 17 million housing deficit. The problems of housing are largely related to issue of affordability; housing is expensive and incomes are too low (UN-Habitat 2011).

To improve the quality of life and welfare of their citizens, governments of developing countries across the globe attach high priority to the provisions of adequate and affordable housing (Ibem & Azuh 2011). However, in most cases the housing provided are of poor quality standard, and located in remote areas without considerations of residents' lifestyle (Ibem et al. 2011, Olotuah 2010). This may be as a result of none addressing the concept of sustainability in housing policies of those nations.

Sustainable development is defined as 'development that meets the needs of the present without compromising the ability of the future generations to meet their own needs' ((Bruntland (1987), cited in Ihuah et al. (2014)). Sustainability is concerned with environmental, economic, social and cultural dimensions. Sustainable housing policies deal with the affordability, social justice, cultural and economic impacts of housing, and assist in achieving healthy residential neighbourhoods (UN-Habitat 2012). The need for sustainable housing is essential particularly in developing countries where there is rapid urbanization and population growth.

The aim of this study is to identify critical success factors for mass housing projects in Nigeria. Critical success factors are certain conditions or circumstances which predict the success of a project (Toor & Ogunlana 2009, Musa et al. 2018).

Mass house building project is defined as 'design and construction of large- scale standardized house- units, usually located in the same area and constructed within the same housing scheme. In any case, the quantity of house-units should not be less than 10 house- units' (Ahadzie et al. 2008).

1.2 *Sustainable housing development*

Generally, sustainable houses are those that are designed, built and managed as healthy, durable, safe secure, and affordable for the whole spectrum of incomes, improve health conditions and labour productivity through better living conditions (UN-Habitat 2012).

Sustainable development is concerned with the environmental factor, economic factor, social factor and cultural factor (UN-Habitat 2012). The environmental sustainability of housing is concerned with the impacts of housing on the environment, as well as the impacts of the environment on housing itself. Social sustainability is concerned with creating affordable, healthy, good quality, secure, inclusive and diverse (mixed-tenure and mixed-income) housing. Cultural sustainability takes into consideration protecting cultural heritage, lifestyles and behaviours of occupants, whereas economic sustainability is concerned with integrating houses with economic activities

Based on review of previous studies (Turcotte & Geiser 2010, Ibem & Azuh 2011, Mulliner & Maliene 2011, UN-Habitat 2012, Ihuah et al. 2014, Mukhtar et al. 2017, Oyebanji 2017, Adabre & Chan 2019) attributes that influence sustainable housing development have been identified and presented in Table 1.

Table 1. Respondents' perceptions of the relative importance factors influencing sustainable mass housing projects in Nigeria.

Attributes Influencing Sustainability	Mean	Rank
Good/safe location	4,23	1
Incorporating green design	4.21	2
Provide occupant-neighborhood linkage	4.17	3
Connected to decent, safe and affordable energy, water and sanitation facilities	4.13	4
Preventing use of polluting materials	4.05	5
Ensuring housing affordability	4.02	6
Integrating houses with economic activities	4.13	7
Locate housing in mixed-use neighborhoods	4,05	8
Protecting cultural heritage, values, norms and traditions of occupants	4.02	9
Use of sustainable local construction materials	3.98	10
Providing serviced land	3.97	11
Enhancing the economic fabric of the local neighbourhood	3.91	12
Flexible/adaptable design	3.87	13
Protecting the houses from external pollution	3.79	14
Accessible credit facilities to developers	3.76	15
Generation of employment and income	3.73	16
Access to basic public facilities (well connected to jobs, shops, health- and child-care, education and other services)	3.65	17
Easy to maintain	3.62	18
Using energy and water most efficiently	3.55	19
Providing access to infrastructure,	3.41	20
Low interest rate	3.36	21
Improving housing finance	3.30	22
Stable macro-economic environment	2.98	23
Long term loan repayment period	2.93	24
Equipped with certain on-site renewable energy generation and water recycling capabilities	2.88	25

2 RESEARCH METHODOLOGY

The study was conducted by means of a questionnaire survey. Initially, literature review was carried out to identify factors that influence sustainability of mass housing projects, in which twenty five factors were identified. The list of the factors was presented to six experts on housing and sustainability for validation. Subsequently, these factors were used to develop survey questionnaire. The respondents were construction professionals with knowledge in sustainability and housing. They include architects, builders, quantity surveyors and real estate managers. The respondents were asked to indicate their opinions on the influence of the factors on housing sustainability, using five-point Likert Scale ranging from 1-5, where 1 represents very low influence and 5 very high influence. A total of 390 questionnaires were administered to the respondents and 243 were returned completed, representing 62% response rate.

The study was conducted in four out of the 36 states of Nigeria. The states were Kaduna, Kano, Bauchi and Gombe, which were chosen for this study because of their similarities in culture and geographical location.

The data collected were analysed using mean and factor analysis, with the aid of SPSS Software version 21.

3 DATA ANALYSIS AND PRESENTATION OF RESULTS

This section analyses the data obtained from the field survey and presents the results. The reliability coefficient of the data collected was calculated using Cronbach's alpha. It was found to be 0.78, which is considered as very good (Kline 2011).

3.1 *Ranking of important attributes that influence sustainability of mass housing projects*

Analysis of the data collected from the field survey reveals that, the mean scores of the twenty five attributes of sustainable housing range from 2.88 to 4.23 as shown in Table 1. The results in Table 1 also indicate that nine attributes have mean scores greater than 4.00 (suggesting very high influence), thirteen attributes have mean scores from 3.00- 4.00 (high influence), while the remaining three attributes have a mean score value of less than 3.00 (average).

Among the attributes with very high influence on housing sustainability, safe location has been ranked first by the respondents, followed by incorporating green design, ranked second, provide occupant-neighborhood linkage (third), connected to decent, safe and affordable energy, water and sanitation facilities (fourth), preventing use of polluting materials (fifth), ensuring housing affordability (sixth), integrating houses with economic activities (seventh), locate housing in mixed-use neighborhoods (eighth), protecting cultural heritage, values, norms and traditions of occupants (ninth).

3.2 *Factor analysis for attributes that influence sustainability of mass housing projects*

Factor analysis has been conducted on the twenty five attributes of sustainable housing, in order to determine the factor grouping that can be used to represent the relationship among the variables.

The Bartlett test of sphericity is 2392.657 and the associated significance level was < 0.05 (p 0.000), suggesting that the population correlation matrix is not an identity matrix. The value of the Kaiser-Meyer-Olkin (KMO) statistic is 0.867, which according to Kaiser (Norusis 1992) is satisfactory for factor analysis.

The factor analysis produced six-factor solution with eigenvalues greater than 1.000, explaining 60 % of the variance, as shown in Table 2. The remaining factors together accounted for 40% of the variance. The factor grouping based on varimax rotation is shown

Table 2. Total rotated factor variance explained for factors influencing sustainable mass housing projects.

	Initial Eigenvalues			Rotation Sums of Squared Loadings		
Component	Total	% of variance	Cumulative %	Total	% of variance	Cumulative %
1	6.691	26.763	26.763	3.212	12.847	12.847
2	2.478	9.912	36.675	2.734	10.937	23.784
3	2.068	8.271	44.947	2.597	10.390	34.174
4	1.590	6.358	51.305	2.539	10.157	44.331
5	1.181	4.723	56.028	2.003	8.010	52.342
6	1.006	4.023	60.051	1.927	7.710	60.051
7	.803	3.213	63.264			
8	.770	3.079	66.343			
9	.722	2.888	69.230			
10	.700	2.802	72.032			
11	.663	2.651	74.683			
12	.626	2.503	77.187			
13	.600	2.400	79.587			
14	.594	2.376	81.963			
15	.536	2.145	84.108			
16	.513	2.054	86.162			
17	.494	1.975	88.137			
18	.481	1.925	90.062			
19	.464	1.854	91.916			
20	.419	1.675	93.591			
21	.372	1.489	95.080			
22	.353	1.413	96.493			
23	.327	1.308	97.801			
24	.283	1.131	98.931			
25	.267	1.069	100.000			

Extraction Method: Principal Component Analysis.

in Table 3. It can be noted that, each variable belongs to only one of the factors, with the loadings on each factor exceeding 0.50.

Table 3. Rotated factor matrix (factor loadings) of CSFs for sustainable mass housing projects in Nigeria.

Factor Attributes	Component					
	1	2	3	4	5	6
Incorporating green design	.734					
Use of sustainable local construction materials	.719					
Preventing use of polluting materials,	.714					
Protecting the houses from external pollution,	.697					
Connected to decent, safe and affordable energy, water and sanitation facilities,	.690					
Good/safe location	.654					
Using energy and water most efficiently,		.762				
Equipped with certain on-site renewable energy generation and water recycling capabilities,		.710				
Flexible/adaptable design,		.700				
Stable macro-economic environment		.570				
Providing serviced land,			.738			

(*Continued*)

Table 3. (*Continued*)

Factor Attributes	Component					
	1	2	3	4	5	6
Access to basic public facilities (well connected to jobs, shops, health- and child-care, education and other services),			.732			
Provide occupant-neighborhood linkage,			.649			
Locate housing in mixed-use neighborhoods			.624			
Providing access to infrastructure,			.613			
Integrating houses with economic activities				.787		
Ensuring housing affordability				.780		
Easy to maintain				.754		
Generation of employment and income,				.729		
Enhancing the economic fabric of the local neighbourhood					.699	
Improving housing finance					.677	
Accessible credit facilities to developers					.595	
Low interest rate,						.659
Long term loan repayment period						.600
Preserving cultural heritage, values, norms and traditions of occupants						.599

The six factors groupings are interpreted based on their attributes as follows:
Factor 1: Green design and healthy environmental
Factor 2: Energy efficiency and adaptable design
Factor 3: Social cohesion and availability of basic facilities
Factor 4: Housing affordability and economic activities integration
Factor 5: Effective housing finance
Factor 6: Preserving cultural heritage

4 CONCLUSION

The concept of sustainability as relates to housing development plays a significant role in meeting present housing needs especially in developing countries. Sustainable housing will assist in the production of adequate housing which are safe, secure, affordable, connected to essential services and facilities, and culturally adequate.

This study was conducted to identify critical success factors for sustainable mass housing projects in Nigeria. Twenty five attributes for sustainable housing were identified from the literature, which were used to develop the survey questionnaire for this study. The perceptions of the participants of the survey revealed that the nine most important attributes contributing sustainability of mass housing projects were safe location, incorporating green design, providing occupant-neighborhood linkage, connecting to decent, safe and affordable energy, water and sanitation facilities, preventing use of polluting materials, ensuring housing affordability, integrating houses with economic activities, locating housing in mixed-use neighborhoods, and protecting cultural heritage, values, norms and traditions of occupants.

Factor analysis was conducted to identify the relationships among the variables. Six factor groupings emerged which explained 60% of the variance. The factors were interpreted based on their attributes as green design and healthy environmental, energy efficiency and adaptable design, social cohesion and availability of basic facilities, housing affordability and economic activities integration, effective housing finance and lastly preserving cultural heritage

There is need for housing policy makers in Nigeria and other developing countries to incorporate these identified factors/attributes in housing policy formulation and implementation in order to address housing challenges facing those countries.

REFERENCES

Adabre, M. A. & Chan, A.P.C. 2019. Critical success factors (CSFs) for sustainable affordable housing. *Building and Environment* 156: 203-214.

Ahadzie, D.K. Proverbs, D.G. & Olomolaiye, P.O. 2008. Critical success criteria for mass house building projects in developing countries. *International Journal of Project Management* 26:675–687.

Arku, G. 2006. The housing and economic development debate revisited: Economic significance of hous-ing in developing countries. *Journal of Housing and Built Environment* 21: 377–395.

Bruntland, H. 1987. *Our Common Future. Report for the World Commission on Environment and Development*. Oxford University Press, New York.

Ibem, E.O. Anosike, M.N. & Azuh, D.E. 2011. Challenges in public housing provision in the post-independence era in Nigeria. *International Journal of Human Sciences* 8(2):421–443.

Ibem, E.O. & Azuh, D.E. 2011. Framework for evaluating the sustainability of public housings programmes in developing countries. *Journal of Sustainable Development and Environmental Protection* 1(3): 24–39.

Ihuah, P.W. Kakulu, I.I. & Eaton, D. 2014. A review of critical project management success factors (CPMSF) for sustainable social housing in Nigeria. *International Journal of Sustainable Built Environment* 3(1):62-71.

Kline, R. B. 2011. *Principles and practice of structural equation modeling (3rd Ed.)*. New York: The Guilford Press.

Mukhtar M, M. Amirudin R. B. Sofield, T. & Mohamad, I. 2017. Critical success factors for public housing projects in developing countries: A case study of Nigeria. *Journal of Environment, Development and Sustainability* 19(5):2029-2066.

Mulliner, E. & Maliene, V. 2011. Criteria for sustainable housing affordability. *Environmental Engineering* 3:966-973.

Musa M, M. Amirudin R.B. Dalhatu, A. & Musa, M.A. 2018. Critical success factors for construction project management in Nigeria. *Advance Science Letters* 24(5):3809-3813.

Norusis, M.J. 1992. *SPSS for Windows*. Professional statistics, Release 5, SPSS Inc., Chicago.

Olotuah, A. O. 2010. Housing development and environmental degeneration in Nigeria. *The Built and Human Environment Review* 3:42–48.

Oyebanji, A.O. Liyanage, C. & Akintoye, A. 2017. Critical Success Factors (CSFs) for achieving sustainable social housing (SSH). *International Journal of Sustainable Built Environment* 6(1):216–227

Toor, S.R. & Ogunlana, S.O. 2009. Construction professionals' perception of critical success factors for large-scale construction projects. *Construction Innovation* 9(2):149–167.

Turcotte, D. A. & Geiser, K. 2010. A framework to guide sustainable housing development. *Housing and Society* 37(2): 87–117.

UN-Habitat. 2010. *A practical guide for conducting housing profiles*. Human Settlements Programme Nairobi: United Nations.

UN-Habitat. 2011. *Affordable Land and Housing in Africa*. Nairobi: United Nations Human Settlements Programme.

UN-Habitat. 2012. *Sustainable Housing for Cities: A Policy Framework for Developing Countries*. Nairobi: United Nations Human Settlements Programme.

Smart construction and construction management

Sustainable Buildings and Structures: Building a Sustainable Tomorrow – Papadikis et al. (Eds)
© 2020 Taylor & Francis Group, London, ISBN 978-0-367-43019-1

Design for construction waste management

J. Xu & W. Lu
Department of Real Estate and Construction, The University of Hong Kong, Hong Kong, China

ABSTRACT: Construction waste is an urging problem to be addressed from the very beginning of a project. This paper argued that instead of working at the construction stage to handle construction waste as most current research and practice have done, construction waste management (CWM) should be planned at the design stage by adopting the ideas of design for manufacturing and assembly (DfMA), as well as design for disassembly (DfD). It further proposed the concept of design for CWM (DfCWM) and its five principles, including standardization, simplification, modularity, reusability, and recyclability. By adopting DfCWM and its principles, construction waste can be reduced at the design process of a project, reused during construction, and recycled when demolished. With the proposition of DfCWM, this paper will bring a paradigm shift opportunity for CWM and design industry by combining the advanced technologies of industrialized construction and construction materials.

1 INTRODUCTION

Construction waste, or construction and demolition waste (CDW), is any substance, matter or thing generated as a result of construction work and abandoned (HKEPD 2018); also defined as surplus materials arising from site clearance, excavation, construction, refurbishment, renovation, demolition and road works (Kofoworola & Gheewala 2009). Similar design strategies, as well as construction technologies and materials are employed, such as cast in-situ dominated while prefabrication emerging, and concrete and rebar composite structures dominated while full steel structures being increasingly seen, lead to similar taxonomy in construction waste management (CWM) systems. It was reported that CDW accounts for 30% to 40% of the total amount of waste in China (Huang et al. 2018). The vast amounts of solid waste produced from construction activities pose a potential crisis to the sustainable development of the construction industry and the whole society. Numerous research looked at how to managing construction waste at the construction stage. However, only a few studies have looked at solving the problem from the design stage (Lu et al. 2017). The fact is, a considerable proportion of construction waste is foredoomed from the design plan. Preventive management of construction waste needs cross-disciplinary collaboration (Xu et al. 2018) and should be considered as early as the initial concept design.

Building/infrastructure is the artificial and aesthetics product composed of a series of designed component combinations. A particularly critical step for construction projects is the selection of the building's components and materials at the design stage (Flager et al. 2009), which involves the assessment of the sustainable, economic, and efficiency-related objectives of a project (Inyim et al. 2014). Hundreds of thousands of combination options lead to the designer be exhausted and poor decision-making facing potential design strategies. Thus, designers usually defer decisions to later stages of the design process (Basbagill et al. 2012). Comprehensively considering the component combination requires adequate and multi-subjects knowledge about engineering, materials, building equipment, etc., which is difficult for designers to handle all these issues integrated and optimized. Usually, designers decide building component and material combination mainly depending on their professional experience or personal intuition. Thus designers generally prefer to source unfamiliar knowledge and information through informal interactions with their colleagues (Aurisicchio et al. 2010). However, a volume of dynamic factors, such as multi-variables affecting component adoption and multi-objective affecting whole lifecycle benefits, make the decision-making hard to balance all interests. The highly fragmented design process leads to a lack of consideration in sustainability, and to be specific, construction waste management.

With the rising of design concepts such as design for manufacturing (DfM) (Ulrich et al. 1993), design for assembly (DfA) (Boothroyd & Knight 1993), design for manufacturing and assembly (DfMA) (Boothroyd 1994), and design for disassembly (DfD) (Boothroyd & Alting 1992), a thinking and practicing revolution in design industry is taking place. Such changes will have direct or indirect impacts on the downstream industry chain players, including raw material supplier, manufacturers, logistics suppliers, construction companies, construction managers, facility managers, demolition companies, and environment agencies and organizations. However, the CWM issue remains untouched by designers. Though there are design for sustainability (DfS) and green building design concepts and practices, they are not directly linked with or lead to the reduction of construction waste (Oktay 2004, Lu et al. 2019). Therefore, the proposition of design for construction waste management (DfCWM) from a life-cycle perspective is needed and will bring subversive changes to CWM. Although, DfCWM is not calling an independent and repeated design plan specifically for CWM, a principle to bear in mind when design happens. It can be jointly and simultaneously considered with DfMA, DfD, and other objectives.

This paper aims to propose the idea of DfCWM to shift the paradigm of CWM. It calls for attention to CWM from the design stage with a life-cycle design thinking. It will also suggest several principles for DfCWM. The rest of the paper is organized as follow: after a literature review of the status of CWM and construction-related design concepts at Section 2; the concept and principles of DfCWM will be presented at Section 3; Section 4 will discuss the potential impact of DfCWM on CWM followed by a discussion and conclusion part at Section 5.

2 LITERATURE REVIEW

Among various DfX (X stands for excellence), DfMA and DfD are the most related to construction management and especially CWM. In traditional departmental responsibility system and sequential engineering, the architectural design process is disconnected from the component manufacturing process, which makes the building less manufacturable, assemblable, and maintainable. For example, the problems of design changes, long construction development cycle, high construction cost, and unstable product quality are endless. DfMA is seen as a critical design philosophy and method for improving the productivity of the construction industry (Gao et al. 2018, Chen & Lu 2018). DfMA focuses on both manufacturability and assimilability, and aims to achieve a comprehensive optimization of these two goals through the practice of a series of guidelines. This vocabulary, carrying the genes produced by the manufacturing industry, brings new possibilities and directions to the fragmented, inefficient, and excessively wasteful construction industry.

Design for Disassembly (DfD) are examples of better construction waste management methods (Rios et al. 2015). Its practice is to simplify the deconstruction process and procedures through planning and design. Deconstruction is the process of demolishing a building, but restoring the use of the removed material. The deconstruction process has fundamentally changed the traditional waste management process. The DfD process is therefore an essential strategy for protecting raw materials (Ciarimboli & Guy 2007, Rio et al. 2015). DfD mainly applies five principles to make this design philosophy feasible: 1) documentation of appropriate materials and deconstruction methods; 2) designing accessible connections and joints for ease of disassembly; 3) separate non-recyclable, non-reusable and non-removable items such as mechanical, electrical and plumbing (MEP) systems; 4) Design simple structures and forms that allow for standardization of components and dimensions; 5) Design that reflects labor practices, productivity and safety (Ciarimboli & Guy 2007).

Although various academic researchers and industry practitioners have input plethora of efforts to CWM research, none was found to introduce DfMA or DfD with CWM. Only several research investigated CWM at the design stage. Osmani et al. (2008) summarized that design changes, design and detailing complexity, design and construction detail errors, unclear or unsuitable specification, and poor coordination and communication at the design stage would cause extra construction waste. It was also proven that implementing waste reduction management at the design stage can effectively reduce construction wastes (Ding et al. 2018). Parametric design integrated with modular coordination is also simulated to lower the waste generation (Banihashemi et al. 2018). Emerging technologies are also adopted to CWM. For example, building information modeling (BIM) is adopted for waste management collaboration, waste-driven design process, waste analysis, innovative technologies integration, and improved waste management documentation (Lu et al. 2017, Akinade et al. 2018). Yet, how to managing construction waste from design processes remained unanswered.

3 DESIGN FOR CONSTRUCTION WASTE MANAGEMENT

3.1 Concept of design for construction waste management

Design for construction waste management (DfCWM) should aim to minimize waste generation from the design process when considering the whole life-cycle of the project. It should bear the 3R principles of waste management, i.e., reduce, reuse, and recycle, and advanced design principles, especially standardization, simplification, and modularity, in mind. Though standardization, simplification, and modularity in existing design fashion will directly contribute to the reduction of construction waste (Osmani et al. 2008), the reuse and recycling of construction materials are largely unattended. Based on the aim, principles, and the background, a tentative definition is given as:

Design for construction waste management (DfCWM) is a design methodology which plans for life-cycle CWM from the design stage with the aim of minimizing waste generation by applying the principles of standardization, simplification, modularity, reusability, and recyclability.

3.2 Principles of design for construction waste management

As highlighted at the definition of DfCWM, there are five principles, i.e., standardization, simplification, modularity, reusability, and recyclability. They will be further explained as follows.

3.2.1 Standardization

It was pointed out that the lack of rudimentary protocol and standards at design would lead to waste (Ioannidis et al. 2014). Although it was talking about biomedical research, the cause-effect is almost the same for the construction industry. By making standardization the first principle of DfCWM, this paper appeals for uniformed standards of construction components and methods. Standardization will ease not only the design and construction processes but also the maintenance, operation, and demolition and therefore reduce the waste generated by special components during manufacturing and assembly. The principle of standardization also keeps in line with the requirements of buildability (Wong & Lam 2008) and the trends of industrialized and smart construction (Larsson et al. 2014).

3.2.2 Simplification

As construction waste generation is closely associated with design, detailing, and construction complexity (Liu et al. 2015), simplification is a must route for waste reduction. Keeping the components, joints, decorations, and their combination simple will reduce unnecessary materials wastes and ease the works at manufacturing, assembly, and disassembly. For example, simple plane structures can help save steel, concrete, shuttering, and supporting materials compared to hetero-structures. Simplification is the method towards standardization and the result of it at the same time. Associated with standardization, simplification will also lay a foundation for prefabrication and construction automation.

3.2.3 *Modularity*

Modular integrated construction (MiC) is also an emerging paradigm shift initiative. It borrows the manufacturing business model to the construction industry where components are prefabricated offsite and then transported to construction sites for assembly (Wuni et al. 2019), which shares a similar idea with DfMA. MiC, which can be measured by modularity, will reduce the construction waste produced at construction sites. Modularity is the degree of connections to which structural elements are powerfully connected among themselves and relatively separated from others (Baldwin & Clark 2000). Design for modularity will mitigate the works at construction sites and therefore reduce construction waste.

3.2.4 *Reusability*

With standardization, simplification, and modularity directly contribute to construction waste reduction, reusability calls for reusing existing structures and materials as much as possible. By doing so, fewer new materials will be added to the construction works and existing materials are fully used. Actually, existing structures and materials from demolition sites of the project or other places can be used for new construction works. For example, the demolished concrete can be segregated to obtain required sizes of aggregate for new construction (Husain & Assas 2013). Furthermore, if a building is carefully disassembled rather than demolished, there is a great chance that more components and equipment might be reused (Addis & Schouten 2004). Therefore, designers should think in advance how a building or infrastructure can be disassembled for reuse at the primary stage.

3.2.5 *Recyclability*

Supplemented for reusability, some structured and components that cannot be reused and have to be demolished can be recycled for other usages. Recyclability encourages the adoption of degradable, non-toxic, and recyclable materials and equipment when design. Many efforts have been paid to the recycling of demolished concrete, masonry rubble, and localized cutting by blasting of concrete (Hansen 1992, Poon 1997, Rao et al. 2007, Addis 2012). Apart from concrete and masonry rubbles, the steel, timber, MEP (mechanical, electrical and plumbing) equipment, and furniture can also be recycled. Designers should choose recyclable and durable materials and equipment during design processes to avoid irreversible transformation from material to waste.

3.3 *The relationships among the five principles*

The relationships among the five principles are illustrated in Figure 1. Standardization, simplification, and modularity are closed linked with each other, together they contribute to the reduction of construction waste. It is difficult to set one of them apart from the others, for example, standardization will inevitably lead to simplification and modularity, so on so forth. The three principles are also linked with reusability and recyclability. On one hand, standardized and simplified modular components will ease the implementation of reusability and recyclability. On the other hand, to achieve the reuse and recycling of components and

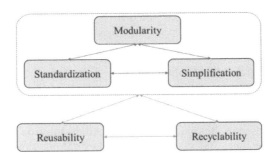

Figure 1. The relationships among five principles.

equipment will promote the adoption of standardization, simplification, and modularity in turn. Reusability and recyclability are not independent with each other, either. Thus, these five principles are unified under the same objective of construction waste management but emphasize on different aspects and strategies.

4 DISCUSSION AND CONCLUSION

This paper proposed a paradigm shift concept to ask for the planning of CWM at the design stage. The concept of DfCWM is not trying to add burden and responsibility to designers but ask for designers and developers to keep CWM in mind at the initial design stage for sustainable buildings and infrastructure. Such considerations will eventually add value to the design and the project by mitigating construction complexities and difficulties and protecting the environment at the same time. Moreover, by reducing construction waste generation, the money will be saved since an increasing fee will be charged for landfill and waste disposal. The reuse and recycle of structures, materials, and equipment can also help save a big fortune. Therefore, DfCWM will not only contribute to the environment but also has its economic value.

With the five principles of DfCWM, i.e., standardization, simplification, modularity, reusability, and recyclability, advanced construction technologies and ideas can also be integrated. Uniformed standards, simplified and modularized components, and simplified manufacturing, assembly, and disassembly methods will provide a perfect setting for technologies including BIM, prefabrication, automated machines, and construction robotics. Besides, the principle of reusability and recyclability calls for the utilization of fresh outcomes of advanced construction materials. To conclude, the concept of DfCWM and its principles bring together the industrialized construction technologies and advanced construction materials for a smarter and more sustainable construction industry.

REFERENCES

Addis, B. 2012. *Building with reclaimed components and materials: A design handbook for reuse and recycling.* Routledge.
Addis, W. & Schouten, J. 2004. *Design for reconstruction-principles of design to facilitate reuse and recycling.*
Akinade, O.O. Oyedele, L.O. Ajayi, S.O. Bilal, M. Alaka, H.A. Owolabi, H.A. & Arawomo, O. O. 2018. Designing out construction waste using BIM technology: Stakeholders' expectations for industry deployment. *Journal of Cleaner Production* 180:375-385.
Aurisicchio, M. Bracewell, R. & Wallace, K. 2010. Understanding how the information requests of aerospace engineering designers influence information-seeking behaviour. *Journal of Engineering Design* 21(6):707-730.
Baldwin, C.Y. & Clark, K. B. 2000. *Design rules: The power of modularity.* MIT press.
Banihashemi, S. Tabadkani, A. & Hosseini, M.R. 2018. Integration of parametric design into modular coordination: A construction waste reduction workflow. *Automation in Construction* 88:1-12.
Basbagill, J. Flager, F. Lepech, M. & Fischer, M. 2013. Application of life-cycle assessment to early stage building design for reduced embodied environmental impacts. *Building and Environment* 60:81-92.
Boothroyd, G. 1994. Product design for manufacture and assembly. *Computer-Aided Design* 26(7): 505-520.
Boothroyd, G. & Alting, L. 1992. Design for assembly and disassembly. *CIRP annals* 41(2): 625-636.
Boothroyd, G. & Knight, W. 1993. Design for assembly. *IEEE Spectrum* 30(9): 53-55.
Chen, K. & Lu, W. 2018. Design for Manufacture and Assembly Oriented Design Approach to a Curtain Wall System: A Case Study of a Commercial Building in Wuhan, China. *Sustainability* 10(7): 2211.
Ciarimboli, N. & Guy, B. 2007. *Design for Disassembly in the built environment: a guide to cloosed-loop design and building.* Pennsylvania State University.
Ding, Z. Zhu, M. Tam, V.W. Yi, G. & Tran, C.N. 2018. A system dynamics-based environmental benefit assessment model of construction waste reduction management at the design and construction stages. *Journal of Cleaner Production* 176:676-692.
Flager, F. Welle, B. Bansal, P. Soremekun, G. & Haymaker, J. 2009. Multidisciplinary process integration and design optimization of a classroom building. *Journal of Information Technology in Construction (ITcon)* 14(38):595-612.

Gao, S. Low, S.P. & Nair, K. 2018. Design for manufacturing and assembly (DfMA): a preliminary study of factors influencing its adoption in Singapore. *Architectural Engineering and Design Management* 14(6): 440-456.

Hansen, T. C. 1992. *Recycling of demolished concrete and masonry*. CRC Press.

Hong Kong Environmental Protection Department (HKEPD). 2018. An Overview on Challenges for Waste Reduction and Management in Hong Kong.

Huang, B. Wang, X. Kua, H. Geng, Y. Bleischwitz, R. & Ren, J. 2018. Construction and demolition waste management in China through the 3R principle. *Resources, Conservation and Recycling* 129:36-44.

Husain, A. & Assas, M.M. 2013. Utilization of demolished concrete waste for new construction. *World Academy of Science, Engineering and Technology* 73:605-610.

Inyim, P. Rivera, J. & Zhu, Y. 2014. Integration of building information modeling and economic and environmental impact analysis to support sustainable building design. *Journal of Management in Engineering* 31(1):A4014002.

Ioannidis, J.P. Greenland, S. Hlatky, M.A. Khoury, M.J. Macleod, M.R. Moher, D. Schulz, K.F. & Tibshirani, R. 2014. Increasing value and reducing waste in research design, conduct, and analysis. *The Lancet* 383(9912):166-175.

Kofoworola, O. F. & Gheewala, S.H. 2009. Estimation of construction waste generation and management in Thailand. *Waste Management* 29(2): 731-738.

Larsson, J. Eriksson, P.E. Olofsson, T. & Simonsson, P. 2014. Industrialized construction in the Swedish infrastructure sector: core elements and barriers. *Construction Management and Economics* 32(1-2):83-96.

Liu, Z. Osmani, M. Demian, P. & Baldwin, A. 2015. A BIM-aided construction waste minimisation framework. *Automation in Construction* 59:1-23.

Lu, W. Webster, C. Chen, K. Zhang, X. & Chen, X. 2017. Computational Building Information Modelling for construction waste management: Moving from rhetoric to reality. *Renewable and Sustainable Energy Reviews* 68:587-595.

Lu, W. Chi, B. Bao, Z. & Zetkulic, A. 2019. Evaluating the effects of green building on construction waste management: A comparative study of three green building rating systems. *Building and Environment* 155: 247-256.

Oktay, D. 2004. Urban design for sustainability: A study on the Turkish city. *The International Journal of Sustainable Development & World Ecology* 11(1): 24-35.

Osmani, M. Glass, J. & Price, A.D. 2008. Architects' perspectives on construction waste reduction by design. *Waste Management* 28(7): 1147-1158.

Poon, C. S. 1997. Management and recycling of demolition waste in Hong Kong. *Waste Management & Research* 15(6):561-572.

Rao, A. Jha, K.N. & Misra, S. 2007. Use of aggregates from recycled construction and demolition waste in concrete. *Resources, Conservation and Recycling* 50(1): 71-81.

Rios, F.C. Chong, W. K. & Grau, D. 2015. Design for disassembly and deconstruction-challenges and opportunities. *Procedia Engineering* 118:1296-1304.

Ulrich, K. Sartorius, D. Pearson, S. & Jakiela, M. 1993. Including the value of time in design-for-manufacturing decision making. *Management Science* 39(4): 429-447.

Wong, F. W. & Lam, P.T. 2008. Benchmarking of buildability and construction performance in Singapore: is there a case for Hong Kong? *International Journal of Construction Management* 8(1): 1-27.

Wang, Z. Li, H. & Zhang, X. 2019. Construction waste recycling robot for nails and screws: Computer vision technology and neural network approach. *Automation in Construction* 97: 220-228.

Wuni, I. Y. Shen, G.Q. & Mahmud, A.T. 2019. Critical risk factors in the application of modular integrated construction: a systematic review. *International Journal of Construction Management* 1-15.

Xu, J. Lu, W. Xue, F. Chen, K. Ye, M. Wang, J. & Chen, X. 2018. Cross-boundary collaboration in waste management research: A network analysis. *Environmental Impact Assessment Review* 73:128-141.

Behaviour and big data in construction waste management: A critical review of research

M.W.W. Lee & W. Lu
The University of Hong Kong, Hong Kong Special Administrative Region, China

ABSTRACT: In recent years, there has been a more widespread application of big data analytics to researches on marketing. Nevertheless, the use of big data which is a more reliable tool in comparison to traditional research methods is still at its infant stage in the academia. This review contributes to the understanding of the current status of construction waste management in Hong Kong, the way in which big data analytics had been applied to construction waste management behavioural research as well as related areas of big data research which are not yet addressed by existing literature.

1 INTRODUCTION

Big data analytics have been gaining popularity in both academic and industrial researches on marketing and consumer behaviours in recent years. This can be due to the fact that big data is capable of generating a fuller picture of the subject matter under investigation as well as minimising sampling bias which is one of the main issues in traditional research. However, its application to academic research on construction waste management ("CWM") is still at its infant stage. Contextualised in Hong Kong, by reviewing previous literature on CWM behaviours which had applied big data analytics, this paper aims at identifying key issues not being addressed by previous studies. The concepts of construction waste ("CW"), current status of CWM in Hong Kong and big data analytics are introduced in Sections 2, 3 and 4 respectively. A review of previous literature on CWM behaviour conducted by way of big data analytics is presented in Section 5. Section 6 provides a discussion of the research gaps identified from the literature mentioned in Section 5. Finally, Section 7 gives a conclusion of this paper.

2 CW IN HONG KONG

CW includes any abandoned surplus materials generated from construction activities such as excavation, construction, site clearance, demolition, refurbishment, etc. (EPD 2015, EPD 2017a). In general, CW can be classified into two categories, namely, inert and non-inert wastes. While inert waste includes earth, soil, concrete and asphalt which can be recycled for reuse in other construction activities (EPD 2017a) such as serving as fill materials in reclamation projects and site formation (Lu & Tam 2013, Lu et al. 2016), non-inert waste comprises timber, bamboo, packaging waste, plastic as well as other organic materials that can only be landfilled (EPD 2015). Previous studies also reported that the amount of CW generated by demolition activities is more than 10 times that from construction activities associated with new developments (Chen & Lu 2017). Therefore, CW is sometimes referred as "construction and demolition waste" ("C&DW").

While the construction industry accounts for only approximately 4.5% of Hong Kong's GDP (C&SD 2017), CW takes up about one-quarter of the total solid waste deposited in landfills in the year of 2015 (EPD 2017b). This implies that CW has also taken up a considerable amount of landfill space which in turn exerts tremendous pressure on Hong Kong, a compact city where land resources are extremely scarce (Hao et al. 2008, Lu & Tam 2013, Yuan et al.

2013, Lu et al. 2018, Lu et al. 2019). CW being dumped at landfills not only depletes valuable land in Hong Kong, but its anaerobic degradation also causes air pollution and contamination to ground water and soil (Lu et al. 2015a, Lu et al. 2015b). Statistics also showed that the government had spent more than HK$200 million per year on treatment of CW (Lu & Tam 2013). In response to the significant impact of CW both environmentally and financially, the government had introduced a number of measures to manage CW which will be elaborated in the next section.

3 CURRENT STATUS OF CWM BEHAVIOURS IN HONG KONG

The implementation of the Construction Waste Disposal Charging Scheme ("CWDCS") in 2006 to encourage waste producers to minimise disposal costs through reduction, on-site sorting and recycling of CW so as to preserve landfill space (EPD 2017a) is among the most robust attempt made by the government in recent years to alleviate the adverse impacts generated by CW. Under the CWDCS, except those which can be reused or recycled, all CW shall be disposed of at the government's waste reception facilities (i.e. public fill reception facilities, sorting facilities, landfills or outlying islands transfer facilities, depending on the relative composition of inert and non-inert CW). The current charges levied on the main contractor for each tonne of waste disposed of are as follows: - HK$71 for every tonne of inert waste dumped at public fill reception facilities (which only accept inert waste); HK$175 for every tonne of mixed inert and non-inert waste accepted by sorting facilities (which only accept waste with more than 50% by weight of inert materials); and HK$200 per tonne of mixed inert and non-inert waste dumped in landfills and outlying islands transfer facilities (EPD 2017a).

The waste being disposed of at sorting facilities would undergo further processing involving both mechanical sorting and handpicking for the purpose of segregating inert materials from non-inert waste (Lu 2013). The inert materials being sorted out will then be temporarily stored at the public fill reception facilities for future reuse in other projects, in particular reclamation projects, whereas the non-inert waste will be dumped in landfills. Since Hong Kong is lacking in local reclamation projects in recent years, the vast majority of the inert materials received by public fill reception facilities have been exported to Taishan, a county-level city in Mainland China, for reclamation purpose (Audit Commission 2016).

Throughout the years, the government has also tried to minimise on-site CW by using prefabrication, which was found to be more effective in waste reduction compared with traditional methods, in public housing projects in Hong Kong, especially in the construction of buildings under the Home Ownership Scheme (Lu 2013).

In response to the problem of illegal dumping of CW, the government had adopted an enhanced trip-ticket system ("TTS") since 2004 under which all the final destinations of CW from public works, i.e. the particular government waste reception facility which the relevant CW had gone to, can be tracked (Lu & Yuan 2012, Lu & Tam 2013, Lu et al. 2016). However, this did not stop the problem of fly-tipping which will be further elaborated in Section 5 of this paper. Many previous studies criticised that the existing transaction costs associated with improper management of CW are not high enough to motivate contractors to improve their CWM practices (Lu et al. 2016).

4 CONCEPT OF BIG DATA ANALYTICS

Big data is best known for its volume, variety, velocity, veracity as well as value which are collectively referred to as "5Vs". "Volume" means big data is huge in quantity which is often measured in terabytes; "variety" refers to the fact that big data can be structured, unstructured, semi-structured or a combination of them (Chen & Lu 2017); "velocity" stands for the high speed of data generation; "veracity" refers to the quality of big data which has significant impact on the results of analyses (Big Data & Framework 2019); "value" refers to the value of the hidden pattern(s) identified from big data analyses (Mayer-Schönberger & Cukier 2013).

Big data analytics are employed to deal with data which is of huge volume (Chen & Lu 2017). In comparison to traditional methods of data analysis, the results obtained from big data analyses are capable of providing a fuller picture of the subject matter being studied which can in turn minimise the potential bias associated with small sampling data (Chen & Lu 2017). By uncovering unknown correlations and hidden patterns, big data analytics offer the benefit of extracting new insights or creating new values for more accurate prediction and better decision-making (McAfee & Brynjolfsson 2012), thereby capable of changing markets, organizations as well as the relationship between governments and the general public (Mayer-Schönberger & Cukier 2013). Data mining which is a relatively new technique employed to discover useful information from a large amount of data and converting it to organized knowledge is one of the major techniques applied in big data analyses (Lu et al. 2015b).

5 APPLICATION OF BIG DATA TO RESEARCH ON CWM BEHAVIOUR

While the vast majority of previous research on CWM behaviour had adopted traditional research methods such as questionnaire surveys (Lu et al. 2018), sorting and weighting of on-site CW (Lu et al. 2018), direct observation (Poon et al. 2001) and interview with employees of construction companies (Lu et al. 2018) which are often being criticised for having small samples (Chen & Lu 2017), the application of big data to research on CWM behaviour is still at its infant stage. This section aims at critically evaluating past studies on various areas of CWM behavioural research so as to identify gaps for further research by means of big data analytics.

5.1 *CWM behaviour in general*

CWM performance is most commonly measured by waste generation rate (WGR). In a study undertaken by Lu et al. (2015b), WGRs of a total of 4227 projects belonging to five categories (building, civil, demolition, foundation and "maintenance and renovation" ("M&R")) were calculated using big data, with the top 15% being regarded as "not so good", the bottom 15% being benchmarked as "good" and the remaining 70% in between being marked as "average". It was found that demolition works, with their median inert and non-inert WGRs significantly higher than the other four types, were amongst the most wasteful category of project. Building and M&R projects had generated the least amount of waste, with more non-inert waste than inert waste. The amount of inert waste generated by foundation and civil works were greater than that of non-inert waste. The study also recommended that the government should encourage the use of best CWM practices by awarding contractors performing well in CWM. The major limitation of this study is that no follow-up case studies or interviews had been conducted to further explore the findings from big data analyses.

5.2 *Difference in CWM behaviours between public and private sectors*

Lu et al. (2016) had conducted a comparative study of CWM behaviours of the same pool of contractors engaging in both public and private projects using big data. Results of big data analyses suggested that the private projects had only outperformed the public projects in managing CW generated from building and M&R works whereas the public projects performed better in CWM in all other types of projects, including but not limited to civil, demolition and foundation works. In order to gain a better understanding about the phenomenon of the same pool of contractors behaving differently in public and private projects, case studies by way of participant observations on construction sites and interviews were being conducted. The results from big data analyses were subsequently supplemented by the qualitative data obtained from the case studies which suggested that public projects were subject to higher social scrutiny whereas private projects which emphasised on efficiency and timely completion tend to put less emphasis on CWM. It is also worth noting that the public sector with a large number of projects enabled inert CW generated from a project to be reused in other projects whereas the private sector with less projects needs to incur relatively high cost in searching for prospective customers of inert CW and thus would still prefer simply dumping CW at the government's waste reception facilities. By way of combination of quantitative analysis using big data analytics and qualitative analysis employing follow-up case studies and interviews, this study had discovered the reasons behind the difference in behavioural patterns of the same group of contractors when undertaking projects of different nature.

5.3 *CWM behaviour in demolition activities*

It is commonly understood that demolition waste generation ("DWG") is much affected by the duration of demolition (Yuan et al. 2011), the technologies available (Hendriks & Janssen 2001) and the relevant contractor's capability in waste management (Mcdonald & Smithers 1998). Chen & Lu (2017) suggested that big data can be utilised in substantiating the relationship between DWG in Hong Kong and its attributing factors which can in turn identify those factors with significant impact on DWG. By conducting regression analysis of big data, Chen & Lu (2017) found that DWG which is measured by weight of waste was largely and positively affected by the external factors of contract sum and duration of project. It was also observed that demolition works in the New Territories generated less waste than those on Hong Kong Island and Kowloon which can be attributed to the fact that the amount of land available on Hong Kong Island and Kowloon is much less than that in the New Territories. With scarce land resources and more urgent need of redevelopment, contractors undertaking demolition works on Hong Kong Island and Kowloon are lacking in the incentive to invest resources in CWM. It is also worth noting that private demolition projects generated 10 times more waste than public works which is consistent with the findings of Lu et al. (2016). Since the public sector had outperformed the private sector in CWM, Chen and Lu (2017) opined that public projects should take the lead in environmental protection. Similar to Lu et al. (2015b), this study could be further improved by using follow-up interviews to explore the phenomenon gleaned from big data analyses.

5.4 *Effects of green building rating system on CW minimisation*

In a study conducted by Lu et al. (2018) by way of triangulating big data with qualitative data from case studies and semi-structured interviews with green building ("GB") practitioners, it was found that CWM had strongly influenced the demolition stage of GB projects, but had little effect on foundation and building works which could be explained by the fact that the Hong Kong BEAM Plus, a commonly used green building rating system ("GBRS"), had granted more GB credits to demolition waste reduction items compared with other CWM-related items. It was also observed that the amount which can be saved by reusing C&DW was negligible in comparison to the huge labour cost incurred in managing the waste.

Chen et al. (2018) had conducted another study on the impact of GB on CWM by comparing the WGRs of a group of BEAM-driven buildings with those of a group of ordinary buildings using big data analytics. Similar to the results of Lu et al. (2018), it was observed that Hong Kong BEAM Plus had failed to incentivise CWM behaviour which can be attributed to the low total weighting given to CWM-related items (except those at the demolition stage) as can be seen from the fact that only 2.18% credits had been allocated to waste minimization at the foundation and building stages. Therefore, for the purpose of promoting CWM behaviour in GB projects, Chen et al. (2018) suggested that the existing GBRS under Hong Kong BEAM Plus should be revised by including "foundation waste minimisation" as an additional item with higher weighting. However, unlike Lu et al. (2018), this study had neither supplemented its findings with follow-up case studies nor interviews.

5.5 *Effectiveness of the TTS enhanced in 2004*

The usefulness of the TTS being introduced in 1999 and enhanced in 2004 in preventing illegal dumping of CW is highly questionable. It is worth to note that the number of prosecution cases in relation to illegal dumping on private land had drastically increased from 44 in 2010 to 96 in 2017 (EPD, 2013, EPD 2016, EPD 2018) whereas the total number of public reports of fly-tipping of CW on public land received by various government departments had escalated from 6,153 in 2010 to 10,507 in 2017 (EPD 2013, EPD 2016, EPD 2018).

According to Lu (2019) who had applied big data analytics to records of dump trucks compiled by the EPD and Civil Engineering and Development Department to develop a model on predicting illegal dumping behaviour, it was observed that long queuing time at the government's waste reception facilities was a major factor contributing to fly-tipping and freelance waste haulers were more prone to commit illegal dumping behaviours. This study was among the first attempt to apply big data analytics in developing a model for crime detection in Hong Kong. However, the number of offending trucks being used to train the model was only six which was less than sufficient and more offending trucks should have been used.

6 DISCUSSION

Based on the literature discussed in Section 5, certain gaps for further studies have been identified which will be further elaborated in this section.

Both studies on the effects of GBRS on waste minimisation have observed that the low weighting of CWM-related items (except those at the demolition stage) under the prevalent GBRS of Hong Kong BEAM Plus was the major barrier to promotion of better CWM practices in GB projects. It is therefore suggested that future studies should be conducted to explore in detail the relative potential of each CWM-related item (e.g. "prefabrication" which has a long history of adoption in public projects) in contributing to waste reduction so as to identify the particular CWM-related item(s) whose weighting should be increased in the long run.

In relation to the problem of fly-tipping, in addition to conducting big data analyses based on dump trucks' records, it is worthwhile to explore the opportunity costs of committing illegal dumping behaviours as well as the behavioural patterns of offenders other than dump truck drivers such as logistics companies who had helped contractors to dispose CW in unauthorised areas.

It is commonly known that big data analytics has a relatively long history of application in the marketing industry, in particular the analyses of consumer behaviours. Although it appears that marketing and construction are two entirely different and unrelated disciplines, it is believed that certain concepts in marketing behavioural analytics could still be applied to academic research on CWM, especially in relation to future analyses on the government's efforts in launching and promoting CWM practices in Hong Kong. As recommended by Lu et al. (2015b), the government is obligated to take the lead in protecting the environment by using award schemes to incentivise contractors to adopt better CWM practices.

Since big data sometimes does include personal data which is subject to the protection of the Personal Data (Privacy) Ordinance of Hong Kong, special care should also be taken in handling information with personal data in order to avoid data leakage and any kind of unauthorised use.

7 CONCLUSION

Although academic research on CWM using big data is still at its infant stage, previous studies on the different behavioural patterns exhibited by the same group of contractors undertaking projects of different nature, CWM of demolition activities which are among the most wasteful type of project in Hong Kong, effects of GBRS on CWM behaviour as well as illegal dumping have already been conducted. It is also worth noting that big data analyses seldom standalone – most previous researches tended to use qualitative data obtained from follow-up interviews to explain the phenomena identified from big data analyses. This paper has also identified certain gaps for further research, including the future directions in conducting research on CWM in GB projects, fly-tipping as well as promotion of better CWM practices. It is believed that the concerted endeavours of different stakeholders, including the government, contractors, construction practitioners ranging from frontline staff to senior management as well as the academia, are of utmost importance to the long-term improvement of CWM behaviour in Hong Kong.

REFERENCES

Audit Commission. 2016. *Management of abandoned construction and demolition materials, Chapter 4 of the Director of Audit's Report No. 67.*

Big Data and Framework. 2019. *The Four V's of Big Data: The 4 Characteristics of Big Data.*

Census and Statistics Department (C&SD). 2017. *Building, Construction and Real Estate Sectors.*

Chen, X. & Lu, W. 2017. Identifying factors influencing demolition waste generation in Hong Kong. *Journal of Cleaner Production* 141: 799–811.

Chen, X. Lu, W. Xue, F. & Xu, J. 2018. A cost-benefit analysis of green buildings with respect to construction waste minimization using big data in Hong Kong. *Journal of Green Building* 13(4): 61–76.

Environmental Protection Department (EPD). 2013. *Measures to tackle fly-tipping of construction and demolition waste and illegal land filling.*

EPD. 2015. *What is construction waste?*

EPD. 2016. *Actions to combat illegal land filling and fly-tipping of construction and demolition waste.*

EPD. 2017a. *Construction Waste Disposal Charging Scheme.*

EPD. 2017b. *Monitoring of solid waste in Hong Kong: Waste statistics for 2015.*

EPD. 2018. *Enforcement Against Illegal Land Filling and Fly-tipping of Construction Waste.*

Hao, J.L. Hills, M.J. & Tam, V.W.Y. 2008. The effectiveness of Hong Kong's Construction Waste Disposal Charging Scheme. *Waste Management & Research* 26(6): 553–558.

Hendriks, C.F. & Janssen, G.M.T. 2001. Construction and demolition waste: General process aspects. *HERON* 46(2): 0046–7316.

Lu, W. & Tam, V.W.Y. 2013. Construction waste management policies and their effectiveness in Hong Kong: A longitudinal review. *Renewable & Sustainable Energy Reviews* 23(C): 214–223.

Lu, W. & Yuan, H. 2012. Off-site sorting of construction waste: What can we learn from Hong Kong? *Resources, Conservation & Recycling* 69(C): 100–108.

Lu, W. & Yuan, H. 2013. Investigating waste reduction potential in the upstream processes of offshore prefabrication construction. *Renewable and Sustainable Energy Reviews* 28(C): 804–811.

Lu, W. 2013. *Construction waste – Hong Kong Style.*

Lu, W. 2019. Big data analytics to identify illegal construction waste dumping: A Hong Kong study. *Resources, Conservation & Recycling* 141: 264–272.

Lu, W. Chen, X. Ho, D.C.W. & Wang, H. 2016. Analysis of the construction waste management performance in Hong Kong: the public and private sectors compared using big data. *Journal of Cleaner Production* 112 (P1):521–531.

Lu, W. Chen, X. Peng, Y. & Liu, X. 2018. The effects of green building on construction waste minimization: Triangulating 'big data' with 'thick data'. *Waste Management* 79: 142–152.

Lu, W. Chen, X. Peng, Y. & Shen, L. 2015b. Benchmarking construction waste management performance using big data. *Resources, Conservation & Recycling* 105(PA): 49–58.

Lu, W. Chi, B. Bao, Z. & Zetkulic, A. 2019. Evaluating the effects of green building on construction waste management: A comparative study of three green building rating systems. *Building and Environment* 155: 247–256.

Lu, W. Peng, Y. Webster, C. & Zuo, J. 2015a. Stakeholders' willingness to pay for enhanced construction waste management: A Hong Kong study. *Renewable and Sustainable Energy Reviews* 47: 233–240.

Mayer-Schönberger, V. & Cukier, K. 2013. *Big data: a revolution that will transform how we live, work, and think.* Boston: Houghton Mifflin Harcourt.

McAfee, A. & Brynjolfsson, E. 2012. Big Data: The Management Revolution. *Harvard Business Review* 90(10): 60–68.

Mcdonald, B. & Smithers, M. 1998. Implementing a waste management plan during the construction phase of a project: a case study. *Construction Management and Economics* 16(1): 71–78.

Poon, C.S. Yu, A.T.W. & Ng, L.H. 2001. On-site sorting of construction and demolition waste in Hong Kong. *Resources, Conservation & Recycling* 32(2): 157–172.

Yuan, H. Lu, W. & Hao, J.J. 2013. The evolution of construction waste sorting on-site. *Renewable and Sustainable Energy Reviews* 20(C): 483–490.

Yuan, H. Shen, L. Hao, J.J. & Lu, W. 2011. A model for cost–benefit analysis of construction and demolition waste management throughout the waste chain. *Resources, Conservation & Recycling* 55(6): 604–612.

Sustainable Buildings and Structures: Building a Sustainable Tomorrow – Papadikis et al. (Eds)
© 2020 Taylor & Francis Group, London, ISBN 978-0-367-43019-1

Construction waste cross jurisdictional trade under PESTEL context

Z. Bao, W. Lu & B. Chi
Department of Real Estate and Construction, Faculty of Architecture, University of Hong Kong, Pokfulam, Hong Kong

C.S. Chin & J. Hao
Department of Civil Engineering, Xi'an Jiaotong-Liverpool University, Suzhou, China

ABSTRACT: The increasing consciousness of various detrimental effects brought about by construction waste calls for successful construction waste management, which needs to properly manage the inert portion of construction waste, termed as construction waste material (CWM). One desirable strategy to manage CWM is to boost CWM cross jurisdictional trade. To better govern the CWM cross jurisdictional trade, the authors in this paper propose a central argument that CWM trading must consider different political, economic, social, technological, environmental, and legal (PESTEL) scrutiny before the implementation to achieve a win-win situation. This paper also lists a series of PESTEL factors that should be taken into scrutiny although the factors may not be exhaustive enough. The research deliverables of this paper could provide a significant reference for jurisdictions that have been long suffering from the ingrained CWM issues, and further the progress to tackle the general construction waste issues could move a leap forward.

1 INTRODUCTION

Construction is widely regarded as a pillar industry not only for it materializing the prosperity of the built environment but also maintaining the national economy. However, construction also receives its fair share of criticism. Construction by nature is hostile to the natural environment. It can lead to a series of negative impacts, including land depletion and deterioration, energy consumption, solid waste generation, dust and gas emission, noise pollution and consumption of non-renewable natural resources (Shen et al. 2007, Lu & Yuan 2011). Amid the contributors leading to the environmental degradation, the solid waste generated from construction activities, termed as construction waste, has raised worldwide attention due to its prodigious volume and potential adverse effects on the environment if not properly managed.

Construction waste, or sometimes called construction and demolition (C&D) waste, refers to the solid waste arising from construction, renovation and demolition activities (Roche & Hegarty 2006, Lu et al. 2019). Every year, a huge amount of C&D waste is generated due to mounting construction activities for urbanization and urban renewal (Lu et al. 2011). Consequently, the issues arising from C&D waste calls for construction waste management around the world. In such economies as the UK, Hong Kong, Singapore, Malaysia, and Australia, construction waste is normally classified as either inert or non-inter depending on whether or not it has stable chemical properties (EPD 1998). For non-inert construction waste, such as timber, bamboo, vegetation, paper, or plastics, the common practice for dealing with it is to dispose it into landfills (Lu et al. 2011). It is a type of solid waste. Landfilling not only gives rise to negative social-economic impacts, but also leads to environmental degradation due to anaerobic decay of the disposed materials and thus the production of carbon dioxide, methane and leachate (Lu et al. 2015). It also rapidly exhausts invaluable landfill space for compact cities. Contrastingly, inert construction waste, such as soil, earth, slurry, rocks and broken concrete account for the vast majority of all construction waste. In Hong Kong, the inert portion accounts for 93% of all construction waste in 2017, more than ten times as non-inert portion (EPD 2018). Arguably, the key to successful construction waste management is to properly manage the inert waste due to their predominant quantity. The inert portion usually as public fill can be used for different purposes, such as recovery and recycling into rocks, or other aggregate for concrete, site formation, and land reclamation. Therefore, they are called "construction waste materials (CWM)". It is the focus of this study.

"Waste material is just a misplaced resource" (Kaseva & Gupta 1996). Reducing, reusing, or recycling (also known as "3R") inert waste is the typical ways to manage the CWM. In addition to producing the recycled materials, sharing CWM amongst different sites/projects is often promoted. However, the sharing cases are only few and far between. The buffer of a municipality is often too small. The windows of opportunity for one site to generate CWM and another site to receive CWM (e.g. for backfilling), even with proper compensation, often mismatch. People start to turn to bigger areas, e.g. multiple municipalities, hoping they will provide more opportunities to share/trade due to little buffer. On the one hand, some regions cannot fully consume its generated CWM through land reclamation, site formation, or the recycling industry. Consequently, the rapid build-up of huge amounts of inert CWM has been a looming crisis since finding sufficient public fill space to contain it is extremely burdensome particularly for some compact cities like Hong Kong. On the other hand, some regions may need vast amounts of inert CWM as filling materials for site formation or land reclamation under some circumstances. If the generated inert CWM within the region is short of supply, they will turn to purchase virgin filling materials or other sources. Considering the aforementioned dilemmas, sharing/trading cross different regions is elevated as a desired strategy. Given the fact that regions here are the entities to have the official power to make legal decisions and judgements, they are called "jurisdictions" and that is the reason why this study is termed as "cross-jurisdictional CWM trading".

Actually, there already have been some *ad hoc* CWM cross jurisdictional trade/sharing cases around the world. For example, in China, Shenzhen disposed around 20% of its CWM to nearby cities in 2015. Hong Kong has transported its CWM to Taishan China for land reclamation over the past years. Besides, in Singapore, all of its CWM is transported to Semakau, an island located 8 km south of Singapore. However, the current extent of CWM cross jurisdictional trade is still too modest, which cannot fully overcome the long-standing CWM problems as aforementioned. The hindrance to CWM cross jurisdictional trade is a lack of such a systematic way to govern it, which is the research gap this study intends to fill. Therefore, in this paper, the central argument the authors would like to propose in this paper is that when it comes to different jurisdictions, CWM sharing/trading must consider different political, economic, social, technological, environmental, and legal (PESTEL) scrutiny. If it can pass the PESTEL scrutiny, then the trade is worth of encouraging. The research deliverables are of benefits for jurisdictions that have been long suffering from the ingrained CWM issues around the globe to tackle the dilemmas so that a win-win situation could be achieved between the jurisdictions of importing and exporting CWM.

2 RESEARCH METHODS

This paper is mainly to propose a central argument that CWM cross jurisdictional trade should be governed by a PESTEL scrutiny to achieve a win-win situation and thus PESTEL is inherently the analytical method this paper adopts. By definition, PESTEL analysis is an acronym for political, economic, social, technological, environmental, and legal and it is widely used as a tool to analyse and monitor the macro environmental factors that may have a profound impact on an organisation's performance (B2U 2016). The original form of PESTEL was first created by Aguilar as ETPS (economic, technical, political, and social) (Aguilar 1967). Then Arnold Brown Institute of Life Insurance used it in strategic evaluation of trends in a form of STEP and subsequently, it was further modified as STEPE and employed for the analysis of the external environment or scanning for environment change (Yüksel 2012). Finally, the legal dimension was added into this method in the 1980s (Richardson 2006), which represents the formal advent of PESTEL analysis at this moment.

At the early stage, PESTEL was firstly widely applied in the marketing since before the implementation of any kind of strategy or tactical plan, it is essential to conduct a situational analysis (Oxford 2016). With the increasing awareness of its usefulness, it was widely applied in other fields (Katko 2006, Richardson 2006, Shilei & Yong 2009). Particularly for this study, the authors would like to argue that before any implementation of CWM cross jurisdictional trade, a PESTEL scrutiny will be conducted to guide the decision making so that a win-win situation could be achieved rather than the predatory dumping of CWM.

3 PESTEL SCRUTINY FOR CWM CROSS JURISDICTIONAL TRADING

Generally, for CWM cross jurisdictional trade, PESTEL scrutiny should fully consider the situations in both the exporter and importer of CWM, which can be illustrated in Figure 1. As shown in Figure 1, for either the exporter or importer, the PESTEL, five dimensions are inter-related and interacted to impact the CWM cross jurisdictional trading. Furthermore, the PESTEL conditions in exporter will affect the PESTEL conditions in importer, and vice versa. Additionally, the process is in a dynamic manner, which will change from time to time. Therefore, the process of PESTEL scrutiny for CWM cross jurisdictional trading is rather sophisticated, and thus the authors will not discuss how the PESTEL will be conducted due to the space constrain. Instead, the authors would like to elaborate what exactly political, economic, social, technological, environmental, and legal factors should be taken into consideration when talking about CWM cross jurisdictional trade to achieve the objective of a win-win situation in the following space.

Based on the literature review and the authors' understandings, a list of PESTEL factors which the authors conceive should be takin into scrutiny before the implementation of CWM cross jurisdictional trade are summarized below (See Table 1) based on literature review, which will be also elaborated individually.

Political factors: Political factors will ascertain the extent to which jurisdictional government and government policy have impact on CWM trade. For example, if the jurisdiction stability is not satisfactory, the CWM trade may not be able to proceed towards the initial objective. Also, the special tariff is too high charged for the trade, it may hinder the boost of the CWM trade between the jurisdictions. Similarly, other political factors listed in the Table 1 include government involvement in trade unions and agreements, bilateral/multilateral relationship and trade control.

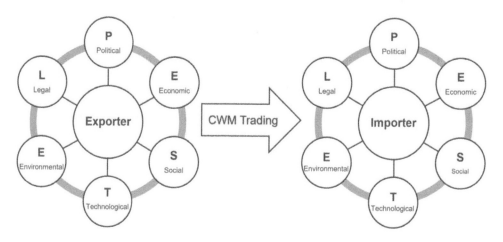

Figure 1. CWM trading under PESTEL contexts.

Table 1. PESTEL factors for CWM cross jurisdictional trade.

PESTEL	Factors
Political factors	Jurisdiction stability; Special tariffs; Government involvement in trade unions and agreements; Bilateral/multilateral relationship; Trade control
Economic factors	Gross domestic product trend; Interest rate; Inflation rate; Jurisdictional government deficits; Availability of credit; Unemployment trend
Social factors	Health consciousness; Attitude towards jurisdictional government; Ethical concerns; Racial equality
Technological factors	Technological change; Access to new technology; Level of innovation; R&D activity; Technology incentives; Technological awareness
Environmental factors	Environmental policies; air and water pollution; climate change; Recycling standards; Support for waste minimisation
Legal factors	CWM disposal laws; Health and safety laws; Discrimination laws; Employment laws; CWM import-export laws

Source (Song et al. 2017, Bing et al. 2016, Ziout et al. 2014)

Economic factors: Economic factors may impact on the economy and its performance, which will affect the jurisdiction in turn. For example, a series of factors in Table 1, such as gross domestic product trend, interest rate, inflation rate, and jurisdictional government deficits will help estimate whether the importer truly needs the traded CWM to earn economic benefits. Also, another economic factor, availability of credit, will help evaluate whether the exporter will execute the agreements, e.g. compensation if it exists.

Social factors: Social factors can determine the social environment and emerging trends. For example, if the health consciousness of the public in the jurisdiction importing the CWM is very high, the CWM trade could receive a lot of protest, which may hinder the CWM trade. Other social factors like ethical concerns and racial equality also cares whether the application of traded CWM is ethical as the unethical phenomena do exist in some jurisdictions where the traded waste is only applied to where the majority of residents nearby are people of colour.

Technological factors: Technological factors consider impact of the current technology level and its development on the CWM cross jurisdictional trade. For example, a list of factors in Table 1, technological change, access to new technology, level of innovation, R&D activity, technology incentives, and technological awareness will help evaluate whether the level of technology in the jurisdiction of importing the CWM has the ability or potential to eliminate the adverse effects brought about by the traded CWM.

Environmental factors: Environmental factors focus on the influence of surrounding environment. These factors, such as air and water pollution and climate change could help assess whether the environment of the jurisdiction importing the CWM has enough resilience to bear the environmental degradation caused by the traded CWM. Other factors include environmental policies and support for waste minimisation as seen in Table 1.

Legal factors: Legal factors determine the legislations in both jurisdictions allow the CWM cross jurisdictional trade and their impact. For example, the most widely recognized legislation internationally is the Basel Convention. If the traded CWM contains the composition where the Basel Convention bans by explicit ordinances, the trade must be forbidden as it is illegitimate. Besides the import-export laws, other laws, such as CWM disposal laws, health and safety laws, discrimination laws, employment laws are also vital to determine the legitimacy of the trade.

4 CONCLUSIONS

Various issues arising from construction waste call for the successful construction waste management. Arguably, the key to successful construction waste management is to properly manage the inert portion of construction waste, termed as construction waste material (CWM) due to the predominant quantity in contrast to the non-inert portion. One desirable strategy of managing CWM is to boost CWM cross jurisdictional trade considering the facing dilemmas. However, the current extent of CWM cross jurisdictional trade is still too modest, which is far from tacking the dilemmas. The reason behind is a lack of systematic way to govern it. Consequently, the authors in this paper propose a central argument to help govern it that CWM trading must consider different political, economic, social, technological, environmental, and legal (PESTEL) scrutiny before the implementation to achieve a win-win situation. If it can pass the PESTEL scrutiny, then the trade is worth of encouraging. This paper also lists a series of PESTEL factors that should be taken into scrutiny although the factors may not be exhaustive enough. The final decision on whether the CWM trade is encouraged would be a comprehensive evaluation of PESTEL factors and how exactly to evaluate it will be the research focus in the future study. The research deliverables of this paper could provide a significant reference for jurisdictions that have been long suffering from the ingrained CWM issues around the globe to tackle the dilemmas, and further the progress to tackle the general construction waste issues could move a leap forward.

REFERENCES

Aguilar, F. J. 1967. Scanning the business environment. Macmillan.

Bing, X. Bloemhof, J.M. Ramos, T.R.P. Barbosa-Povoa, A.P. Wong, C.Y. & van der Vorst, J.G. 2016. Research challenges in municipal solid waste logistics management. *Waste management* 48: 584-592.

Business-To-You. 2016. Scanning the Environment: A PESTE Analysis. [Online], https://www.business-to-you.com/scanning-the-environment-pestel-analysis/(Access 12 July 2019)

EPD. 1998. Monitoring of Solid Waste in Hong Kong. Environmental Protection Department, Hong Kong. [Online], https://www.wastereduction.gov.hk/sites/default/files/msw1998.pdf (Access 12 July 2019)

EPD. 2018. Monitoring of Solid Waste in Hong Kong. Environmental Protection Department, Hong Kong. [Online], https://www.wastereduction.gov.hk/sites/default/files/msw1998.pdf (Access 12 July 2019)

Kaseva, M.E. & Gupta, S. K. 1996. Recycling-an environmentally friendly and income generating activity towards sustainable solid waste management. Case study—Dar es Salaam City, Tanzania. *Resources, Conservation and Recycling* 17(4): 299-309.

Katko, T. S. 2006. Road safety fatalities, management, and policy in Finland, 1970-2003. Public Works Management & Policy 11(2): 126-138.

Lu, W. & Yuan, H. 2011. A framework for understanding waste management studies in construction. *Waste Management* 31(6):1252:1260

Lu, W. Chen, X. Peng, Y. & Shen, L. 2015. Benchmarking construction waste management performance using big data. *Resources, Conservation and Recycling* 105: 49-58.

Lu, W. Chi, B. Bao, Z. & Zetkulic, A. 2019. Evaluating the effects of green building on construction waste management: A comparative study of three green building rating systems. *Building and Environment* 155: 247-256.

Lu, W. Yuan, H. Li, J. Hao, J. J. L. Mi, X. & Ding, Z. 2011. An empirical investigation of construction and demolition waste generation rates in Shenzhen city, South China. *Waste Management* 31(4): 680-687.

Oxford College of Marketing. 2016. What is a PESTEL analysis? [Online], https://blog.oxfordcollegeof marketing.com/2016/06/30/pestel-analysis/ (Access12 July 2019)

Richardson Jr, J.V. 2006. The library and information economy in Turkmenistan. *IFLA journal* 32(2):131-139.

Roche, T. & Hegarty, S. 2006. Best practice guidelines on the preparation of waste management plans for construction and demolition projects. *Department of the Environment, Community and Local Government*: Dublin, Ireland.

Shen, L. Y. Li Hao, J. Tam, V.W.Y. & Yao, H. 2007. A checklist for assessing sustainability performance of construction projects. *Journal of Civil Engineering Management* 13(4): 273-281.

Shilei, L. & Yong, W. 2009. Target-oriented obstacle analysis by PESTEL modeling of energy efficiency retrofit for existing residential buildings in China's northern heating region. *Energy Policy* 37(6): 2098-2101.

Song, J. Sun, Y. & Jin, L. 2017. PESTEL analysis of the development of the waste-to-energy incineration industry in China. *Renewable and Sustainable Energy Reviews* 80: 276-289.

Yüksel, I. 2012. Developing a multi-criteria decision making model for PESTEL analysis. *International Journal of Business and Management* 7(24): 52.

Ziout, A. Azab, A. & Atwan, M. 2014. A holistic approach for decision on selection of end-of-life products recovery options. *Journal of Cleaner Production* 65: 497-516.

Sustainable Buildings and Structures: Building a Sustainable Tomorrow – Papadikis et al. (Eds)
© 2020 Taylor & Francis Group, London, ISBN 978-0-367-43019-1

A new strategy framework for construction waste recycling in China

H. Guo & W. Lu
Department of Real Estate and Construction, Faculty of Architecture, The University of Hong Kong,
Hong Kong

ABSTRACT: Construction waste recycling (CWR) is a key focus within academia and industry, as it is directly related to the effectiveness of construction waste management (CWM), and to a great extent, affects the sustainable development of green buildings and the environment protection. To support practical and feasible waste recycling strategies for governments, questionnaire survey was conducted in Sichuan, China, and factor analysis technique was used to extract the critical factors for waste recycling activities. Results show that the factors can be condensed into four clusters, including policy cluster, economy cluster, technology cluster, and awareness cluster, according to which a new strategy framework for CWR is proposed for policy recommendation.

1   INTRODUCTION

With the growing of urban population and quickening of urbanization worldwide, the construction industry produces an increasingly huge amount of construction waste each year. Nowadays, the industry has become one of the largest solid waste producers worldwide (Lu & Yuan 2010). According to statistics, there is 40% of the municipal solid waste generation coming from the construction industry, which causes serious damage of land and groundwater, also greatly affects the sustainability of the environment. However, only 5% of waste generation can be recycled in China, while more than half of waste is recycled and reused in developed countries such as the US, Germany, Netherlands, and Japan. In this situation, how to improve construction waste management (CWM) has become an urgent problem.

Construction waste recycling (CWR), as one of the most important approaches for CWM, has drawn attention from researchers for decades. Contractors have been focused as the principal of CWM for directly waste production, while the collaborative nature of stakeholders has been ignored, which can cause the insignificant effectiveness and efficiency of CWM.

Much of the earlier work of CWM considered contractor-centered point of view. The pioneering study of Lu & Yuan (2010) proposed some critical factors of CWM, based on which, researchers have further suggested that to improve waste management in the construction industry, contractors should take effective measures to reduce waste generation and reuse waste on-site. Yuan et al. (2011) noted that waste management behaviors of contractors could help improve their economic benefits in the construction process. Similarly, Marzouk & Azab (2014) encourage contractors to reuse and recycle waste by evaluating a range of CWM methods. Apart from encouraging contractors to conduct waste management, much research has also focused on their different management strategies. For example, waste minimization contracts, waste segregation, maximization reuse of sustainable materials, efficient logistic management (Ajayi et al. 2017). In addition, factors which can have an influence on CWM behaviors include economic viability and governmental supervision (Wu et al. 2017). The chief focus of these studies is on how to encourage and help contractors perform better on construction sites, but most of them have ignored an essential fact that only good behaviors of contractors cannot achieve the optimal waste management.

In recent years, some studies, however, have taken a different view by paying attention to stakeholders-based CWM. In an attempt to go beyond the traditional perspective, researchers have considered other stakeholders sharing the responsibility of CWM. Wang et al. (2015) hold the view that waste minimization design as one of the key strategies for effective waste management should be highly valued by decision makers. Following his lead, Li et al. (2015) demonstrated that designers' attitude could influence waste minimization effect in the long-term. Furthermore, Bakshan et al. (2017) raised the awareness of construction stakeholders (such as owners, governments, and even social pressure) who have influence on CWM.

The 3R principle (reduction, recycle and reuse") is being promoted globally. However, compared with waste reduction, CWR research is still in its infancy. In order to support practical and feasible waste recycling strategies for governments, this study considers stakeholders-based CWM and uses factor analysis technique to extract the critical factors for waste recycling activities based on a survey study in Sichuan, China.

2 THEORETICAL MODEL

According to the characteristics of CWM, the behaviours of various stakeholders are not independent, but can influence and interact with each other to determine the effect of CWR. Based on the features of waste management in China, the theoretical model of CWR is constructed as shown in the Figure 1.

In this model, governments act as the role of guiding CWR activities of other units including project owners, designers, contractors, and recyclers, among which, owners as investors who own authority to manage designers and contractors, while recyclers are viewed as a third-party who are responsible for waste recycling. All the stakeholders can act as a whole to generate a comprehensive impact on CWR.

Figure 1. Theoretical model for CWR.

3 RESEARCH METHOD

3.1 Questionnaire survey

The questionnaire survey was formed based on the Likert scale system. For the CWR activity in China, each statement was provided with five alternative answers: strongly agree, agree, not sure, disagree and strongly disagree, represented with 5, 4, 3, 2, 1 scores, respectively. It can be seen from Table 1 that 13 factors affecting the recycling of waste are listed. The questionnaire was then set out in April 2017 by site investigation in Sichuan, China, and 108 valid responses, the valid rate was 72.00%. As supported by the literature (Peterson 1994), Cronbach's coefficient alpha was tested to determine the internal consistency among the factors, the value of Cronbach's α is 0.647 (>0.5), showing an acceptable level for further analysis.

3.2 Factor analysis technique

The technique of factor analysis can be used to reduce or regroup individual factors identified from a larger number to a smaller and more critical set (Jiang & Wong 2016). In this study, 13 factors were entered into SPSS 19.0 for factor analysis. The results show that the value of KMO is 0.683, and the significance level of Bartlett test of Sphericity is 0.000, suggesting that the population correlation matrix is not an identity matrix and the sample is acceptable for factor analysis (Kaiser 1974).

Table 1. Selected factors for waste recycling.

Stakeholders	Code	Activities
Government	F1	Policy support for recycling
	F2	Fund support for recycling
	F3	Punishment to illegal dumping
	F4	Landfill charge
Owner	F5	Environmental awareness
	F6	Waste recycling management willingness
Designer	F7	Green design concept
	F8	Green design capability
Contractor	F9	Waste management plan
	F10	Division of responsibility for waste management
	F11	Waste on-site sorting
Recycler	F12	Market share of recycler
	F13	Recycling costs and market demand for recycled materials

4 RESULTS AND DISCUSSIONS

4.1 *Data analysis process and results*

The statistical analysis in the previous section has led to the identification of 13 factors for CWR activities, in fact, many factors in Table 1 are interrelated, factor analysis was conducted in order to lower the interrelation of factors and to obtain a "concise" list of critical factor clusters (Lu et al. 2008). After deleting two factors (F4, F8) for their lower communalities (<0.70), the principal component analysis generated a four-factor solution with eigenvalues larger than 1.0, explaining 85.71% of the variance. Then, the factor grouping based on the varimax rotation is shown in Table 2. According to the suggestion by previous researchers (King & Teo 1996, Lu et al. 2008), only the factors with loading greater than 0.50 were selected to evaluate the factor patterns. It can be noticed that 10 factors are selected except F11 for its repeated distribution to two categories, and for the other factors, each of them belongs to only one of the four clusters generated by factor analysis.

4.2 *Discussions*

For further interpretation, each cluster shown in Table 2 will be given a name, which can be seen as a critical factor (CF) comprised of a cluster of CWR factors. The four clusters are policy (CF1), economy (CF2), technology (CF3), and awareness (CF4). Furthermore, the CFs can be summarized in a new strategy framework for CWR as shown in Figure 2.

Table 2. Results of factor analysis.

Cluster	CFs	Factors
1	Policy	F1 Policy support for recycling
		F12 Market share of recycler
2	Economy	F2 Fund support for recycling
		F3 Punishment to illegal dumping
		F9 Waste management plan
3	Technology	F10 Division of responsibility for waste management
		F13 Recycling costs and market demand for recycled materials
		F5 Environmental awareness
4	Awareness	F6 Waste recycling management willingness
		F7 Green design concept

Figure 2. Strategy framework for CWR.

Strategy 1: Policy
In fact, the quality of recycled materials is in full compliance with most engineering standards, and their environmental indicators are much higher than natural materials. However, the demand for recycled materials does not possess a certain degree of market share. In this circumstance, strategies are recommended to carry out quality certification for recycled materials, so as to encourage the use of sustainable construction materials.

Strategy 2: Economy
Subsidies are one of the most effective incentives for waste recycling. In order to promote the development of waste recycling market, economical strategies are suggested to issue subsidies for projects that fulfill recycling requirements. For these construction enterprises, subsidies can include tax reductions, financing offer, and other preferential treatments. Meanwhile, penalty for illegal dumping can be increased to restrain the behavior of contractors.

Strategy 3: Technology
Construction is the process that determine whether waste can be recycled, waste management plans, on-site waste sorting, and other activities are critical to CWR. Thus, technological strategies are recommended to strengthen waste management training for related enterprises, clarify the responsibility division for waste management, and develop effective waste management plans.

Strategy 4: Awareness
Construction enterprises with good green image is not only the requirements of external environment, but also a significant approach to improve internal competitiveness. Compared with owners and contractors, the impact of designers on waste recycling activities has always been ignored. In fact, the importance of green design has been discussed by previous studies. Governments may consider establishing a corporate green image evaluation system and providing public with a green image standard for waste recycling.

5 CONCLUSIONS

This research shows that critical strategies affecting CWR in terms of policy, economy, technology, and awareness. Specifically speaking, they are policy support for waste recycling, economic measures for waste recycling (both subsidy and penalty), technologic means for waste recycling, and green awareness for waste recycling. The study provides governments with reference for CWR policies, which can not only promote the development of CWR, but also affect the sustainable development of green buildings and the environment protection in Sichuan, China. For future direction, the model proposed can also be applied in other regions of China for CWR strategy development, and further studies are suggested to verify these strategies through simulation techniques so that more reliable policy recommendations can be ensured.

REFERENCES

Ajayi, S. O. Oyedele, L.O. Bilal, M. Akinade, O.O. Alaka, H.A. & Owolabi, H.A. 2017. Critical management practices influencing on-site waste minimization in construction projects. *Waste Management* 59: 330-339.

Bakshan, A. Srour, I. Chehab, G. El-Fadel, M. & Karaziwan, J. 2017. Behavioral determinants towards enhancing construction waste management: A Bayesian Network analysis. *Resources, Conservation and Recycling* 117: 274-284.

Jiang, W. & Wong, J. K. 2016. Key activity areas of corporate social responsibility (CSR) in the construction industry: a study of China. *Journal of Cleaner Production* 113: 850-860.

Kaiser, H. F. 1974. An index of factorial simplicity. *Psychometrika* 39: 31-36.

King, W. R. & Teo, T. S. 1996. Key dimensions of facilitators and inhibitors for the strategic use of information technology. *Journal of Management Information Systems* 12: 35-53.

Li, J. Tam, V.W. Zuo, J. & Zhu, J. 2015. Designers' attitude and behaviour towards construction waste minimization by design: A study in Shenzhen, China. *Resources, Conservation and Recycling* 105: 29-35.

Lu, W. Shen, L. & Yam, M.C. 2008. Critical success factors for competitiveness of contractors: China study. *Journal of Construction Engineering and Management* 134: 972-982.

Lu, W. & Yuan, H. 2010. Exploring critical success factors for waste management in construction projects of China. *Resources, Conservation and Recycling* 55: 201-208.

Marzouk, M. & Azab, S. 2014. Environmental and economic impact assessment of construction and demolition waste disposal using system dynamics. *Resources, Conservation and Recycling* 82: 41-49.

Peterson, R. A. 1994. A meta-analysis of Cronbach's coefficient alpha. *Journal of Consumer Research* 21: 381-391.

Wang, J. Li, Z. & Tam, V.W. 2015. Identifying best design strategies for construction waste minimization. *Journal of Cleaner Production* 92: 237-247.

Wu, Z. Ann, T. & Shen, L. 2017. Investigating the determinants of contractor's construction and demolition waste management behavior in Mainland China. *Waste Management* 60: 290-300.

Yuan, H. Shen, L. Hao, J.J. & Lu, W. 2011. A model for cost–benefit analysis of construction and demolition waste management throughout the waste chain. *Resources, Conservation and Recycling* 55: 604-612.

Sustainable Buildings and Structures: Building a Sustainable Tomorrow – Papadikis et al. (Eds)
© 2020 Taylor & Francis Group, London, ISBN 978-0-367-43019-1

Developing information requirements for BIM-enabled quantity surveying practice: An adaptation approach

J. Wang & W.S. Lu
Department of Real Estate and Construction, The University of Hong Kong, Pokfulam, Hong Kong

ABSTRACT: Numerous studies have reported the benefits of building information modelling (BIM) to enhance the accuracy and efficiency of quantity surveying (QS) practice. Yet, its real-life application is largely limited. A significant barrier lies in the formulation of BIM information requirements to facilitate evaluate/enrich designers' BIM product for estimation. This paper aims to propose a process model to formulating QS BIM information requirements. It advocates the adaptation approach to reference and adapt the prevailing BIM guidelines for local use to ensure the guidelines' development efficiency and local applicability. It does so by presenting the findings of a case study investigating the BIM adoption practice of a Hong Kong QS consultant. The study outcomes will inform QS firms, especially the early adopters of a locality, to devise BIM strategies and enhance BIM adoption to a higher maturity level.

1   INTRODUCTION

Numerous studies have reported the benefits of building information modeling (BIM) for quantity surveying (QS) tasks. With a visualized representation of design information, BIM facilitates understanding the designer's intentions and preparing the conceptual estimate (Matipa et al. 2008). BIM integrates information from multiple professionals into a single platform, providing more reliable, detailed and organized design information for QS than that in 2D drawings (Kaner et al. 2008, Arayici et al. 2011, Monteiro & Martins 2013). Embedded with predefined rules, BIM has the potential to automatically produce quantity takeoff, cost estimations, and bills of quantities (Sabol 2008, Ma et al. 2011, Wu et al. 2014), and timely updated the design documents based on design changes along with the project process (Raphael & Priyanka 2014). In a nutshell, BIM can offer tremendous opportunities for quantity surveyors to simplify, speed up, automatize, and perhaps the most importantly, increase the accuracy of the quantity surveying tasks.

Despite many benefits, the adoption of BIM in real-life applications is largely lagging. The barriers can be multifaceted, such as the lack of training, software toolkits and so on. This paper focused on two critical barriers. First, there lacks a comprehensive set of rules to inform how BIM-enabled QS practice can be unfolded (Wu et al. 2014), e.g., to govern its procedures, deliverables, and collaboration among different professionals (Lu et al. 2018). The other barrier is rooted in the relationship and collaboration between quantity surveyors and other designers. Using BIM for QS requires upfront collaborations between quantity surveyors and designers, e.g., providing more detailed design information, inputting the correct coding, zoning and ensuring that each BIM object contains essential information for quantity take-off. The reality is that quantity surveyors cannot clearly convey their requirements to designers, let alone having such a BIM handed over for subsequent QS tasks.

This has given rise to the need for a set of clearly-defined requirements to support BIM-enabled QS practice. While the QS professionals are shifting towards a BIM-based practice, i.e., estimate based on geometric and non-geometric information in BIM, they need to ask for a BIM product containing specific information from designers, or modify the designer's BIM to the specific QS requirements by themselves. To ensure a smooth BIM process, it is important to have a set of BIM requirements with the appropriate amount of details and preferable presentation formats to specifically target QS practice (Cavka et al. 2017). Currently, several organizations have developed their own information requirements for QS BIM practice, e.g., Singapore's Quantity Surveying BIM Attribute Requirements (QSBAR).

Developing QS BIM requirements can be rather challenging. For one thing, the majority organizations are either late adopters or small- and medium-size companies; they lack sufficient resources to develop BIM requirements on their own, e.g., BIM experience, expertise, and professionals. Furthermore, the requirements are usually embedded into various sources, e.g., standard methods of measurement (SMM), contracts, BIM technical manuals, etc. To efficiently solve the problems, QS organizations tend to follow the existing information requirements and adapt them to suit the specific local requirements, e.g., local business targets, SMM, etc. In this regard, the development of BIM requirements can be translated into a series of challenges of identifying local QS needs, collect existing BIM requirements, and adapt the BIM requirements based on the local QS needs.

The primary aim of this paper is to formulate a process model to characterize the BIM requirements that support QS tasks. The proposed process model highlights an adaptation thinking – instead of developing the BIM requirements from a blank, this process can be enhanced by referencing and adapting the high-quality BIM requirements for local QS practice. The remainders of the paper are organized as follows. Subsequent to this introduction presents the research method which involves a case study to investigate the development of QS BIM information requirements by a Hong Kong quantity surveying consultant. Based on the investigation, a four-step process model was proposed in Section 3. Discussions and conclusion are drawn in the last section.

2 RESEARCH METHODS

The research methods of this study involved a case study to investigate the development of BIM requirements by a leading QS consultant in Hong Kong (denoted as *Firm A*). Amid the BIM trend in Hong Kong, Firm A joined the early BIM adopter group to explore using BIM for QS practice. The firm set up a short-term target to achieve BIM-enabled pre-tender practice, including preparing the BIM model specifically for QS use, BIM-based quantity takeoff, cost estimate, and bills of quantities, and a long-term target to adopt BIM for its full business. The firm also actively explored BIM technological and process solutions for their own practice. However, they encountered difficulties hindering further BIM adoption. The reality was that architects did not offer BIM models or merely provide a visual representation; it was rather time-consuming to develop and enrich the BIM model with sufficient information for QS tasks. To solve the problem, the management head decided to devise a set of QS BIM information requirements to convey specific QS requirements to designers and facilitate quantity surveyors to develop and enrich BIM models for QS tasks.

During the past two years, the authors work closely with Firm A to help the firm developed BIM strategies. The engagement, in turns, enabled the researchers to closely observe, investigate, and reflect from the real-life BIM adoption practice. Data was collected from various sources. First, the data was collected and analyzed to understand the overall BIM requirement development process, including the minutes of meeting, BIM strategic plan, emails, and organizational manual and technical guidelines. After that, interviews will be arranged with BIM experts, senior quantity surveyor and project team leaders within the organization to further refine the process model and understand the advantages and problems of developing BIM requirements following such a process model.

3 LOCALIZING BIM GUIDELINES – AN INVESTIGATION FROM A HK QS CONSULTANT

To ensure smooth BIM-based quantity takeoff and tender document (e.g., bills of quantity) preparation, Firm A aimed to develop a guideline specifying its BIM information requirements.

The proposed guideline will serve for two purposes – conveying the detailed QS requirements to designers or BIM modelers, and facilitating QS to enrich the hand-over BIM models for the subsequent measurements and tender document preparation. The authors' investigation suggested that Firm A adopted a four-step process model to develop its BIM information requirements (See Figure 1), including:

1. Specifying the requirements to unfold QS tasks in the local organization environment by understanding the existing QS practice and its local operating environments;
2. Collecting and organizing prevailing QS BIM information guidelines from leading organizations in the globe;
3. Analyzing the alignment between the collected BIM guidelines and local specific QS requirements; and; and
4. Devising a localized version of BIM requirements by rejecting/adapting/adopting the collected BIM guidelines, followed by reviewing and finalizing the requirements.

3.1   *Specifying local QS requirements*

The first step was to specify the requirements that QS tasks can unfold in the locality. To do so it required to collect different sources of data to develop an in-depth understanding of the existing process to conduct quantity take-off and prepare tender documents based on 2D drawings. The data were collected via mixed methods, with desktop studies to acquire knowledge, project engagement to closely observe the process, and expert interview probing into the detailed QS requirements. The data sources are summarized in Table 1.

A deeper probe into the data uncovered that the QS requirements can be grouped under three categories – industry-level, organization-level, and project-level requirements. The industry-level requirements comprised local codes and guidelines, e.g., HKSMM4, which set out local-accepted requirements on how building works should be measured, grouped and presented. They also specified the codes to represent each measured building components. Notably, the industry level requirements vary in different countries/regions; this should be carefully considered when adapting the BIM requirements from overseas organizations. The organization-level requirements were aligned to the firm's specific business targets and BIM objectives, e.g., BIM for quantity takeoff and tender document preparation. It determined the scopes of the QS BIM requirement. The project-level requirements regulated how different stakeholders should collaborate, and how BIM and related information should be exchanged. This required the QS BIM information requirement guidelines should contain a certain level of flexibility such that it can be tailored to different projects.

3.2   *Collecting and organizing prevailing guidelines*

In the second step, the prevailing BIM requirements were collected from the global organizations that lead the QS BIM practice. The data collection mainly focused on documented and

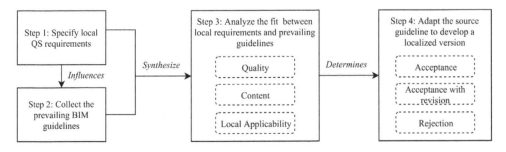

Figure 1.   A four-step process model to adapt and develop local BIM information requirements.

Table 1. Overview of collected data.

Data collection methods	Source of data	Description
Desktop studies	Hong Kong Standard Methods of Measurement 4 (HKSMM4)	• Providing the mutually accepted rules for measurement and a standard format to present the measured work
	Organization's training materials	• Used for training the newly-employed quantity surveyors; • Revealing the general process and common practice to unfold QS tasks in the firm
	Technical solutions	• The in-house applications that support data management and document preparation during the lifecycle of the firm's business
	Organization's technical guidelines	• Guiding quantity surveyors to use the in-house technical solutions
Project engagement	Contract documents	• Setting out the detailed contract requirements that QS should fulfill
	Schedules of different materials	• The detailed arrangement of measured quantities of different building objects; • Generated by applications and revised/grouped by quantity surveyors
	Bills of quantities	• One of the most important tender documents prepared by quantity surveyors. • Presenting an itemized list of the component parts of the building, and sets out the quality and quantity of all the component parts necessary for the construction of the works
	Project models and drawings	• The BIM models and drawings of the projects, • helping to understand how the design information is presented and integrated into BIM and drawings and utilized for QS tasks
Interviews	Interviews with the senior quantity surveyors Interviews with the BIM managers	• To develop a deeper understanding of the existing process and the detail information requirements that support QS tasks

published guidelines from the organizations sharing similar QS practice with the Hong Kong QS consultant, such as the UK, Singapore, Australia, and New Zealand. This is because information requirements are largely localized products that are shaped by QS practice procedures. Referencing the guidelines from those regions sharing similar QS practice could incur fewer adaptation efforts. Table 2 summarized some guidelines and information requirements used as reference by Firm A. Notably, the outcome of this step was an organized set of documented information requirements, which will serve as the basis for analysis, adaptation, and consolidation into a clear set of requirements to inform BIM-based design handover and QS tasks.

3.3 Assessing the referencing BIM requirements and devising BIM requirements for local QS practice

Once the referencing guidelines were collected, the next step went to analyze and assess the guidelines to formalize recommendations to develop the BIM requirements for local use. The assessment was unfolded in three dimensions, i.e., the quality, content, and applicability of referencing BIM requirements. The standards and thresholds of each dimension were determined by an expert panel consists of BIM experts and senior quantity surveyors. The collected guidelines were first scanned to determine its quality and inclusion for local guideline development. A detailed analysis was then conducted to critically analyze the selected referencing guidelines, especially their scopes of applications, structural framework, and presentation

Table 2. Global prevailing BIM information requirements.

Name of files	Issuing years, bodies, and countries	Description
BIM for cost managers: requirements from the BIM model	2015, Royal Institution of Chartered Surveyors (RICS), UK	• Providing a guidance note for quantity surveyors for delivering BIM-based cost consultancy services • Developed in conjunction with other protocol documents in the UK, such as New Rules of Measurement.
Quantity Surveying BIM Attribute Requirements	2018, Singapore Institutes of Surveyors and Valuers, Singapore	• A set of documents including detailed cost breakdown structure of the elemental attributes and geometry required within a BIM model; • Act as a guide for design consultants to model BIM to quantity surveyors' requirements; • Aligned to the SMM of Singapore.
Product Data Templates	2016, Chartered Institutes of Building Service Engineering (CIBSE), UK	A recommended template for manufacturers to generate a set of information about a given manufacturer's product.
BIM Object/Element Matrix	2013, NAPSPEC, Australia and New Zealand	• Defining a large number of objects and elements and their properties concerned by different professional activities including cost estimate; • Providing standards for BIM creation and information exchange at various stages of BIM develop and use. • Developed based on Uniformat/OmniClass classification and LOD at different stages in the building's lifecycle

formats. The selected guidelines will be further analyzed to identify its alignment with the local QS requirements, and applicability to the local QS practice. The assessment of these three dimensions will provide the basis for inspiring decision making about which referencing guidelines are relevant and for identifying which recommendations can be directly applied or adapted.

3.4 Devising a localized BIM guideline

The acquired knowledge and information from referencing information requirements were consolidated in the last step, which adapted the referencing BIM requirements into one suitable to use in the local context. In the case studies, local BIM requirements were consolidated and developed during a number of internal meetings and penal discussions amongst the management head, the senior quantity surveyors, the team leader of trial BIM projects, the BIM center directors, and in-house technicians. Five decisions were documented during this process, including:

- **Rejection** of the whole referencing guidelines. This decision was made in considering the poor quality of the guidelines per se or the huge incompatibility between the guideline and local QS practice;
- **Acceptance** of the referencing guidelines or a part of the guidelines. The decision was made in considering that both the quality and the compatibility of the referencing guideline, or a part of it, reaching the pre-defined thresholds, and thus can be directly used in the local QS practice and its specific local industry, organization, and project requirements.
- **Acceptance with revision**. This is an intermediate decision. The decision was made in considering that the quality or the compatibility of the referencing guidelines reach a lower threshold. The referencing guidelines could be useful for the local practice subjected to a certain

level of adaptation to their frameworks or contents based on the local industry, organization, and project requirements. The adaptation may involve adding new data and rephrasing certain expressions to better reflect the firm's context.

Following the decisions to adapt the referencing guidelines, a draft document was produced to consolidate the study outcomes. Notably, the document was presented at the elemental level to suit the object-based parametric presentation of BIM, i.e., the information of BIM is represented, included and organized into BIM objects as their attributes. The drafted BIM requirement was then evaluated by both external panels, including staff in the firms who did not participate in the guideline development, research institutes, designers, BIM modelers, owners, and other parties with no conflict of interests to ensure the applicability and readability of the requirements.

4 CONCLUSIONS

Adaptation of prevailing practice guidelines for local use has been advanced as an efficient means to improve the applicability of the developed guidelines. Yet, such an adaptation process is still unclear and remained challenging, as the prevailing practice, mainly developed in other countries/regions, are developed with embeddedness of local factors such as specifications and standard methods of measurements. It needs a solid process to identify the embedded contextual factors and adapt the referencing guidelines accordingly for specific local needs.

This paper provides valuable information for guideline adaptation and development for the local use. It aims to formulate a process model by investigating the QS BIM information requirement development and adaptation process by a Hong Kong quantity surveying consultant, also an early explorer of BIM in QS practice in Hong Kong. It identified four essential steps, namely specifying the local requirements underpinning the QS practice, collecting the prevailing guidelines from leading organizations/countries as the reference, analyzing the referencing guidelines, and adapt/develop a localized version of the BIM requirements.

The adaptation approach helps with the efficient development of a BIM guideline for local use. It eases the intensive learning process for BIM information guideline development, and facilitate the preparation of an applicable guideline especially for the organizations that are still struggling to explore BIM. However, this does not mean that the adaptation approach significantly reduces time and efforts. Indeed, the adaptation process still incurs lots of time and efforts. For one thing, the real-life QS practice is very complex; the local QS specification is complicated and can vary in individual's interpretations. Also, the guideline development needs to overcome the different information grouping principles underpinning QS and BIM practice – the former clusters information from different types of building elements based on local standards while the latter dispersed information in each BIM object. Therefore, it may involve an iterative process to evaluate and revise the draft BIM requirements to truly suit the real-life QS practice.

REFERENCES

Arayici, Y. Coates, P. Koskela, L. Kagioglou, M. Usher, C. & O'reilly, K. 2011. Technology adoption in the BIM implementation for lean architectural practice. *Automation in construction* 20(2):189-195.

Cavka, H.B. Staub-French, S. & Poirier, E.A. 2017. Developing owner information requirements for BIM-enabled project delivery and asset management. *Automation in construction* 83:169-183.

Kaner, I. Sacks, R. Kassian, W. & Quitt, T. 2008. Case studies of BIM adoption for precast concrete design by mid-sized structural engineering firms. *Journal of Information Technology in Construction* 13 (21): 303-323.

Lu, W. Lai, C.C. & Tse, T. 2018. *BIM and Big Data for Construction Cost Management*. London: Routledge.

Ma, Z. Wei, Z. & Zhang, X. 2013. Semi-automatic and specification-compliant cost estimation for tendering of building projects based on IFC data of design model. *Automation in Construction* 30:126-135.

Matipa, W. M. Kelliher, D. & Keane, M. 2008. How a quantity surveyor can ease cost management at the design stage using a building product model. *Construction Innovation* 8(3):164-181.

Monteiro, A. & Martins, J.P. 2013. A survey on modeling guidelines for quantity takeoff-oriented BIM-based design. *Automation in Construction* 35:238-253.

Raphael, V. & Priyanka, J. 2014. Role of building information modelling (BIM) in quantity surveying practice. *International Journal of Civil Engineering and Technology* 5(12):194-200.

Sabol, L. 2008. *Challenges in cost estimating with Building Information Modeling.* IFMA world workplace 1-16.

Wu, S. Wood, G. Ginige, K. & Jong, S.W. 2014. A technical review of BIM based cost estimating in UK quantity surveying practice, standards and tools. *Journal of Information Technology in Construction* 19:534-562.

Sustainable Buildings and Structures: Building a Sustainable Tomorrow – Papadikis et al. (Eds)
© 2020 Taylor & Francis Group, London, ISBN 978-0-367-43019-1

Influence of scene crisis perception on the coping behavior of public during shopping mall fire

X. Gu, Y. Chen & Y. Peng
School of Public Administration, Zhejiang University of Finance & Economics, Hangzhou, P.R. China

M. Liu
Hangzhou Fire Rescue Detachment Fuyang District Brigade, P.R. China

ABSTRACT: As commercial buildings usually have a high number of potential fire hazards, high population density, and complex layout resulting in difficulty of direction identification, it is important for the public to timely perceive the fire crisis and take effective behaviors in coping shopping mall fire. Based on the situational theory, this research constructs the fire scene of shopping mall, and further surveyed the crisis perception and coping behavior of people in the shopping mall with questionnaire and analyzed their relationship with multiple regression analysis. The results demonstrate that crisis perception of fire scene in shopping malls can positively affects the behavior, including alerting policies, organizing emergency evacuation, notification to others, and escaping from emergency exits, and negatively affects escaping from elevators, but has no significant effects on the behavior e.g., waiting for rescue, and jumping off buildings. The results provide references for improving performance of fire drills through enhancing crisis perception and learning of positive coping behaviors.

1 INTRODUCTION

Because of high frequency and high loss, fire is one of the main disasters that threaten public safety, and socioeconomic development. With rapid urbanization, the number of commercial buildings is increasing. As commercial buildings usually have a high number of potential fire hazards, high population density and complex layout resulting in difficulty of direction identification, fire hazards occurred in commercial buildings usually results in high losses. According to the statistics of the Fire and Rescue Bureau of the Emergency Management Department in 2018, a total of 237,000 fires were reported nationwide, resulting in 1,407 deaths, 798 injuries and 3.675 billion yuan of direct property losses, including 17,740 fires in shopping malls and other crowded places (China Fire Almanac 2018). Shopping malls are highly crowded and have a large flow (Li et al. 2005), the casualties and losses caused by each fire accident are much higher than those of other places (China Fire Almanac 2013). From 2013 to 2018, electric fire in shopping malls accounted for more than 30%, followed by inadvertent use of fire and production operations (China Fire Almanac 2013-2018), which made it difficult to completely avoid potential fire hazards in shopping malls. Therefore, it is particularly important for the public to timely perceive the crisis scene of fire and adopt the positive response behavior during the mall fire.

Crisis perception refers to people's psychological feelings and perceptions of crisis, which affects their emergency response, prevention and response to crisis. The existing study of crisis perception can be divided into two categories, one is to explore the influencing factors of crisis perception, and the other is to explore the impact of crisis perception on behavior. For example, Holbrook et al. (2018) adopted web-based network sampling to explore the role of mainstream media reporting in public crisis perception. Wang et al. (2017) used questionnaires to investigate the impact of institutional trust, industry trust, and corporate trust on consumer-product crisis perception. Bowen et al. (2017) compared the impact of American and German consumers' crisis perception on crisis response under the corporate scandal based on the cross-cultural perspective.

In the study of crisis perception of natural disasters and accident disasters, it mainly involves public environmental crisis perception (Sun 2016), traffic crisis perception (Long et al. 2017, Li 2016, Yang et al. 2017), coal mine crisis perception (Blignaut 1979), and building safety crisis perception (Han et al. 2019). In environmental crisis, the more serious the public's perception, the lower the trust on the central government and local government, while the use of the Internet will further reduce the trust on the central government (Sun 2016).

In traffic crisis, the drivers' perception of the crisis is affected by driving experience, attention span, and risk type (Long et al. 2017) while the pilots' risk perception is affected by the flight environment, pilot characteristics, and organizational culture (Li et al. 2016). Traffic signal sensitivity can help cyclists had better perceive traffic crises, but the multi-task use of audiovisual organs will reduce the public's perception of crisis (Yang et al. 2017). In building safety crisis, workers' crisis perception can be significantly affected by length of service, position, frequency, and severity of accidents (Han et al. 2019). In emergent crisis, the public is accustomed to assessing crisis events based on personal experience.

The public's crisis perception will not only affect their behavior, but also affect the evolution of crisis events. However, the existing research on fire response behavior mainly explored the impact of age, gender, education, fire experience, and fire drill on public response behavior (Xiao et al. 2002, Song 2008, Bryan 1983, Wood 1972, Yan 2005, Zhao 2013, Wu 2016). Few researches have investigated the influence of crisis perception on fire coping behavior. Based on the situational theory, this study constructs a fire scene in shopping malls, designs a questionnaire based on existing research, and explores the impact of public scenario crisis perception on fire coping behaviors. This study can provide reference for improving the public fire-crisis perception ability, optimizing fire department publicity and education, and conducting fire drills in different scenarios.

2 RESEARCH DESIGN

In order to realize the research aim, this study designed the questionnaire based on existing studies, which was further improved through expert interviews. The formal questionnaire includes basic information including gender, age, education, type of workplace building, control variables including fire experience, fire knowledge, fire drills, and the number of visiting the shopping mall, key independent variable crisis perception, and dependent variable coping behavior. Crisis perception and coping behavior were identified based on the constructed fire scenario. Likert 7-level scale was adopted to represent the public's behavioral inclination. A general fire scenario and a severe fire scenario was constructed based on the situational theory, which descried the two scenarios with texts. In the formal survey, each respondent was randomly invited to fill in a questionnaire of coping behavior in either severe or general mall fire scenarios. Finally, the data of the questionnaire were coded and sorted out. Descriptive statistics and multiple regression analysis were carried out by SPSS software to explore the impact of crisis perception on fire coping behavior. The main research variables are described in Table 1.

3 RESEARCH FINDINGS AND DISCUSSIONS

3.1 Sample and preliminary analysis

Five hundred and ninety-four questionnaires were distributed, including 546 electronic questionnaires and 48 paper questionnaires. After eliminating the invalid samples, 572 valid

Table 1. Measurement of involved variables.

Classification	Variable	Measure
Dependent variable	Coping behavior	Possibility of adopting coping behaviors (waiting for rescue, alerting policies, organizing emergency evacuation, notification to others, escaping from emergency exits, following crowd, jumping off the building, escaping from elevator) (1-7, Extremely low =1, Extremely high =7)
independent variable	Fire scenario	0= General fire; 1= Severe Fire
	Crisis perception	Crisis perception of the fire scene (1-7), The crisis is extremely low =1, extremely high =7
Social demographic characteristics	Gender	Female =0; Male =1
	Age	Under 18 years old =1, 18-25 years old =2, 26-35 years old =3, 36-45 years old =4, 45-60 years old =5, 60 years old or older =6
	Education	Primary school or below =1, Junior high school =2, High school =3, Junior College =4, Bachelor degree or above =5
Fire safety experience	Fire experience	No =0; Yes =1
	Fire knowledge	Never know=1, Learn a little from daily life =2, Received fire propaganda =3, Received professional training =4
	Number of fire drills	Never=0, Once or twice =1, 3-4 times=2, 5-7 times=3, 8-10 times=4, More than 10 times =5
	Time distance from the last fire drill	1-3days=1, 4-7days=2, 8-30days=3,1-3 months=4, 3 months-1year=5, Did not participate in 1 year =6
	Familiarity with firefighting equipment	Specific number
Shopping mall familiarity	workplace building type	Non-commercial building =0, Commercial building =1
	Number of visits to the mall per month	Once or twice =1,3-4 times =2,5-7 times =3,8-10 times =4, more than 10 times =5
	Time distance from the last trip to the mall	1-3 days =1, 4-7 days =2, 8-30 days =3, 1-3 months =4, more than 3 months =5
	Overlook of Emergency Exit in Shopping Malls	1-4, Pay attention every time =1, Never pay attention =4

samples were valid, with an effective rate of 95.8%, including 329 questionnaires with severe fire scene and 243 questionnaires with general fire scenes. The sample distribution is shown in Table 2.

3.2 *Analysis of the impact*

In this paper, all variables in the questionnaire were imported into the multivariate regression model for analysis, the results are shown in Table 3. Based on the regression analysis, this study investigated the impact of crisis perception on different fire coping behaviors.

From the regression results in Table 3, it can be found that crisis perception positively affected the coping behaviors of alerting policies, notification to others, and escaping from emergency exits at 0.01 significance level, and positively affects organizing emergency evacuation at 0.05 significance level while negatively affects escaping from elevator at 0.01 significance level. Crisis perception has no significant effect on behaviors such as waiting for rescue and jumping off a building.

When the public perceived the fire crisis, in order to effectively deal with the fire crisis, it was necessary to take active coping behaviors to resolve the crisis and reduce their own casualties. The higher crisis perception, the more threat of fire that individual can feel, which made them alert policies, notify others and escape from emergency exits and other positive behaviors to ensure their safety. The perception of crisis in the fire scene also affected the

Table 2. Sample distribution.

	Category	General fire scenario		Severe fire scene	
		Sample size	Percentage	Sample size	Percentage
Gender	Female	130	22.73%	199	34.79%
	Male	113	19.76%	130	22.73%
Age	Under 18 years old	3	0.52%	7	1.22%
	18-25 years old	180	31.47%	235	41.08%
	26-35 years old	52	9.09%	52	9.09%
	36-45 years old	3	0.52%	10	1.75%
	45-60 years old	5	0.87%	23	4.02%
	60 years old or older	0	0.00%	2	0.35%
Education	Primary school or below	2	0.35%	8	1.40%
	Junior high school	4	0.70%	14	2.45%
	High school	15	2.62%	25	4.37%
	Junior College	31	5.42%	35	6.12%
	Bachelor degree or above	191	33.39%	247	43.18%
Workplace building type	Commercial building	27	4.72%	40	6.99%
	Cultural and educational building	98	17.13%	101	17.66%
	Medical building	8	1.40%	17	2.97%
	Hotel building	3	0.52%	9	1.57%
	Traffic building	2	0.35%	6	1.05%
	industry building	14	2.45%	14	2.45%
	else	91	15.91%	142	24.83%
Fire experience	Yes	233	40.73%	305	53.32%
	No	10	1.75%	24	4.20%

identification of the risk of coping behavior. If the judgment of the severity of fire is higher, the public would think that the behavior of escaping from elevator is more dangerous, thus avoiding or reducing such risk behaviors. In addition, compared with those in the general fire scene, the public in the severe fire scene were less inclined to inform others and escape from the emergency exits, but tended to follow the crowd to escape.

When the public was aware of crisis in a serious fire, the higher the crisis perception, the more urgent their coping behaviors will be. Therefore, compared with general fires, it was easier for the public to escape with the crowd, but lacked self-coping behavior judgment and decision-making. In fact, escaping with the crowd was not a dangerous behavior in itself. Its threat lies in the lack of correct crisis perception. The public with weak cognitive decision-making ability is prone to follow the crowd without thinking, instead of making correct decisions and coping behaviors by themselves. When the crowd took positive behaviors such as escaping from emergency exits, it can help the victims escape from the fire, otherwise, it would result in casualties of the victims. However, crisis perception has no significant impact on passive coping behaviors such as waiting for rescue and radical behaviors such as jumping off a building.

When the public was faced with fire and felt that the fire was dangerous, people were more likely to escape (Slovic et al. 2007) rather than passively waiting for being rescued and letting the fire invade. This maybe the reason waiting for rescue was not significant in the regression models. However, jumping off buildings was extremely dangerous, and it posed a serious threat to the safety of life. Whether the public's perception of the fire scene crisis is high or low, they were not inclined to jump off buildings radically to survive in order to avoid fire damage.

Table 3. Multivariate regression results.

Coping behavior	Waiting for rescue	Alerting policies	Organizing emergency evacuation	Notification to others	Escaping from emergency exit	Following the crowd	Jumping off building	Escaping from elevator
Fire scenario	-0.065 (0.158)	-0.175 (0.116)	-0.128 (0.152)	-0.28** (0.126)	-0.193* (0.105)	0.24* (0.138)	0.133 (0.113)	-0.066 (0.1)
Crisis perception	-0.072 (0.063)	0.364*** (0.046)	0.14** (0.06)	0.272*** (0.05)	0.321*** (0.042)	0.098* (0.055)	-0.072 (0.045)	-0.143*** (0.04)
Gender	-0.2 (0.159)	-0.293** (0.117)	-0.24** (0.152)	-0.132 (0.127)	-0.165 (0.106)	-0.418** (0.139)	-0.016 (0.114)	0.055 (0.1)
Age	0.086 (0.11)	0.158* (0.081)	0.367 (0.106)	0.007 (0.088)	0.124* (0.073)	0.001 (0.096)	-0.21** (0.079)	-0.11 (0.07)
Education	-0.116 (0.1)	0.167** (0.074)	0.067 (0.096)	0.157** (0.08)	0.151** (0.067)	0.26** (0.088)	-0.01 (0.072)	-0.096 (0.063)
Fire experience	-0.257 (0.329)	-0.571** (0.242)	-0.38 (0.315)	-0.21 (0.262)	-0.498** (0.218)	-0.084 (0.288)	0.142 (0.235)	0.163 (0.208)
Fire knowledge	0.214 (0.147)	0.118 (0.108)	0.278** (0.141)	0.263** (0.117)	0.126 (0.097)	0.185 (0.128)	-0.249** (0.105)	-0.103 (0.093)
Number of fire drills	-0.107 (0.081)	0.062 (0.059)	-0.013 (0.078)	0.073 (0.064)	0.136** (0.054)	0.049 (0.071)	-0.025 (0.058)	-0.077 (0.051)
Time distance from the last fire drill	-0.238** (0.102)	0.191** (0.075)	0.123 (0.098)	0.166** (0.081)	0.174** (0.068)	0.11 (0.089)	-0.139* (0.073)	-0.142** (0.064)
Familiarity with firefighting equipment	0.006 (0.034)	0.053** (0.025)	0.105** (0.032)	0.055** (0.027)	0.051** (0.022)	0.003 (0.029)	0.014 (0.024)	-0.028 (0.021)
workplace building type	-0.015 (0.031)	0.035 (0.023)	-0.002 (0.03)	0.014 (0.025)	0.039* (0.021)	0.001 (0.027)	-0.021 (0.022)	-0.057** (0.02)
Number of visits to the mall per month	-0.057 (0.074)	-0.111** (0.054)	-0.049 (0.071)	-0.114* (0.059)	-0.177*** (0.049)	-0.073 (0.065)	0.082 (0.053)	0.103* (0.047)
Time distance from the last trip to the mall	-0.009 (0.072)	-0.053 (0.053)	-0.031 (0.069)	-0.042 (0.058)	-0.023 (0.048)	-0.107* (0.063)	0.04 (0.052)	0.059 (0.046)
Overlook of Emergency Exit in Shopping Malls	-0.15 (0.114)	0.025 (0.084)	-0.033 (0.109)	0.147 (0.091)	0.06 (0.076)	0.193* (0.099)	-0.018 (0.081)	-0.095 (0.072)
Constant	5.309*** (0.992)	1.636** (0.729)	1.748* (0.951)	1.702** (0.791)	1.839** (0.659)	1.95** (0.867)	3.974*** (0.71)	4.542*** (0.627)
R^2	0.034	0.204	0.081	0.113	0.212	0.093	0.050	0.098
Adj.R2	0.01	0.184	0.058	0.092	0.192	0.070	0.026	0.075
F	1.396	10.181	3.490	6.121	10.684	4.060	2.103	4.321
sig.	0.150	0.000***	0.000***	0.000***	0.000***	0.000***	0.010**	0.000***
N	572	572	572	572	572	572	572	572

* The values in the table are non-standardized coefficients, * $P < 0.1$, ** $P < 0.05$, *** $P < 0.001$, and the standard error values of the variables in the model are in parentheses.

4 CONCLUSION

Based on existing studies, this paper developed the questionnaire and used multiple regression analysis to investigate the influence of crisis perception on fire coping behaviors. The research found that crisis perception has significant positive impacts on the behavior of alerting policies, organizing emergency evacuation, notification to others, and escaping from emergency exits. Crisis perception has a significant negative impact on the risk-taking behavior of escaping from the elevator. The results of this study show that fire departments need to arrange fire scenarios to simulate different levels of complexity in order to improve the effectiveness of fire drill in shopping malls. Before the drill, fire accident video and other ways can be used to enhance crisis perception, so as to enhance the public's cognition and learning of positive coping behaviors in fire drills.

However, the sample lacks elderly people, which yet is one of the vulnerable groups in fire hazards. Future studies should expand the sample and investigate the impact of crisis scene perception on coping behaviors for the elderly people. In addition, the general and severe fire scenarios were constructed with text description, which results in difficulty of obtaining more effective feedback. Future study can explore constructing the scenario with pictures.

REFERENCES

Blignuat, C.J.H. 1979. The perception of hazard II. The contribution of signal detection to hazard perception. *Ergonomics* 22(11): 1177-1183.

Bowen, M. Freidank, J. Wannow, S. & Cavallone, M. 2017. Effect of perceived crisis response on consumers' behavioral intentions during a company scandal – an intercultural perspective. *Journal of International Management* 24(3): 222-237.

Bryan, J. L. 1983. A review of the examination and analysis of the dynamics of human behavior in the fire at the MGM Grand Hotel, Clark County, Nevada as determined from a selected questionnaire population. *Fire Safety Journal* 5(3):233-240.

China fire almanac: 2013. *China personnel publishing house.*

China fire almanac: 2014. *Yunnan people's publishing house.*

China fire almanac: 2015. *Yunnan people's publishing house.*

China fire almanac: 2016. *Yunnan people's publishing house.*

China fire almanac: 2017. *Yunnan people's publishing house.*

China fire almanac: 2018. *Yunnan people's publishing house.*

Han, Y. Feng, Z. Zhang, J. Jin, R. & Aboagye-Nimo, E. 2019. Employees' safety perceptions of site hazard and accident scenes. *Journal of Construction Engineering and Management* 145(1).

Holbrook, T.J. & Kisamore, J.L. 2018. The effects of media slant on public perception of an organization in crisis. *Social Influence* 1-13.

Li, W.Q. Chen, H. & Li, Y. 2016. Influencing factors and processing mechanism of pilot risk perception. *Psychological Science* 6:1385-1390.

Li, G.F. Wang, J. & Li, H. 2005. Discussion on fire hazard and fire safety countermeasures of modern shopping malls. *Fire Science and Technology* (b03):109-112.

Long, S. Chang, R. & Shuang, L. 2017. Effects of driving experience and hazard type on young drivers' hazard perception. *Chinese Journal of Ergonomics* 11-16.

Slovic, P. Peters, E. Grana, J. Berger, S. & Dieck, G.S. 2007. Risk perception of prescription drugs: results of a national survey. *Therapeutic Innovation & Regulatory Science* 41(1):81-100.

Song, Y. 2008. Research on behavioral reliability of evacuation in building fire. *Hunan University of Science and Technology.*

Sun, W. 2016. Public environment crisis perception, Internet use and government trust - analysis based on CGSS2010 data. *Journal of Fujian Administrative College* 3:34-43.

Wang, X. Chao, G. & Wan, G. 2017. The role of macro-level trust in consumers' perception of product crisis. *Management Review* 2.

Wood, P. G. 1972. The behavior of people in fires. *Fire Research Station.*

Wu, D. 2016. Chongqing underground commercial building fire evacuation behavior influence factor research. *Chongqing University.*

Xiao, G. Wen, L. & Chen, B. 2002. Reliability model of human behavior in building fire. *Journal of Northeastern University (natural science edition)* 23(8):761-764.

Yan, W. Chen, B. & Zhong, M. 2005. Research on the effect of different floors on evacuation psychology and behavioral response of college students in the case of fire. *China Safety Production Science and Technology* 1(4):32-37.

Yang, C. Y. & Wu, C.T. 2017. Primary or secondary tasks? Dual-task interference between cyclist hazard perception and cadence control using cross-modal sensory aids with rider assistance bike computers. *Applied Ergonomics* 59: 65-72.

Zhao, M.G. 2013. Study on the human behavior in metro fire. *Applied Mechanics and Materials* 353-356:1456-1460.

Sustainable Buildings and Structures: Building a Sustainable Tomorrow – Papadikis et al. (Eds)
© 2020 Taylor & Francis Group, London, ISBN 978-0-367-43019-1

Challenges of off-site construction: A critical review

L. Hou & Y.T. Tan
School of Engineering, RMIT University, Melbourne, Australia

ABSTRACT: Off-site Construction (OSC) has been recognised as a promising construction method to tackle challenges of constructing structurally and functionally intricate contemporary buildings. Considering the current scale of OSC is not commensurate with traditional construction methods, to understand OSC-related opportunities and bottlenecks, a significant amount of research works has been carried out over the past decade. It is also perceived that formulation and utilisation of state-of-the-art technologies, applications and methods have gained growing attention for researchers that are striving for improving OSC performance in construction projects. Therefore, a rational understanding of technology development, bottlenecks and improvement is certainly of pivotal importance, which became the motive of this literature review study. This study, methodically, identified a total of 1212 publication records using Web of Science, followed by a literature screening process which remained 834 publications for further examination and analysis. Next, a co-citation network, co-occurrence keywords network and a timeline view of co-occurrence keywords were generated to visually present the solicited knowledge domains and research trends. With the fundamental challenges presented, this paper examined a wide range of information and communication technologies (ICT) that can be tentatively applied in OSC process. Last but not least, this paper also presented an in-depth reflection on the gaps-in-knowledge and resolutions for technological evolvement.

1 INTRODUCTION

The construction industry is an information-dependent industry given its practice being consistently dealing with drawings, cost analysis sheets, budget reports, risk analysis, charts, contract documents, planning schedules (prior to construction), logistics and inventory management, progress monitoring, quality assurance (during construction), facilities and assets maintenance and rehabilitation (post-construction).

In general, most of the firms within the construction industry have been bedeviled with difficulty in delivering value to their customers on schedule. Other criticisms of the industry also include lack of stability, susceptibility to fluctuation in demand cycle, uncertainty in production, unspecific project, product demand and most specifically divergent skills etc. The industry has grown to the stage of handling large and complex projects with modern construction methods such as Off-site Construction (OSC). OSC is a promising technique which has potential to overcome multitudinous challenges that range across construction productivity, efficiency, quality, safety, sustainability and so on (O'Connor et al. 2015, Svajlenka & Kozlovska 2017). However, due to inadequate supply chain control, lack of stability, fluctuation in demand, design, supply, production and demand uncertainty, workforce upskilling and many other requirements, the market share of OSC is far less than conventional construction methods (Goulding et al. 2015, Dave et al. 2017).

Emerging technologies may introduce effective solutions to overcome the limitations of OSC, for instance, the use of information and communication technologies (ICT) can streamline the ways of collecting, analysing and communicating on project data, making it possible for effective supply chain management and project quality assurance (Rohani et al. 2014,

Chen et al. 2017, Zhong et al. 2017). In order to make sense of essential rationales around ICT and its applications and understand exactly what and how the technologies could play a role in OSC projects, it is worth soliciting the significant body of knowledge from the contemporary literature in the areas of, for instance, ICT, Building Information Modelling (BIM), smart sensing and tracking, 3D printing, and so on. Besides, understanding technology limitations and uncertainties provided the difference between OSC and traditional construction is also important in this kind of study which aims to derive insights around technology opportunities and future development. To sum up, the objective of this paper is to identify existing intellectual lessons on technologies-advanced OSC and to formulate possible resolutions to guide prospective researchers to bridge the gaps-in-knowledge.

2 METHODOLOGY

Scientometric and bibliometric analysis is an effective quantitative method to analyse the intellectual landscape of a knowledge domain and identify the research topics of interest (Chen 2016). Therefore, this paper mainly capitalised on a scientometric and bibliometric tool named CiteSpace 5.2.R4 to analyse: 1) the research hotspots and knowledge domains; 2) the co-citation network of the literature; and 3) the co-occurrence keywords network (Chen 2017). In addition, with a possibility of making use of advanced technologies in the project lifecycle, this study analysed applicable technologies in detail across the project stages, and then discussed some broader application points, such as supply chain and Just-In-Time (JIT) methodologies. Importantly, this study proposed viable technological solutions to enhance information flow, improve supply chain control and mitigate project uncertainties, via contrasting promising BIM-enabled 3D printing against traditional OSC workmanship in various factors.

3 ANALYSIS OF THE KNOWLEDGE DOMAINS AND KEY CLUSTERS

In CiteSpace, document co-citation analysis can be conducted to analyse knowledge domains and demonstrate the state of cited references (Chen 2016). In this study, a document co-citation network containing 331 nodes and 900 links was generated and presented in Figure 1, which presents each node denoting a cited reference and its size representing a co-citation frequency. The Figure 1 network only displays 50 most frequently-cited articles. Modularity Q and Silhouette are two important metrics that reveal the overall structural properties of the

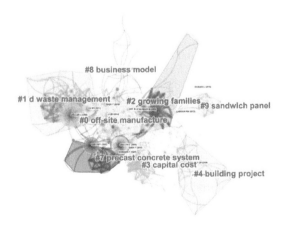

Figure 1. An overview of the co-citation network.

network. For instance, a network's Modularity Q measures the extent to which the network can be organised into multiple components/modules (Chen et al. 2010). The higher the Modularity Q, the more reliable/credible the network is (Chen et al. 2010). Our network was a credible network with the Q value well above 0.71.

According to the network, significant clusters were identified based on the keywords of the publications cited within each cluster, using the Log-Likelihood Ratio (LLR) algorithm – a method that can select the best labels for related clusters. As can be seen from the figure, most important clusters in the co-citation network relate to the analysed literature that includes the keywords of off-site manufacture/production, waste management, risk factor, off-site production, production output, modular construction, product architecture model, and precast concrete system etc. Cluster #1 refers to the research works around OSC's benefits on reducing construction waste such as timber, concrete, plasterboard, bricks, asbestos, vegetation, rock and soil, indicating an important research direction in the OSC domain. Compared to conventional construction, OSC workmanship is likely to reduce material waste and environmental pollution. Underpinned by novel technology applications such as data mining and internet of things (IoT), OSC practitioners could better oversee the usage of diverse construction materials (Lu et al. 2015, Chen & Lu 2017). Cluster #2 primarily focusses on formulating effective OSC strategies and overcoming potential uncertainties, which could be multifold, overarching and encountered at all stages (Arashpour et al. 2016, Li et al. 2016). Clusters #7 and #9 aggregate some material, information and process innovation research that sheds light on panelised walls, optimised fabrication platforms, double-skin facade systems, integrated environmental systems and the like (Said et al. 2017, Kilaire & Stacey 2017).

4 FUNDAMENTAL CHALLENGES

The construction industry is an information-dependent industry given its practice being consistently dealing with drawings, cost analysis sheets, budget reports, risk analysis, charts, contract documents, planning schedules (prior to construction), logistics and inventory management, progress monitoring, quality assurance (during construction), facilities and assets maintenance and rehabilitation (post-construction). In general, most of the firms within the construction industry have been bedeviled with difficulty in delivering value to their customers on schedule. According to Wang et al. (2018), one of the main bottlenecks is the delivery of prefabricated components, with the prolonged construction period and increased storage and labour requirements. Other criticisms of the industry also include lack of stability, susceptibility to fluctuation in demand cycle, uncertainty in production, unspecific project, product demand and most specifically divergent skills etc.

The industry has grown to the stage of handling large and complex projects with modern construction methods such as OSC. Unfortunately, it is quite unclear of OSC efficiencies associated with the practice of tracking of prefab materials, equipment and tradesmen in both the off-site and on-site settings, which often gives rise to uncertainties such as scheduling, costing and resources of projects (Arashpour et al. 2016). In addition, with the increasing global competition there is need for the construction industry to be in alliance with technology players and influence the productivity of its industry and projects (notice that this should be done in technology R&D). Apparently, ICT have been perceived to be an essential ingredient for project success as they allow integrated business processes across various functional units and project stages.

On the one hand, OSC seems to have emanated from internal initiatives that aim at improving overall effectiveness of project delivery and management, by utilising multiple managerial and technical factors such as integrated project delivery (IPD), BIM and lean construction. The focus of OSC, on the other hand, has to be extended to adding values to the organisation. Jin et al. (2018) reckoned that OSC seeks to improve performance via the better use of internal and external capabilities (e.g. technologies) in order to form seamlessly-coordinated OSC workmanship. However, OSC is not an isolated concept, and the inter-connected relationship between OSC and emerging technologies need to be made clear (Jin et al. 2018).

In order for organisations to implement effective design, there needs to be a good understanding of the OSC design process inclusive of fabrication, assembly, cost, quality, safety, sustainability, reliability, so on so forth (Jaillon & Poon 2010). Currently, design experiences primarily derived from sporadic OSC projects have demonstrated huge potential of cost savings and efficiency attainments (Khalfan & Maqsood 2014). In line with this is the need to investigate into the technologies that support project stakeholders to perform early stage design and planning activities. In addition, as OSC practitioners will need the knowledge and skills to move between off-site and on-site working environments, the state-of-the-art knowledge in this instance must be examined to meet the changing demand.

As well, digital capabilities relating to the operation of prefabs, auxiliary equipment, workforce health, safety and wellbeing, quality control, waste disposal and management, team dynamics and all other sorts of factors that could be underpinned by the implementation of innovative techniques and technologies deserve a thorough examination. While dealing with an increasing amount of sizeable, complex and fast-track projects, the present-day construction industry is confronting with a growing demand of assuring the function and reliability of the built structure and its multifold asset types. Whether intensive and routine maintenance and repair activities of an OSC project could be realistically performed lies in the understanding of information captured as part of the routine technical-administrative procedures of most OSC projects.

Finally yet importantly, sustainable asset management of OSC structures for long-term use requires a good understanding of asset conditions such as tear, wear and functional failure. Traditional auditing which mainly relies on close distance visual checks are quite onerous and sometimes may not suffice the requirements of operation and maintenance safety, efficiency and other considerations. BIM-enabled construction 3D printing is the latest technology introduced in the construction industry, studies on this method of construction is still scarce compared to the studies carried out on traditional modular construction. The knowledge concerning the potential, limits of application and process performances of both these technologies in conjunction with modular construction is essential as they are the future drivers of enhanced construction practices throughout the world. To overcome this lack of available understanding, this paper formulates a list of comparison points between BIM (and peripheral technologies)-enabled construction 3D printing and traditional OSC against various factors such as design complexity, structure property, material type and so on.

5 SUMMARY

OSC techniques are conducive to buildability, sustainability, waste control, resource utilisation and labor productivity. Nowadays, both developed and developing countries have started their journey to research and develop their OSC techniques and applications across a wide range of areas. This study explored the OSC literature published within the last decade and identified ten key research areas. The recent research works start to leverage emerging technologies to design, construct, operate and maintain OSC projects. Overall, the investigation on synthesised literature has led to important cues which can help researchers understand challenges that may hamper the development of OSC techniques. The uniqueness of this study is also perceived from a perspective that connects the traditional OSC industry to innovative technological paradigms, and to collaborators including architects, engineers, building professionals, property developers and builders; along with suppliers of timber and engineered wood, software, manufacturing equipment, prefabricated systems, and building materials. This paper will enhance the sector's confidence on addressing the encountering problems associated with conventional OSC project delivery and technological evolvement and make use of the lessons learnt from this study into the futuristic OSC work practice (Table 1).

Table 1. BIM/peripheral technologies-supported 3D printing.

Factors	BIM and 3D Printing enabled OSC	Traditional OSC
Complex design, structures, materials	One of the main features of 3D printing methods in construction is that it enables the design and subsequent creation of complex structures	Production of complex structures is difficult
Speed of construction	High speed of construction (especially in case of contour crafting) – operation construction 3D printing methods operate at an unrelenting and steady pace as the on-site construction is machine-driven	The speed of construction/assembly on site is subject to variations depending upon the transportation system, labour capability and the supply chain system
Compatibility with large-scale/ mass production	The in-situ approach of 3D printing requires a relatively less detailed pre-project planning as changes can still be made on-site	Thorough and a high level of pre-planning must be as accurate as possible since changes to the components cannot be made once manufactured off-site

REFERENCES

Arashpour, M. Wakefield, R. Lee, E.W.M. Chan, R. & Hosseini, M.R. 2016. Analysis of interacting uncertainties in on-site and off-site activities: Implications for hybrid construction. *International Journal of Project Management* 34(7):1393-1402.

Chen, C. 2016. *CiteSpace: a practical guide for mapping scientific literature*. Nova Science Publishers, Incorporated.

Chen, C. 2017. Science Mapping: A Systematic Review of the Literature. *Journal of Data and Information Science* 2(2): 1-40.

Chen, C. Ibekwe-SanJuan, F. & Hou, J. 2010. The structure and dynamics of cocitation clusters: A multiple-perspective cocitation analysis. *Journal of the Association for Information Science and Technology* 61(7): 1386-1409.

Chen, X. & Lu, W. 2017. Identifying factors influencing demolition waste generation in Hong Kong. *Journal of cleaner production* 141: 799-811.

Chen, K. Xu, G. Xue, F. Zhong, R. Y. Liu, D. & Lu, W. 2017. A physical internet-enabled building information modelling system for prefabricated construction. *International Journal of Computer Integrated Manufacturing* 31(4-5): 349-361.

Chen, C. Ibekwe-SanJuan, F. & Hou, J. 2010. The structure and dynamics of cocitation clusters: A multiple-perspective cocitation analysis. *Journal of the American Society for information Science and Technology* 61(7): 1386-1409.

Dave, M. Watson, B. & Prasad, D. 2017. Performance and perception in prefab housing: An exploratory industry survey on sustainability and affordability. *Procedia engineering* 180: 676-686.

Goulding, J.S. Pour Rahimian, F. Arif, M. & Sharp, M.D. 2015. New offsite production and business models in construction: priorities for the future research agenda. *Architectural Engineering and Design Management* 11(3): 163-184.

Jaillon, L. & Poon, C.S. 2010. Design issues of using prefabrication in Hong Kong building construction. *Construction Management and Economics* 28(10): 1025-1042.

Jin, R. Gao, S. Cheshmehzangi, A. & Aboagye-Nimo, E. 2018. A Holistic Review of off-site Construction Literature Published between 2008 and 2018. *Journal of Cleaner Production* 202: 1202-1219.

Khalfan, M. & Maqsood, T. 2014. Current state of off-site manufacturing in Australian and Chinese residential construction. *Journal of Construction Engineering*.

Kilaire, A. & Stacey, M. 2017. Design of a prefabricated passive and active double skin façade system for UK offices. *Journal of Building Engineering* 12:161-170.

Li, C. Z. Hong, J. Xue, F. Shen, G.Q. Xu, X. & Mok, M.K. 2016. Schedule risks in prefabrication housing production in Hong Kong: a social network analysis. *Journal of cleaner production* 134:482-494.

Lu, W. Chen, X. Peng, Y. & Shen, L. 2015. Benchmarking construction waste management performance using big data. *Resources, Conservation and Recycling* 105: 49-58.

O'Connor, J.T. O'Brien, W.J. & Choi, J.O. 2015. Standardization strategy for modular industrial plants. *Journal of Construction Engineering and Management* 141(9): 04015026.

Rohani, M. Fan, M. & Yu, C. 2014. Advanced visualization and simulation techniques for modern construction management. *Indoor and Built Environment* 23(5): 665-674.

Said, H. M. Chalasani, T. & Logan, S. 2017. Exterior prefabricated panelized walls platform optimization. *Automation in Construction* 76: 1-13.

Svajlenka, J. & Kozlovska, M. 2017. Modern method of construction based on wood in the context of sustainability. *Civil Engineering and Environmental Systems* 34(2): 127-143.

Wang, Z. Hu, H. & Gong, J. 2018. Framework for modeling operational uncertainty to optimize offsite production scheduling of precast components. *Automation in Construction* 86: 69-80.

Zhong, R. Y. Peng, Y. Xue, F. Fang, J. Zou, W. Luo, H. Thomas Ng, S. Lu, W. Shen, G.Q.P. & Huang, G.Q. 2017. Prefabricated construction enabled by the Internet-of-Things. *Automation in Construction* 76: 59-70.

Sustainable Buildings and Structures: Building a Sustainable Tomorrow – Papadikis et al. (Eds)
© 2020 Taylor & Francis Group, London, ISBN 978-0-367-43019-1

Real-time 3D reconstruction using SLAM for building construction

T. Qian & C. Zhang

Xi'an Jiaotong-Liverpool University, Suzhou, China

ABSTRACT: 3D reconstruction in construction site is now receiving remarkable attention in construction management. Currently, researchers mainly focus on off-line reconstruction using image-based algorithms or directly obtain point clouds using laser scanner. Both methods are time-consuming. This research proposes a real-time 3D reconstruction system using simultaneous localization and mapping (SLAM). This system reaches 96.8% accuracy in measuring the outline length of a small cabinet. Moreover, the accuracy, data collecting time and data processing time are evaluated through comparing the results of off-line image-based method Structure from Motion (SfM) and laser scanner-based method. The new system achieves satisfactory accuracy rate and efficient data collecting and processing, however, many details in the appearance of objects were missing and there were many noise points in the 3D reconstruction model.

1 INTRODUCTION

3D reconstruction is a process of capturing the geometry and appearance of an object. Point cloud models, mesh models and geometric models are three main models to represent recon-structed models, where the point cloud models are the basis. Point cloud is a set of data points including XYZ coordinates data and color data, which form the external surface of an object. 3D reconstruction is widely used in construction site, for example, 3D reconstruction of as-built model in construction site can be applied in construction progress tracking, geometry quality inspection, construction safety management etc. (Wang et al. 2019). There are two main 3D reconstruction methods to obtain 3D point cloud models, which are the image-based ranging algorithms method and the laser scanner-based method. Image-based ranging algo-rithms processes 2D images to get 3D point cloud model, while 3D laser scanner directly gen-erates point cloud model by depth information based on round trip time of the laser beam/ ray. Both two methods can only generate 3D point clouds off-line. Off-line generating point cloud models causes delay of detecting errors in the collecting data step. For example, engin-eers generate a model and find that there are some missing parts. They have to recollect the data of the missing parts. Therefore, it is better to generate a point cloud model during the data collecting step. In the present paper, we propose a real-time 3D reconstruction system by using simultaneous localization and mapping (SLAM) and evaluate this system with the image-based algorithm SfM and laser scanner.

2 LITERATURE REVIEW

Laser scanner is widely used in industry to generate 3D models by using Light Detection and Ranging (LiDAR) to scan the scene and generate 3D point clouds. Multiple scans have to be done to cover the whole scene in the same coordinate system. Embedded software is used to register and combine all those scans for the complete point clouds of the whole scene. Image-based 3D reconstruction processes a group of 2D overlapping images to generate 3D point clouds. Image data are obtained from monocular cameras, binocular cameras and video cameras. A monocular camera is the common camera which is used in daily life. The binocular camera gets stereo images which are two monocular images in pair, while the video camera gets video images including a series of monocular images or a series of stereo images.

In the construction domain, Arabic (2007) used 3D terrestrial laser scanner to digitize 3D information of real-world object and terrain down to millimeter detail. Zhang et al. (2013) set up a preliminary system which can assess construction process control with minimum human input. They got robust and accurate result under laboratory conditions with objects of regular shape. Randall (2011) analyzed the construction engineering requirements of laser scanning technology and proposed a multidisciplinary framework to integrate laser scanning technology with the fundamentals of 3D model-based design. Although laser scanner has high precision in dense 3D point cloud reconstruction, it still has a set of limitations. The cost of one laser scanner device is from 10,000 to 130,000 USD (Golparvar-Fard et al. 2011). In addition, the process of data collecting step is time-consuming, which needs several hours to generate one separate point cloud in high resolution. Therefore, it may take at least several weeks to get a large-scale model of a building. Moreover, if a moving object is in the line of sight of the laser scanner's optical sensing instrument, it will create occlusion on the scene. Moreover, with the increasing distance between the scanned objects, the level of details of captured scene will be reduced.

Golparvar-Fard et al. (2011) developed an automatic 3D reconstruction application using SfM with collection of unordered daily construction site photos. They evaluated point cloud data obtained by SfM with point cloud data obtained by terrestrial laser scanning approach. The result showed that the accuracy of using the image-based point cloud models is less than the point cloud model generated by the laser scanner. Furukawa et al. (2009) proposed a fully automated 3D reconstruction and visualization system for architectural scenes using Bundler Structure from Motion (BSfM) package and Patch-Based Multi-View Stereo (PMVS). Their system can generate a 3D model of interiors of a small house. Brilakis et al. (2011) developed a videogrammetric framework using moving camera to progressively reconstruct dense 3D point cloud of construction site mainly using Speeded-Up Robust Features (SURF) and SfM. This framework showed acceptable results only under controlled setting with miniature scale model of a simulated construction site. As it used video data with moving camera, this framework can solve the problem of image occlusion compared with normal image-based reconstruction. From the above review of image-based algorithms, SfM is the most widely used algorithms. The acquisition of image data is easy, efficient and cost-effective. Image data can be obtained by normal consumer camera and the cost of camera is only hundreds of dollars. Golparvar-Fard et al. (2011) concluded that the time consuming in getting SfM point cloud only needs about 6% man-hours compared with getting point cloud using laser scanner. However, the above mentioned algorithms still cannot achieve real-time 3D reconstruction.

To the best of our knowledge, there is limited existing research on real-time 3D reconstruction in construction site. The objective of the present research is to develop an automated real-time 3D reconstruction system using SLAM, which is widely used in real-time robot localization and mapping. Our focus is on using dense point cloud generated by SLAM to do 3D real-time reconstruction.

3 METHODOLOGY

The proposed 3D real-time reconstruction system contains four steps: visual odometry, optimization, loop closing, and 3D point cloud mapping. Figure 1 shows the framework of the methodology.

3.1 *Visual odometry*

The main functionality of visual odometry is to estimate camera motion based on images taken. Images contain a matrix of brightness and color. It is difficult to estimate camera motion directly from a matrix of data. Therefore, some representative points are extracted from the image first and these points will remain the same after changing the positon and

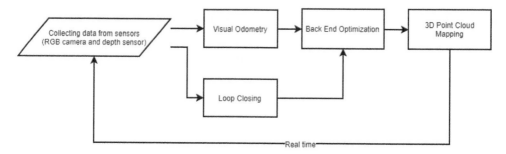

Figure 1.   Framework of 3D real-time reconstruction system.

orientation of the camera. These representative points are called *Features* in visual SLAM. The camera position and orientation will be estimated by using calculating the changes of these *Features*. The corners and edges in the image are taken as features since they are more distinguishable among different images. However, when the camera position is far away from one object, the corner may not be recognized. Therefore, in this research, Oriented FAST and Rotated BRIEF (ORB) is used to detect *Features* based on four characteristics: repeatability, distinctiveness, efficiency and locality (Rublee et al. 2011). The ORB feature is composed of two parts: the key point and descriptor. Its key point is called Oriented Features from Accelerated Segment Test (FAST), which is the FAST corner point (Rosten et al. 2006). Its descriptor is called the Binary Robust Independent Elementary Features (BRIEF) (Calonder et al. 2010). After the ORB features are calculated, a fast library for approximate nearest neighbors (FLANN) is used to match ORB features in each image (Muja et al. 2014). Finally, the camera positon and orientation is calculated using PnP (Perspective-n-Point) with the positon data of the ORB features (Lepetit et al. 2013).

3.2   Back end optimization

The error of camera position estimation will be increased with the increasing number of frames because only two adjacent frames are considered in visual odomerty step. Optimization step aims to reduce the error caused in the visual odemetry step through calculating a fixed number of previous frames. Bundle Adjustment (BA) algorithm is used in calculating the total errors of camera position estimation among previous frames (Triggs et al. 1999). Non-liner optimization library g2o is used to solve the minimum error of BA to get optimized position and orientation of camera which makes bundles of light rays reflected from all feature points converging to the camera's optical center (Kümmerle 2011).

3.3   Loop closing

Similar data are detected in this step among data collected by the camera in similar positions. Loop closing is important for SLAM system as it helps to remove cumulative errors while estimating trajectory over a long period of time. Bag-of-Words (BoW) is used to distinguish the similarity of images. In BoW, each object is represented by one word. A dictionary is made up from all the words describing the objects in the whole scene. The similarity between two images is then calculated by identifying the number's difference of specific objects. The dictionary of this research is generated by the DBoW3 library (2017).

3.4   3D point cloud mapping

In the mapping step, all the Red Green Blue Depth (RGBD) data are combined to generate 3D reconstruction point cloud. As the camera position and orientation of each image has been estimated in the previous three steps, all the coordinates of separate point clouds of RGBD images can be calculated and combined together to get a complete point cloud. Due to the effective range of Kinect V2 is from 0.5m to 4.5m, noisy points are removed where the depth value is out of this range. After that, statistical filtering methods are used to remove isolated noise points. The filter counts the distribution of the distance values of each point from its nearest N point and removes points with a large mean distance. Finally, downsampling is performed using a Voxel Filter. Since there are overlapping fields of view in multiple viewing angles, there are a large number of points in the overlapping area. Voxel filtering guarantees that there is only one point in a certain size cube, so as to remove redundancy.

4 PRELIMINARY RESULT

In a preliminary test, we used a small cabinet to evaluate the real-time SLAM system. Firstly, a 3D point cloud reconstruction model of a small cabinet was generated by using real-time SLAM (Figure 2). A Kinect V2 RGBD camera was used to collect normal RGB and depth images. And then, the same cabinet is reconstructed by using visual SfM proposed by Wu (2013) (Figure 3) with 26 monocular images taken by a camera. In addition, a more precise model was generated by using Leica P40 laser scanner (Figure 4).

a. Front side   b. Left side   c. Right side   d. Back side   e. Top side

Figure 2.   Point cloud of cabinet (SLAM).

a. Front side   b. Left side   c. Right side   d. Back side   e. Top side

Figure 3.   Point cloud of cabinet (SfM).

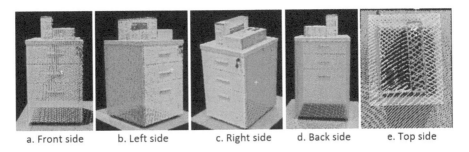

a. Front side   b. Left side   c. Right side   d. Back side   e. Top side

Figure 4.   Point cloud of cabinet (Laser Scanner).

Table 1 presents the appearance evaluation of three point cloud models from five sides. The Point cloud generated by SLAM is blurry in all the five sides due to noise points over the surface. These noise data are hard to be removed as the data are mixed up together. This issue is caused by the fault estimation of camera position and orientation. In SfM model, appearances of left side, right side and back side are totally missing because there are very few features due to the surface of these three sides are flat and with solid color. SLAM model has detailed information on these three sides because we use RGBD camera which has depth information to know that these three sides are flat. As to the laser scanner model, the appearances are very clear in all five sides.

The accuracy is evaluated through measuring the ratio of x/y, y/z and x/z (x, y and z represents length, width and height of the small cabinet respectively). The calculated ratio of each algorithm is compared with the actual ratio (Table 2) and the equation to compute the error is shown as below:

$$e = \left| \frac{R_{cal} - R_{act}}{R_{act}} \right| \tag{1}$$

where e = Error; R_{cal} = calculated ratio; and R_{act} = Actual ratio

Moreover, the data collecting time, data processing time and average accuracy are evaluated and the results are shown in Table 3. Because the processing of SLAM is in real-time, the data collecting and processing time is combined together. The average accuracy of SLAM is 96.8% which is an acceptable result and it only took 2.36% total time of SfM and 0.104% total time of Laser scanner.

Table 1. Appearance evaluation.

	Front side	Left side	Right side	Back side	Top side
SLAM	blurry	blurry	blurry	blurry	blurry
SfM	Some parts missing	Missing	Missing	Missing	Some parts missing
Laser scanner	Clear	Clear	Clear	Clear	Clear

Each column evaluates the appearance of cabinet model in one side and each row evaluates the appearance of cabinet model generated by one algorithm.

Table 2. The error evaluation of models generated by using SLAM, SfM and Laser scanner.

	Actual ratio	Calculated ratio of SLAM/Error	Calculated ratio of SfM/Error	Calculated ratio of Laser scanner/Error
x/y	0.833	0.808/0.03	0.856/0.028	0.842/0.01
y/z	0.8	0.786/0.018	0.784/0.02	0.79/0.012
x/z	0.667	0.635/0.048	0.672/0.007	0.666/0.002

Error is calculated based on Equation 1.

Table 3. Processing time and average accuracy of SLAM, SfM and Laser scanner.

	SLAM	SfM	Laser scanner
Data collecting time	10 s in total	70 s	6900 s
Data processing time		352 s	2700 s
Accuracy	96.8%	98.2%	99.2%

5 CONCLUSION AND FUTURE WORK

This research used SLAM to achieve real time 3D reconstruction in construction site. Currently, only a small-scale experimental environment was evaluated and got relatively precise results. However, the detailed information of the object was missing and there were many noise points which influenced the appearance of reconstruction. In the future, 3D real-time reconstruction in real construction site will be evaluated; we will establish data interaction between Building Information Model (BIM) and real-time reconstructed model to achieve automatic processing monitoring, construction safety management and geometry quality inspection etc. Moreover, we will try to achieve real-time 3D reconstruction using unmanned ground vehicles (UGV) or unmanned ground vehicles (UAV).

ACKNOWLEDGEMENT

We would like to acknowledge the support from the Key Program Special Fund of Xi'an Jiaotong-Liverpool University, project code KSF-E-04.

REFERENCES

Arayici, Y. 2007. An approach for real world data modelling with the 3D terrestrial laser scanner for built environment. *Automation in Construction* 16(6): 816-829.

Brilakis, I. Fathi, H. & Rashidi, A. 2011. Progressive 3D reconstruction of infrastructure with videogrammetry. *Automation in Construction* 20(7): 884-895.

Calonder, M. Lepetit, V. Strecha, C. & Fua, P. 2010. September. Brief: Binary robust independent elementary features. *In European conference on computer vision, Springer, Berlin, Heidelberg* 778-792.

DBoW3. 2017. https://github.com/rmsalinas/DBow3. (Accessed on July 8th, 2019)

Furukawa, Y. Curless, B. Seitz, S.M. & Szeliski, R. 2009. September. Reconstructing building interiors from images. *In 2009 IEEE 12th International Conference on Computer Vision IEEE* 80-87.

Golparvar-Fard, M. Bohn, J. Teizer, J. Savarese, S. & Peña-Mora, F. 2011. Evaluation of image-based modeling and laser scanning accuracy for emerging automated performance monitoring techniques. *Automation in Construction* 20(8):1143-1155.

Ham, Y. & Golparvar-Fard, M. 2013. An automated vision-based method for rapid 3D energy performance modeling of existing buildings using thermal and digital imagery. *Advanced Engineering Informatics* 27(3): 395-409.

Han, K.K. & Golparvar-Fard, M. 2017. Potential of big visual data and building information modeling for construction performance analytics: An exploratory study. *Automation in Construction* 73:184-198.

Kümmerle, R. Grisetti, G. Strasdat, H. Konolige, K. & Burgard, W. 2011. May. G^2o: A general framework for graph optimization. *In 2011 IEEE International Conference on Robotics and Automation (pp. 3607-3613)*.

Lepetit, V. Moreno-Noguer, F. & Fua, P. 2009. Epnp: An accurate o (n) solution to the pnp problem. *International Journal of Computer Vision* 81(2):155.

Muja, M. & Lowe, D.G. 2014. Scalable nearest neighbor algorithms for high dimensional data. *IEEE Transactions on Pattern Analysis and Machine Intelligence* 36(11): 2227-2240.

Randall, T. 2011. Construction engineering requirements for integrating laser scanning technology and building information modeling. *Journal of Construction Engineering and Management* 137(10): 797-805.

Rosten, E. & Drummond, T. 2006. Machine learning for high-speed corner detection. *In European conference on computer vision. Springer, Berlin, Heidelberg* 430-443.

Rublee, E. Rabaud, V. Konolige, K. & Bradski, G.R. 2011. ORB: An efficient alternative to SIFT or SURF. In ICCV 11(1):2.

Triggs, B. McLauchlan, P.F. Hartley, R.I. & Fitzgibbon, A.W. 1999. *Bundle adjustment—a modern synthesis. In International workshop on vision algorithms. Springer,* Berlin, *Heidelberg* 298-372.

Wang, Q. & Kim, M.K. 2019. Applications of 3D point cloud data in the construction industry: A fifteen-year review from 2004 to 2018. *Advanced Engineering Informatics* 39: 306-319.

Wu, C. 2013. Towards linear-time incremental structure from motion. *In 2013 International Conference on 3D Vision-3DV IEEE* 127-134.

Zhang, C. & Arditi, D. 2013. Automated progress control using laser scanning technology. *Automation in Construction* 36:108-116.

Sustainable Buildings and Structures: Building a Sustainable Tomorrow – Papadikis et al. (Eds)
© 2020 Taylor & Francis Group, London, ISBN 978-0-367-43019-1

Social impact assessment for sponge city PPPs: Framework and indicators

J. Guo, W. Li & J. Yuan
School of Civil Engineering, Southeast University, Nanjing, China

ABSTRACT: Sponge City is being widely implemented as a measure of urban stormwater management in China, and public-private partnerships are encouraged to apply in the construction of sponge city projects to solve the financial problems. Considering the complicated social impacts brought by sponge city PPPs, it was necessary to evaluate its all positive and negative social impacts, only then sponge city PPPs could play an important role in Chinese urban stormwater management. This paper proposed the framework for social impact assessment of sponge city PPPs, and 33 SIA indicators were selected according to the framework. These SIA indicators could help understand the content and intension of social impacts of sponge city PPPs better, and provide foundation for further analysis of the importance and mechanisms of sponge city PPPs' social impacts.

1 INTRODUCTION

With the continuous development of urbanization, water ecological crisis is becoming a severe problem faced by many cities around the world with its enormous negative impact on human wealthy and city construction. Under this condition, various urban stormwater management schemes and practices have been developed to solve this problem, such as Low Impact Development, Green Infrastructure and Water Sensitive Urban Design in developed countries (Li & Jensen 2018).

In recent years, China suffered from many urban rainstorm waterlogging and water resource pollution disasters. Due to the excessively rapid urbanization, the outdated urban construction and management modes, which lacked consciousness about sustainable development, have impeded the urban sustainability of China seriously. By gaining experience from advanced urban stormwater management practices, "sponge city" is an initiative highly advocated by Chinese government since 2012. According to Technical Guidance on Sponge City Construction published by Chinese Ministry of Housing and Urban-Rural development, sponge city means "a city that has good 'elasticity' in adapting to environmental changes and responding to natural disasters like a sponge, so that it absorbs water, stores water, soaks water and purifies water when raining, then releases and reuses the water when needed". As the application of Low Impact Development in China, sponge city is taken as a key approach to solve the urban water ecological crisis. About 30 cities were selected to be pilot cities for sponge city construction in China with the total investment more than $1.5 trillion nationwide.

Considering the huge amount of capital needed for sponge city's construction, it is encouraged vigorously by Chinese government to apply public-private partnerships into sponge city construction. Since 2012, there has been a boom of public-private partnerships (PPPs) in Chinese infrastructure construction. In terms of sponge city, public-private partnerships are promoted strongly by financial subsidy and other preferential policies (Wang et al. 2017). At present, most pilot cities have adopted public-private partnerships into the construction of sponge city, which would be the main type of Chinese sponge city projects in future.

While the sponge city itself is an infrastructure project with great scale and deep influence, involving city comprehensive planning, water system renewal, green building reform, green land engineering, road and square transformation, its combination with public-private partnerships intensifies the extent of sponge city's social impact through multipartite stakeholders. There is no doubt that the join of private sector could complicate the potential social impacts and risks of sponge city because of the different demand between public and private sector. So, the social impact of sponge city public-private partnerships projects could be obviously different from general infrastructure projects of urban construction.

Social impact assessment of public infrastructure has always been a valid method to analyze and manage the social consequence caused by infrastructure projects (Vanclay 2002), because it would be possible to estimate both positive benefits and negative impacts of infrastructure projects by social impact assessment. Administrators could make interventions and action plans according to the social impact assessment, mitigating negative impacts and amplifying social benefits. In view of the huge scale of sponge city and diversity of stakeholders of public-private partnerships, it is necessary to establish a particular social impact assessment framework for sponge city public-private partnerships, to make it more significant in field of urban water ecology management and urban sustainability.

This paper aims to evaluate social impacts of sponge city public-private partnerships from the perspective of social change process. Then the social impact assessment framework and indicators are formulated. Through the analysis of social impact assessment indicators, the policy suggestions on how to improve the performance of sponge city PPPs projects in China are built. The research of sponge city's social impact may prepare the administration with an effective intervention to manage these large-scale infrastructure projects.

2 LITERATURE REVIEW

2.1 *Social impact assessment*

It is generally believed that social impact assessment originated from the National Environmental Policy Act (NEPA) of the United States in 1969. At that time, social impact assessment was carried out more often as an integral part of environmental impact assessment (Burdge 2003). Since then, the role of social impact assessment in addressing social changes and mitigating social impacts was recognized gradually (Tang et al. 2008), as a powerful policy tool of decision making.

Scholars have not yet achieved consensus about the definition of social impact assessment. Vanclay proposed that 'social impact assessment is analyzing, monitoring and managing the social consequences of development' (Vanclay 2003). It should be taken as a 'philosophy about development and democracy' (Vanclay 2010), not only a technique or method. Becker defined social impact assessment as "the process of identifying the future consequences of a current or proposed actions, which are related to individuals, organizations and social macro-systems" (Becker 2001). In developed countries, with long-term expansion and enrichment, the practitioners of SIA paid more attention on values and principles about human rights, social equity and social inclusion, and so on. Nonetheless, SIA should still aim to mitigate the negative impact and improve social well-being with structural and operational barriers (Tang et al. 2008). It was difficult for SIA in a developing country to exert its intrinsic significance, given the limitation on data currency and the lack of public participation (Esteves et al. 2012). Overall, the cognition of SIA should not restrict on the traditional concept of "social", but more on the comprehensive knowledge including all artificial social changes and impacts of planned interventions (Vanclay 2002).

2.2 *Urban stormwater management practices*

Urban stormwater management terms got gradually complex and various with the constant development of urbanization on a global scale (Fletcher et al. 2013). These terms were used in different countries with distinctive guidelines, including LID, Sustainable Urban Drainage

Systems (SUDS), Water Sensitive Urban Design (WSUD), Best Management Practices (BMPs) or Green Infrastructure (GI).

The United States first began the exploration of urban stormwater management. BMPs was first proposed by the Clean Water Act in 1972 in U.S, which was aimed to solve the non-point source pollution in urban and rural areas initially, then gradually developing into an integrated framework of mitigating the storm water pollution (Chang et al. 2018). Based on BMPs, LID was frequently applied in North America and New Zealand, originated from Prince George's County, Maryland, USA in the early 1990s. The idea was mainly to control the source of rainwater runoff by decentralized small-scale measures, so as to control water pollution caused by storm water. With the widespread use of LID, a new concept named Green Infrastructure (GI) was raised in 1999, distinguishing from the traditional 'Gray Infrastructure' (Benedict & Mcmahon 2002). The emphasis of GI was on the interconnected network of green spaces including green roofs, trees, and permeable pavement that can infiltrate and accumulate rains (Foster et al. 2011). For now, these terms were interchangeable in most contexts, and their meanings were not restricted on storm water management, but a comprehensive approach for urban ecology system.

Besides, other countries also proposed their stormwater management practices, such as Sustainable Urban Drainage Systems (SUDS) in UK, Water Sensitive Urban Design (WSUD) in Australia and Low Impact Urban Design and Development (LIUDD) in New Zealand. These stormwater management practices had different aims or characteristics according to the diverse needs of these countries, seen in Table 1. However, no matter what the original aims of these practices were, they were all becoming an integrated urban management method involving the whole urban water circle and the macro urban development strategy.

2.3 *PPPs: Social dimension*

Public-private partnerships are becoming more and more prevalent around the world, as effective tools to deliver public services and infrastructure (Ghere 2001). In general, PPPs can be defined as a cooperative arrangement between public and private sectors that distinguished by the following characters. First, the partnerships between public and private sectors are long-term infrastructure contracts (Hodge & Greve 2007). Second, PPPs emphasize on risk sharing, transferring part of risks undertaken by public sector in traditional infrastructure projects(Forrer et al. 2010). Another character of PPPs are public product or services provided by the joint of public and private sectors (Hueskes et al. 2017).

There were very limited literatures related to social impact assessment of PPPs, although it has been admitted that PPPs would cause a variety of impacts on social and environmental

Table 1. Urban stormwater management practices in developed countries.

Practices	Country	Year	Features or goals
BMPs	U.S.	1972	solve the non-point source pollution in urban and rural areas
LID	U.S. and New Zealand	1990s	control the source of rainwater runoff by decentralized small-scale measures, control water pollution caused by storm water
GI	U.S.	1999	an interconnected network of green spaces including green roofs, trees, and permeable pavement that can infiltrate and accumulate rains
SUDS	UK	1990s	consider from perspective of sustainable development, combining quantity, quality and habitat in urban drainage problem (Fletcher et al. 2015)
WSUD	Australia	1990s	an integrated water management framework related to the whole urban water circle, from water supply to sewerage and storm water treatment, not only a storm water management method (Wong 2006)
LIUDD	New Zealand	2000s	achieve urban sustainable development by satisfying both economic and environmental needs, improving resource management in urban catchment, eventually enhancing urban biodiversity (Ignatieva et al. 2008)

systems (Forrer et al. 2010, Lund-Thomsen 2009). It is difficult to accurately evaluate the social dimensions of PPPs' impacts as the they involve more qualitative rather than quantitative indicators (Hueskes et al. 2017). As a policy strongly advocated by many developing countries, the "poor evaluation rigor" and lack of measurable criteria has hindered private capital to enter the PPPs market largely (Hodge & Greve 2007, Lund-Thomsen 2009).

One of the most concerned social issues of PPPs focused on local community acceptance associated with the level of public participation in PPPs preparation period (Hodge & Greve 2007, Chan et al. 2010, Hueskes et al. 2017), which is highly concerned in field of social impact assessment too. Another key feature of PPPs was the relationship governance between the public and private sector, as their complex collaboration in risk, cost and benefit sharing (Asquith et al. 2015). Hence, there is a strong demand for the government to establish a strong control mechanism, with adequate legal framework and stable political environment to monitor the private's behaviors in the whole life cycle of PPPs (Chan et al. 2010).

3   A FRAMEWORK FOR CONCEPTUALIZING SIA OF PPP SPONGE CITY PROJECTS

Considering the complexity of social impact, it is difficult to establish an overall framework that contain all dimensions and indicators of social impact, which has been admitted by many scholars (Vanclay 2002, Padilla-Rivera et al. 2016). However, there are some indispensable advantages of social frameworks or indicators, as they work as useful tools in field of public policymaking and social issues addressing. A comprehensive framework can provide an advanced understanding of the specific and various impacts caused by major infrastructure projects (Rossouw & Malan 2007). This section aimed to construct an integrated framework of PPP sponge city project's social impact assessment, synthesizing existed social impact assessment research and the characteristics of PPP sponge city projects.

Through the concept of SIA by Vanclay (2003) in International Principles, social impact assessment aims at "... the intended and unintended consequences on the human environment of planned interventions (policies, programs, plans, projects) and any social change processes invoked by those interventions...". Thus, the projects as a kind of planned interventions lead to the direct social change processes, the so-called first order changes. In addition, the first order social change process can cause several other, the second- and higher- order social change processes (Vanclay 2002, Burdge 2003). For example, urban planning and resettlement, the social change processes, are set in motion by the construction of sponge city projects (planned interventions) directly, then resettlement can bring about the rural to urban migration process (the second order social change process) and changes in living environment (the higher order social change process). Under the local social settings, these social change processes will result in social impacts, such as type of employment and loss of community cohesion. Different from the social changes, 'social impacts' are the impacts actually experienced by humans (individuals or social groups) in the sense of material (physics) or cognition (perception) (Vanclay 2002).

Usually, SIA is far from limited to a narrow or restrictive concept–"social" (Vanclay 2002, Vanclay 2003). Because the social, economic and environmental domains have the inherent and inextricable interconnections (Dempsey et al. 2011, Vanclay 2002). In this way, social changes will lead to changes in economic and environmental domains, and then have the potential to trigger impacts of these two domains, as the some kind of second and higher order impacts and the cumulative impacts (Arce-Gomez et al. 2015, Vanclay 2003). In other words, there must be comprehensive consideration of impact variables in social, economic and environmental system during the process of SIA in projects, of course, including PPP sponge city projects. Moreover, this is aligned with the urban sustainable development goals of the management practices such as social, economic and environmental sustainability in sponge city projects (Ignatieva et al. 2008, Chan et al. 2018).

On the other hand, people's social experience of changes (i.e. the social impacts) can provoke them to take actions that result in further social change processes (Vanclay 2002). For example, the negative social impacts (experience) such as uncertainty and fear amongst the local residents (being unsure about the role and effects of this innovation and new approach, sponge city PPP projects) can activate the formation of opposition groups in the community, sometimes these groups can easily hinder the projects' success. Therefore, the projects (planned interventions) can affect the local communities, and the affected communities can also have a certain reaction on the projects (Arce-Gomez et al. 2015). In order to receive "a social license to operate" and avoid the protest action, on the one hand, the projects should consider the differences in impact distribution of different groups in community, especially the vulnerable groups' impact burden (i.e. enhance projects' adaptability to the community), and on the other hand, the projects should offer a range of approaches and tools to effectively engage the public, which can assist them in understanding and planning for the changes and help them more easily adapt to the changes that will likely affect their lives (Arce-Gomez et al. 2015, Vanclay 2003, Vanclay et al. 2015).

In reality, the implementation of sponge city PPP projects as a new initiative often remains low public perceptions, and these projects are mostly constructed scatted in large areas and above the ground, it is easier to get people's attention, so the extent of adaptability between the project and local communities has a huge bearing on the amount of people's fears and anxieties generated, and the ultimate success of the project (Vanclay et al. 2015, Wang et al. 2017). Thus, in the conceptualizations of SIA of sponge city PPP projects, there should be great concern with ensuring the development of these projects be generally acceptable and more adaptable to the communities (Vanclay et al. 2015, Vanclay 2003).

To meet the huge investment needs, public private partnerships (PPP) as the innovative arrangement has been facilitated as the main impetus for sponge city projects (Wang et al. 2017). Unlike the traditional procurement approaches, the sponge city projects delivery by PPPs allocate risks between the public and private sectors, and encourage the private sectors' efficiency by linking their payments to specific performance criteria they achieved (Li et al. 2016). Due to the nature of public goods, sponge city projects typically are not easily converted into profits, the returns of private sectors mainly depend on government subsidies and purchases (Dai et al. 2018, Li et al. 2016). Therefore, the investment enthusiasm of private investors in the sponge city projects and whether the projects delivery by PPP can satisfy the social needs or solve the social problem (i.e. generate social benefits) are directly related to the government performance (Dai et al. 2018). In addition, whether the social impacts will be caused by social change processes depends on the characteristics of the local social setting and the extent of mitigation measures that are put in place (Vanclay 2002). If the local government can offer a more robust for sponge city PPP projects and manage properly, and at the same time develop strategies to help the communities cope with changes and promote the acceptance to the projects, then the social changes might not create impacts (Vanclay 2002). Thus,

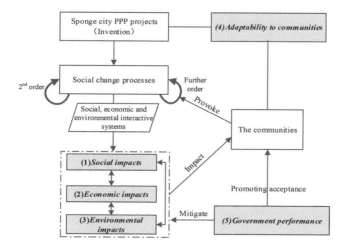

Figure 1. SIA framework of sponge city PPP projects.

good government performance will make much difference to the social impacts of sponge city PPP projects, the significant roles of the government must be taken into account when conducting the SIA of sponge city PPP projects.

According to the analysis above, as shown in Figure 1, based on the principle of SIA and the features of sponge city PPP projects, the following five dimensions should be considered in SIA framework of sponge city PPP projects due to their significant roles in generation and development of social impacts of these projects: (1) Social impacts (SOI); (2) Economic impacts (ECI); (3) Environmental impacts (ENI); (4) Adaptability to communities (AC); (5) Government performance.

4 ESTABLISH THE SIA INDICATOR SYSTEM UNDER THE FRAMEWORK

After building social impact assessment framework of sponge city PPPs, the next step was to form measurable indicators on the basis of the framework. The SIA indicators of sponge city PPPs were introduced with five dimensions shown in Figure 1: economic impacts, environmental impacts, social impacts, adaptability to communities and government performance.

4.1 Social impacts

Indicators of Social impacts dimension emphasized sponge city PPPs' impacts on local society, from the perspective of infrastructure and humanity. Indicators related to infrastructure included *Complement to infrastructure (SOI1), Operational efficiency of infrastructure (SOI2), Daily convenience of local communities (SOI3), Educational, cultural and entertaining facilities of local communities (SOI4)*, where the quantity and efficiency of public infrastructure and services were both taken as the core content of social impact assessment (Wang et al. 2013, Vanclay 2002). Besides, sponge city also imposed impacts on urban population and their activities, where *Proportion of agricultural activities in urban economic activities (SOI5), Proportion of urban population in population (SOI6), Local employment rate (SOI7), Rights protection of vulnerable groups (SOI8), Proportion of poverty population (SOI9)* were used to measure the change of humanity caused by sponge city PPPs (Weidema 2006). In conclusion, Social impact dimension consisted of typical contents that social impact assessment should concentrate on, although some of them were qualitative and difficult to measure in quantitative terms. But they were indispensable to evaluate social impacts of sponge city PPPs comprehensively.

4.2 Economic impacts

Indicators of Economic impacts dimension were considered from three layers: family, city, and country, to measure the change in field of local economic brought by sponge city PPPs entirely. The first layer was family, including three indicators: *Income level (ECI1), Consumption level (ECI2), Engel coefficient (ECI3)*. These indicators mainly measured the economic impacts to families around sponge city PPPs. The second layer was city, including three indicators: *Local GDP (ECI4), Industrial structure (ECI5), Fiscal revenue (ECI6)*. And the third layer was country, including two indicators: *Foreign investment and advanced technology (ECI7), Large trans-regional infrastructure layout (ECI8)* (Vanclay 2003, Colburn & Jepson 2012). Because of the large scale of sponge city project, it was totally worth considering its macro impacts on region and country. From the perspective of country, sponge city projects involved kinds of water treatment technology and affected integrated regional water system, so the application of advanced technology and the layout of trans-regional infrastructure were taken into consideration of the impacts of sponge city projects.

4.3 Environmental impacts

As described above, indicators of Environmental impacts dimension were formulated according to the "Sponge City Construction Performance Evaluation and Assessment Indicators" published by Chinese Ministry of Housing and Construction. Environmental dimension contained four indicators: *Water ecology (ENI1), Water environment (ENI2), Water resource (ENI3), Water safety (ENI4)* (Li et al. 2016). These indicators were all relevant to urban water circulation system. Their meaning has been explained above.

4.4 *Adaptability to communities*

Adaptability to communities (AC) dimension was also given high attention in general social impact assessment, as it paid great attention on the interaction between the project and local communities. Social impact was people's perception to social changes caused by projects to some extent. Therefore, this dimension underlined the attitude and behavior of residents around sponge city PPPs. *Approval rate for projects of local communities (AC1), Information disclosure of projects (AC2), Public participation in project decision making (AC3)* described the role of communities in project planning and management, which would help projects adapt to local society better (Becker et al. 2003). *Public dispute caused by the project (AC4), Risk prevention measures (AC5)* reflected contradiction between project and communities, which could lead to social risk and failure of projects directly (Domínguez-Gómez 2016).

4.5 *Government performance*

Government performance (GP) dimension represented the role of PPPs. Indicators of government performance dimension could be divided into two sectors: one sector presented the political environment for PPPs set by local government, including *Consistency and continuity of PPP policy (GP1), Legal perfection of PPP law (GP2), Coordination between relevant agencies (GP3)*, which represented the healthy market for PPPs (Sachs et al. 2007); the other sector presented the government reputation problems in sponge city PPPs (Chan et al. 2014), including *Government default (GP4), Unreasonable project change (GP5), Rationality of project objectives (GP6)*, which were the common reason for failures of PPPs. Finally, *sharing of benefits and risks (GP7)* between government and the private sector was the most important factor that determined if PPP project could achieve success (Ke et al. 2010).

Table 2. SIA indicators of sponge city PPPs.

	Dimension	Indicator
Social impact assessment of sponge city PPPs	Economic impacts (ECI)	Income level (ECI_1) Consumption level (ECI_2) Engel coefficient (ECI_3) Local GDP (ECI_4) Industrial structure (ECI_5) Fiscal revenue (ECI_6) Foreign investment and advanced technology (ECI_7) Large trans-regional infrastructure layout (ECI_8)
	Social impacts (SOI)	Complement to infrastructure (SOI_1) Operational efficiency of infrastructure (SOI_2) Daily convenience of local communities (SOI_3) Education, culture and sanitation facilities of local communities (SOI_4) Proportion of agricultural activities in urban economic activities (SOI_5) Proportion of urban population in population (SOI_6) Local employment rate (SOI_7) Rights protection of vulnerable groups (SOI_8) Proportion of poverty population (SOI_9)

(Continued)

Table 2. (*Continued*)

	Dimension	Indicator
	Environmental impacts (ENI)	Water ecology (ENI_1) Water environment (ENI_2) Water resource (ENI_3) Water safety (ENI_4)
	Adaptability to communities (AC)	Approval rate for projects of local communities (AC_1) Information disclosure of projects (AC_2) Public participation in project decision making (AC_3) Public dispute caused by the project (AC_4) Risk prevention measures (AC_5)
	Government performance (GP)	Consistency and continuity of PPP policy (GP_1) Legal perfection of PPP law (GP_2) Coordination between relevant agencies (GP_3) Government default (GP_4) Unreasonable project change (GP_5) Rationality of project objectives (GP_6) Sharing of benefits and risks (GP_7)

5 CONCLUSION

This paper constructed a social impact assessment framework for Chinese sponge city public-private partnerships on the basement of social impact assessment researches, taking features of sponge city and public-private partnerships into consideration in the meanwhile. Furthermore, social impact assessment indicators were proposed.

Social impact assessment has always been considered to be a valid way to improve efficiency of infrastructure projects and avoid possible risks or failure, as it would help the decision maker develop specific project management measures based on the advanced analysis of social impacts. Especially for sponge city PPPs, as an innovative project involved in varieties of infrastructure renovation and construction projects, its social impacts characterized by wide geographical area and complex affected populations. Therefore, scientific social impact assessment could play an important role in reducing negative impacts and implement project goals more effectively. Although there are some deficiency using indicator to evaluate the complicated and dynamic social impact (Arce-Gomez et al. 2015), they still were an widely-used tools that could be useful and intelligible for public policymaker as their focus and multi-dimensions on special issues (Padilla-Rivera et al. 2016).

There were also limitations in this research. Although the social impact assessment indicators system has been built, but the significance of different indicator or dimension has not been identified. Measuring the different significance of each dimension and indicator would help understand the internal formation mechanism of social impacts effectively, which could be discussed further in future studies.

REFERENCES

Arce-Gomez, A. Donovan, J.D. & Bedggood, R.E. 2015. Social impact assessments: Developing a consolidated conceptual framework. *Environmental Impact Assessment Review* 50:85-94.

Asquith, A. Brunton, M. & Robinson, D. 2015. Political Influence on Public–Private Partnerships in the Public Health Sector in New Zealand. *International Journal of Public Administration* 38:179-188.

Becker, D. R. Harris, C.C. Mclaughlin, W.J. & Nielsen, E.A. 2003. A participatory approach to social impact assessment: the interactive community forum. *Environmental Impact Assessment Review* 23: 367-382.

Becker, H. A. 2001. Social impact assessment. *European Journal of Operational Research* 128:311-321.

Benedict, M. A. & Mcmahon, E. T. 2002. Green infrastructure: smart conservation for the 21st century. *Renewable Resources Journal* 20:12-17.

Burdge, R. J. 2003. The practice of social impact assessment background. *Impact Assessment and Project Appraisal* 21:84-88.

Chan, A. P. Lam, P.T. Chan, D.W. Cheung, E. & Ke, Y. 2010. Critical success factors for PPPs in infrastructure developments: Chinese perspective. *Journal of Construction Engineering and Management* 136:484-494.

Chan, A. P. Lam, P. T. Wen, Y. Ameyaw, E.E. Wang, S. & Ke, Y. 2014. Cross-sectional analysis of critical risk factors for PPP water projects in China. *Journal of Infrastructure Systems* 21: 04014031.

Chan, F.K.S. Griffiths, J. A. Higgitt, D. Xu, S. Zhu, F. Tang, Y.T. Xu, Y. & Thorne, C.R. 2018. Sponge City in China—A breakthrough of planning and flood risk management in the urban context. *Land Use Policy* 76:772-778.

Chang, N.B. Lu, J.W. Chui, T.F.M. & Hartshorn, N. 2018. Global policy analysis of low impact development for stormwater management in urban regions. *Land Use Policy the International Journal Covering All Aspects of Land Use* 70:368-383.

Colburn, L.L. & Jepson, M. 2012. Social indicators of gentrification pressure in fishing communities: a context for social impact assessment. *Coastal Management* 40:289-300.

Dai, L. Van Rijswick, H.F. Driessen, P.P. & Keessen, A.M. 2018. Governance of the Sponge City Programme in China with Wuhan as a case study. *International Journal of Water Resources Development* 34:578-596.

Dempsey, N. Bramley, G. Power, S. & Brown, C. 2011. The social dimension of sustainable development: Defining urban social sustainability. *Sustainable Development* 19:289-300.

Domínguez-Gómez, J.A. 2016. Four conceptual issues to consider in integrating social and environmental factors in risk and impact assessments. *Environmental Impact Assessment Review* 56:113-119.

Esteves, A.M. Franks, D. & Vanclay, F. 2012. Social impact assessment: the state of the art. *Impact Assessment & Project Appraisal* 30:34-42.

Fletcher, T.D. Andrieu, H. & Hamel, P. 2013. Understanding, management and modelling of urban hydrology and its consequences for receiving waters: A state of the art. *Advances in Water Resources* 51: 261-279.

Fletcher, T.D. Shuster, W. Hunt, W. F. Ashley, R. Butler, D. Arthur, S. Trowsdale, S. Barraud, S. Semadeni-Davies, A. & Bertrand-Krajewski, J.L. 2015. SUDS, LID, BMPs, WSUD and more–The evolution and application of terminology surrounding urban drainage. *Urban Water Journal* 12: 525-542.

Forrer, J. Kee, J.E. Newcomer, K.E. & Boyer, E. 2010. Public–private partnerships and the public accountability question. *Public Administration Review* 70: 475-484.

Foster, J. Lowe, A. & Winkelman, S. 2011. The value of green infrastructure for urban climate adaptation. *Center for Clean Air Policy* 750:1-52.

Ghere, R. K. 2001. Probing the strategic intricacies of public–private partnership: the patent as a comparative reference. *Public Administration Review* 61:441-451.

Hodge, G. A. & Greve, C. 2007. Public–private partnerships: an international performance review. *Public Administration Review* 67: 545-558.

Hueskes, M. Verhoest, K. & Block, T. 2017. Governing public–private partnerships for sustainability: An analysis of procurement and governance practices of PPP infrastructure projects. *International Journal of Project Management* 35:1184-1195.

Ignatieva, M. Stewart, G. H. & Meurk, C. D. 2008. Low Impact Urban Design and Development (LIUDD): matching urban design and urban ecology. *Landscape Review* 12:61-73.

Ke, Y. Wang, S. Chan, A.P. & Lam, P.T. 2010. Preferred risk allocation in China's public–private partnership (PPP) projects. *International Journal of Project Management* 28:482-492.

Li, L. & Jensen, M. B. 2018. Green infrastructure for sustainable urban water management: Practices of five forerunner cities. *Cities* 74:126-133.

Li, X. Li, J. Fang, X. Gong, Y. & Wang, W. 2016. Case studies of the sponge city program in China. *Proceedings of the World Environmental and Water Resources Congress* 295-308.

Lund-Thomsen, P. 2009. Assessing the impact of public–private partnerships in the global south: The case of the Kasur tanneries pollution control project. *Journal of Business Ethics* 90:57.

Padilla-Rivera, A. Morgan-Sagastume, J.M. Noyola, A. & Güereca, L. P. 2016. Addressing social aspects associated with wastewater treatment facilities. *Environmental Impact Assessment Review* 57:101-113.

Rossouw, N. & Malan, S. 2007. The importance of theory in shaping social impact monitoring: lessons from the Berg River Dam, South Africa. *Impact Assessment and Project Appraisal* 25: 291-299.

Sachs, T. Tiong, R. & Qing Wang, S. 2007. Analysis of political risks and opportunities in public private partnerships (PPP) in China and selected Asian countries: Survey results. *Chinese Management Studies* 1:126-148.

Tang, B.S. Wong, S.W. & Lau, M.C.H. 2008. Social impact assessment and public participation in China: A case study of land requisition in Guangzhou. *Environmental Impact Assessment Review* 28: 57-72.

Vanclay, F. 2002. Conceptualising social impacts. *Environmental Impact Assessment Review* 22:183-211.

Vanclay, F. 2003. International principles for social impact assessment. *Impact assessment and project appraisal* 21:5-12.

Vanclay, F. 2010. The triple bottom line and impact assessment: how do TBL, EIA, SIA, SEA and EMS relate to each other? *Tools, Techniques And Approaches For Sustainability: Collected Writings in Environmental Assessment* Policy *and* Management. World Scientific.

Vanclay, F. Esteves, A.M. Aucamp, I. & Franks, D.M. 2015. *Social Impact Assessment: Guidance for assessing and managing the social impacts of projects.*

Wang, P. Lassoie, J.P. Dong, S. & Morreale, S.J. 2013. A framework for social impact analysis of large dams: A case study of cascading dams on the Upper-Mekong River, China. *Journal of environmental management* 117:131-140.

Wang, Y. Sun, M. & Song, B. 2017. Public perceptions of and willingness to pay for sponge city initiatives in China. *Resources, Conservation and Recycling* 122:11-20.

Weidema, B. P. 2006. The integration of economic and social aspects in life cycle impact assessment. *The International Journal of Life Cycle Assessment* 11: 89-96.

Wong, T. H. 2006. Water sensitive urban design-the journey thus far. *Australasian Journal of Water Resources* 10:213-222.

Sustainable Buildings and Structures: Building a Sustainable Tomorrow – Papadikis et al. (Eds)
© 2020 Taylor & Francis Group, London, ISBN 978-0-367-43019-1

Effects of random measurement errors on a linear DEM error model: A case study using TLS point clouds

L. Fan, G. Gong & C. Zhang

Department of Civil Engineering, Xi'an Jiaotong – Liverpool University, Suzhou, P.R. China

ABSTRACT: Terrestrial laser scanning (TLS) has now become an important tool for monitoring ground surface movements during and after construction. To better understand surface changes measured, it is important to have a sound understanding of digital elevation model (DEM) accuracy. A recent study showed that a linear model can be used to represent the digital elevation model (DEM) error in terms of root mean square error (RMSE) for some typical data spacing of TLS point clouds. However, the effects of the measurement noise on that model is not clearly understood. In this study, the measurement noise, as a controlled parameter, is added to data points in a semi-artificial point cloud that is assumed to free of measurement errors to form various new measurement-error-contaminated point cloud datasets. These new datasets were analysed using a statistical resampling technique, with an attempt to quantitatively investigate the effects of random measurement errors on the coefficients of the linear model.

1 INTRODUCTION

In recent years, terrestrial laser scanning (TLS) has been used for a wide variety of applications such as the detection of landslides and terrain surface movements (Schürch et al. 2011, Montreuil et al. 2013, Day et al. 2013), and the characterisation of river bed morphology (Heritage & Milan 2009). In these applications, it is common practice to construct digital elevation models (DEMs) using TLS point clouds because a DEM forms a basic input to many GIS-related tasks. In these applications, an essential part of DEM production is evaluation of DEM accuracy (Fan et al. 2014).

DEM accuracy can be influenced by a few factors, mainly including data density, terrain complexity and interpolation methods (Fisher & Tate 2006, Hu et al. 2009, Guo et al. 2010). Some studies (Krausl et al. 2006, Bater & Coops 2009, Aguilar et al. 2010, Fan & Atkinson 2015) have been carried out to establish the correlations between these factors and DEM accuracy. For example, Krausl et al. (2006) proposed an approach to derive DEM accuracy measures from data points, and discussed the effects of data density and terrain curvature on DEM accuracy from both theoretical and practical perspectives. In Aguilar et al. (2010), a hybrid theoretical-empirical model was proposed for estimation of DEM errors, taking into account terrain slope, data density and interpolation method. Some researchers (Ackermann 1980, Li 1993, Aguilar et al. 2006) proposed a linear model for modelling DEM errors. A more recent empirical study (Fan & Atkinson 2015) confirmed the possibility of using a linear model (Equation (1)) to represent the DEM error in terms of RMSE for some typical data spacing of TLS point clouds. However, that study did not quantitatively investigated the effects of measurement errors on the coefficients of the linear DEM error model, which form the focus of this study.

$$RMSE = KD + E \tag{1}$$

where K is a parameter depending on the characteristics (typically described by surface roughness (Fan & Atkinson 2018) of terrain surfaces, D denotes the data point spacing, E is the propagated error caused by measurement errors.

2 METHODOLOGY

A point cloud that is free of measurement noise would be an ideal study dataset. However, such survey data were not available, and as such a semi-artificial point cloud was created. In this study, the original point cloud representing a bare terrain surface was smoothed using a local polynomial regression of order 2. This processing was used to remove the high frequency spatial variation and the measurement noise, and resulted in a new point cloud that was assumed to be free of noise. The terrain surface considered has an area of 3.5 m by 3.5 m and is inclined at an angle of approximately 16°.

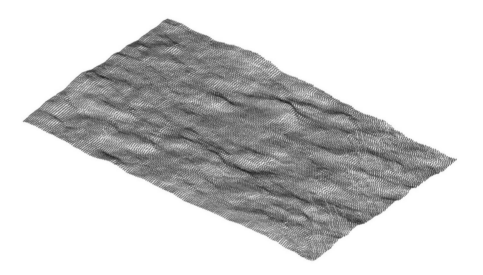

Figure 1. The study data.

Measurement errors were added to the error-free point cloud created to investigate their effects on the linear DEM error model. To characterise the measurement errors, the following assumptions were made: (1) the measurement errors were random variables obeying a normal distribution, and (2) the measurement errors in each direction (i.e. x, y and z) have a mean value of zero and an equal standard deviation (i.e. $\sigma_x = \sigma_y = \sigma_z = \sigma$). The magnitudes of the measurement errors considered ranged from 1 mm to 8 mm, reflecting the millimetric measurement accuracy of typical terrestrial laser scanners.

The method used for the investigation was a statistical resampling method. The procedure is illustrated in Figure 2 and is detailed in the following. In Step 1, 100 check points were selected randomly from the error-free point cloud, and constrained with a minimum distance of 0.3 m between individual check points. This minimum distance was determined with the aid of a variogram for the elevation prediction residuals and was used to avoid spatial correlation between the elevation residuals at the check points. In Step 2, random measurement errors were added to the remaining data points (i.e. without the check points). Using the error-contaminated datasets, the experiments started with a reduction that led to the remaining data points having an equivalent point spacing of 30 mm in Step 3. In Step 4, the thinned point cloud was used to predict the elevation values at the check point locations using a TIN with a linear interpolation. For source data of high density such as TLS data, there may be little justification for using a more complex interpolation method. In Step 5, the predicted elevations were then compared with the reference ones at the check point locations to obtain the elevation residuals, which were then used to calculate the DEM errors in terms of RMSE.

Steps 1 - 5 were run $i = 100$ times for this initial point reduction and the average DEM errors were reported. Once completed, a further reduction of data points (leading to an equivalent spacing of 35 mm) was carried out. Steps 1 - 5 were repeated 100 times again for the second

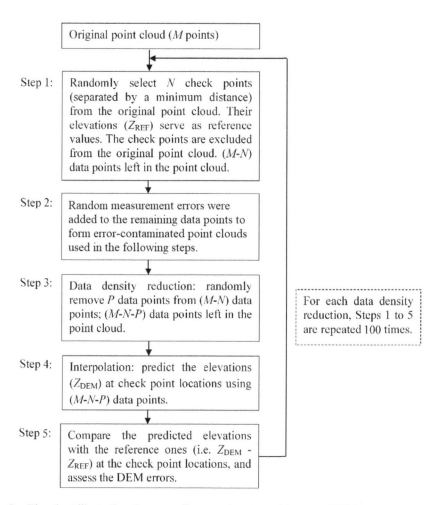

Figure 2. Flowchart illustrating the resampling experiments used to assess DEM accuracy.

reduction. This data point reduction process continued until the equivalent point spacing was 100 mm. This process was implemented in MATLAB® (2017a).

3 RESULTS AND DISCUSSION

The RMSEs for the measurement errors of different magnitudes were shown in Figure 3. As expected, the overall DEM error increased with increasing magnitudes of measurement errors. As such, the approach may be used to infer the relative magnitudes of measurement errors in point clouds representing the same surface but measured by different instruments/techniques. For each measurement error case, it can be observed that the RMSEs increased approximately linearly with the densities (i.e. for the point spacing varying from 30 mm to 100 mm) of the TLS data considered. It should be noted that the linear behaviour would not be valid for all point spacings. When the point spacing is very large, the RMSE will plateau because it is constrained by the population variance of the elevation residuals after any trend is removed (Fan & Atkinson 2015). These linear relationships suggest that the DEM error in terms of RMSE consists of a point-spacing dependent component and a constant component (i.e. Equation (1)). The former depended on the signal spatial variation for a given data point spacing. A higher data density would capture signals in more detail and hence reduce the overall DEM error. The

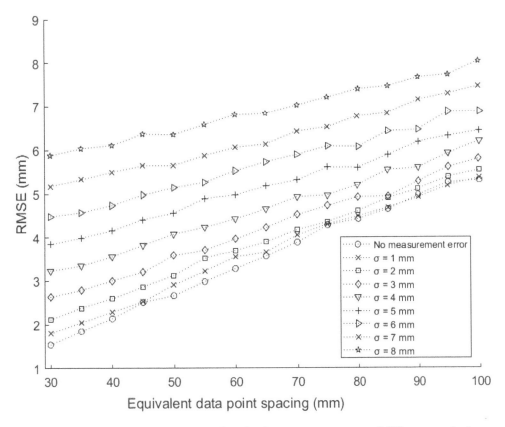

Figure 3.   The RMSEs at different data spacings for the measurement errors of different magnitudes.

constant component was caused mainly by the measurement noise and could not be reduced by a further increase in data resolution because the measurement errors added to the error-free point cloud were spatially random.

As expected, the constant component (i.e. E) increased with increasing the measurement errors applied. Figure 4 illustrates more clearly the effects of the measurement errors on E. The results can be fitted well by a quadratic regression model, suggesting the ratio E/σ also increased with increasing measurement noise. The values of E were found to be smaller than the measurement errors of individual data points. This observation is consistent with previous theoretical studies (Li 1993). The constant component E is essentially the propagated measurement error, and statistically the range of its values should lie between 0 and the measurement error of individual data points.

In Figure 3, the slopes of the linear regression models appeared to decrease with increasing magnitudes of measurement errors. This can be more clearly observed in Figure 5 where the coefficient K of the point-spacing dependent component (i.e. KD) at different magnitudes of measurement errors was quantified using the linear regression models. This suggests that the coefficient K was affected not only by the intrinsic surface complexity but also by the measurement noise for the case studied. The likely reason for this behaviour is that the random measurement errors caused some micro-scale signal spatial variation into noise. Consequently, those micro-scale signals could no longer be detected by the point-spacing dependent part of the linear models.

In this study, three measurement error components (i.e. $\sigma_x, \sigma_y, \sigma_z$) were added to data points separately. The equivalent overall positional error of an individual data point would be equal to $\sqrt{3}\sigma$, based on the law of error propagation with the assumptions of independent and

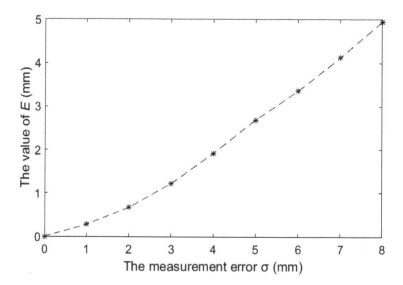

Figure 4. The values of E at the measurment errors of different magnitudes.

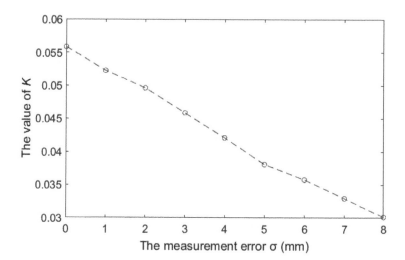

Figure 5. The values of K at the measurment errors of different magnitudes.

identical error components. In this study, the measurement errors were added randomly to individual data points with the same weighting. For future studies, it may be interesting to look into spatially correlated measurement errors (e.g. assuming the measurement error is larger at the local areas where the surface is more complex).

4 CONCLUSIONS

Using the measurement noise as a controlled parameter in statistical resampling experiments, it was found that the constant component E of the linear model increased with increasing random measurement noise, so did the ratio E/σ. On the other side, the coefficient (i.e. K) of the point-spacing dependent component was reduced as the random measurement noise increased, likely due to the change of spatial variations from signals to noises.

ACKNOWLEDGEMENTS

The authors are grateful for the financial support from Xi'an Jiaotong–Liverpool University (RDF-15-01-52 and RDF-18-01-40) and Natural Science Foundation of Jiangsu Province (Grant number BK20160393).

REFERENCES

Ackermann, F. 1980. The accuracy of digital terrain models. *In Proceedings of 37th Photogrammetric Week (University of Stuttgart)* 113-143.

Aguilar, F.J. Aguilar, M.A. Agüera, F. & Sánchez, J. 2006. The accuracy of grid digital elevation models linearly constructed from scattered sample data. *International Journal of Geographical Information Science* 20(2):169-192.

Aguilar, F.J. Mills, J.P. Delgado, J. Aguilar, M.A. Negreiros, J.G. & Pérez, J.L. 2010. Modelling vertical error in LiDAR-derived digital elevation models. *ISPRS Journal of Photogrammetry and Remote Sensing* 65(1): 103-110.

Bater, C.W. & Coops, N.C. 2009. Evaluating error associated with lidar-derived DEM interpolation. *Computers and Geosciences* 35(2): 289-300.

Day, S.S. Gran, K.B. Belmont, P. & Wawrzyniec, T. 2013. Measuring bluff erosion part 1: terrestrial laser scanning methods for change detection. *Earth Surface Processes and Landforms* 38(10): 1055-1067.

Fan, L. & Atkinson, P.M. 2015. Accuracy of digital elevation models derived from terrestrial laser scanning data. *IEEE Geoscience and Remote Sensing Letters* 12(9): 1923-1927.

Fan, L. & Atkinson, P.M. 2018. A new multi-resolution based method for estimating local surface roughness from point clouds. *ISPRS Journal of Photogrammetry and Remote Sensing* 144: 369-378.

Fan, L. Smethurst, J.A. Atkinson, P.M. & Powrie, W. 2014. Propagation of vertical and horizontal source data errors into a TIN with linear interpolation. *International Journal of Geographical Information Science* 28(7):1378-1400.

Fisher, P.F. & Tate, N.J. 2006. Causes and consequences of error in digital elevation models. *Progress in physical geography* 30(4):467-489.

Guo, Q. Li, W. Yu, H. & Alvarez, O. 2010. Effects of topographic variability and lidar sampling density on several DEM interpolation methods. *Photogrammetric Engineering & Remote Sensing* 76:1-12.

Heritage, G.L. & Milan, D.J. 2009. Terrestrial Laser Scanning of grain roughness in a gravel-bed river. *Geomorphology* 113(1–2): 4-11.

Hu, P. Liu, X.H. & Hu, H. 2009. Accuracy Assessment of Digital Elevation Models based on Approximation Theory. *Photogrammetric Engineering & Remote Sensing* 75(1): 49–56.

Krausl, K. Karel, W. Briese, C. & Mandlburger, G. 2006. Local accuracy measures for digital terrain models. *The Photogrammetric Record* 21(116): 342-354.

Li, Z.L. 1993. Theoretical models of the accuracy of digital terrain models: an evaluation and some observations. *The Photogrammetric Record* 14:651–660.

Montreuil, A.L. Bullard, J.E. Chandler, J.H. & Millett, J. 2013. Decadal and seasonal development of embryo dunes on an accreting macrotidal beach: North Lincolnshire, UK. *Earth Surface Processes and Landforms* 38(15):1851–1868.

Schürch, P. Densmore, A.L. Rosser, N.J. Lim, M. & McArdell, B.W. 2011. Detection of surface change in complex topography using terrestrial laser scanning: application to the Illgraben debris-flow channel. *Earth Surface Processes and Landforms* 36(14): 1847–1859.

Sustainable Buildings and Structures: Building a Sustainable Tomorrow – Papadikis et al. (Eds)
© 2020 Taylor & Francis Group, London, ISBN 978-0-367-43019-1

Exploring the influence of information technology on learning performance: A case study of construction management students

S. Zhang & Y. Hua
Suzhou University of Science and Technology, Suzhou, China

S. Galvin
University of South Wales, South Wales, UK

ABSTRACT: With the fast development of information technology (IT) in the last 10 years, the learning approaches and behaviors of students have significantly changed. It is very critical for university teachers harnessing IT effectively to improve the learning performance of the students. Based on literature review on IT and learning performance, this paper conducts questionnaire survey to the construction management students with the aim of investigating the influence of IT on learning performance. It was found that more than half of the university students spend over 6 hours a day on handling IT data. The most popular IT software/tools used by the students in learning are intelligent technology and internet. The most significant influences of IT on learning were accelerating the change of learning style and promoting the collaborative learning with classmates.

1 INTRODUCTION

Information technology (IT) refers to the electronic technology, digital technology, computer technology, communication technology and network technology, which involves information collection, processing and storage (Ren 2006). The rapid development of IT has accelerated the development of modern education. Educational informationization is one of the primary features of educational modernization. It is an effective way to accelerate educational reform, direct educational innovation, optimize educational quality and ensure educational justice in the information era (Shi 2012). IT has changed the traditional learning and teaching model, for example, the blackboard and chalk & talk have been increasingly replaced by multimedia and internet. The learning and teaching approaches are more flexible and convenient in IT environment. In 2011, the Ministry of China initiated the development of Open Course Scheme, which has produced 256 free virtual courses open to the public in eight years. By the end of 2018, the number of visitors to the home page of three largest platforms with Chinese University Virtual Open Course reaches 4,162 million. In 2018, the Ministry of Education officially issued the Education Informatization 2.0 Action Plan. A statistic data showed that from 2013-2017, 8% of the educational investment was spend in the application and development of IT in education by the Chinese government, whereas in 2018, the ratio was increased to 8.5%. All these illustrate the popularity of applying IT in modern education in China.

  With the increasingly dynamic environment of education informationization, the learning behavior and performance of the university students have significantly changed. For example, the passive learning is shifted to active learning, and the individual learning is replaced by cooperative learning. At the same time, the university students are learning knowledge by using a variety of software/tools/methods, such as Wechat, massive open online courses (MOOC), mobile phone (Gao 2017). As a consequence, IT has changed the students' traditional learning behaviors and approaches, which may have significant influences on the learning performance. However, very few studies have been focused on exploring the influence of IT on the learning performance. This paper aim to conduct an empirical study on this topic, which comprises two main objectives: (1) identifying the latest development of IT which have been utilized in the learning process by university students; (2) analyzing influences of IT on the learning performance.

2 LITERATURE REVIEW

2.1 *Application of IT in university education*

At present, the application of IT in university education mainly includes multimedia, digital library, open courses, distance education, virtual laboratory and so on. A survey showed that by 2015, the types of IT used in teaching are: digital media (73.4%), digital library (75.3%), virtual high-quality courses (55.1%), open courses (46.3%), online open courses (24.4%), distance education (29.2%), online general courses (26.4%), virtual laboratory (15.6%), and MOOC (6.3%) (Li et al. 2015). Gu et al. (2016) summarized the new teaching methods derived from the development of IT, which includes MOOC, mini-classroom and digital learning. In fact, many new types of IT have been applied in university education, such as iPad + Apps and Wechat. The iPad + Apps mode refers to download Apps related to learning content through iPad and then learn the material on the platform (Li 2013). For university students, Wechat is not only a chatting software, but also the main way for them to obtain information and to learn (Lv 2013).

2.2 *Evaluation criteria of learning performance*

According to Liu et al. (2010), learning performance refers to the achievement and benefit of a learner who completes an activity in a stated time meeting a certain standard. Generally, the evaluation criteria can be employed to evaluate the students' learning performance. According to Wu (2008), the learning performance criteria can be evaluated in terms of learning ability, learning attitude, learning behavior and learning output. In another study, based on the observation of students' online learning behavior, three criteria were used to evaluate the learning performance, i.e. the online learning behavior change, learning outcome and learner satisfaction (Zhang 2014). When further analyzing the influencing factors of learner satisfaction, it was found that flexibility, quality of online learning material, reliability of online learning system, easy access, and the teachers' attitude to online learning have an impact on the students learning satisfaction level (Huang & Zheng 2011).

2.3 *Influencing factors of learning performance*

Based on literature review, the influencing factors of learning performance are summarized as three categories: learner factors, teacher factors and learning environment factors.

2.3.1 *Learner factors*

If the students are motivated to learn, they are more active and focused on what they learn, which will also improve the learning effectiveness (Li 2008). Hao (2005) found that students' living stress level, emotional condition and learning satisfaction level are the main factors affecting the learning performance. In addition, the mental health level of the university students has a significant impact on the learning performance. Xie (2017) claimed that if the university students try to improve their learning performance, they need to maintain a normal level of mental health, pay attention to the learning process, strive to improve the learning motivation, resolve learning difficulties, optimistically face the unhealthy psychological symptoms, and be willing to communicate with others to construct a comfortable learning environment. According to Pan (2015), three types of factors affect the learning performance. Firstly, the basic characteristic of the learner, such as age, background, etc. Secondly, the learning style of learner, such as the learning interaction, involvement, etc. Thirdly, the external factors, such as teachers' factors, learning media, etc. However, among these factors, the degree of learner involvement is the most critical factor affecting the learning performance.

2.3.2 *Teacher factors*
The teaching methods and strategies adopted by the teachers affect the students' learning styles, and produce different levels of learning efficiency, and as a consequence achieve different levels of learning performance. In addition, the communication ability of teachers is also critical to improve the learning performance. When students encounter difficulties in the learning process, effective communication between the teachers and students can solve the problems in a timely manner, which motivates students to achieve a better learning performance (Wang 2012). The teaching input of teachers is an important factor affecting students' learning output. The more efforts devoted to teaching by the teachers, the students will be more effectively motivated to learning, and consequently the learning performance will be improved (He 2014).

2.3.3 *Learning environment factors*
In modern IT era, internet environment is very important in influencing the learning performance of university students. Many teaching materials can be found easily from the internet. Hence, the online learning has become an integral part of leaning process in addition to face-to-face lecturing (Chen et al. 2016). The cultural environment, learning facilities, learning environment, interpersonal relationship provided by the university have a significant impact on students' learning performance (Qiao 2003).

2.4 *Influence of IT on learning performance*

According to relevant research findings, the application of IT in the university education has both negative and positive influences.

2.4.1 *Positive influence of IT on learning performance*
(1) Constructing an excellent learning environment

The integration of IT and education provides students with a better learning environment. Internet resources are easy to access, so a large amount of reliable and various learning information can be obtained by the students. In addition, IT can stimulate the students' interest, broaden their learning vision, create a virtual real world which is beyond the textbooks and traditional courses, and provide more comprehensive knowledge (Ruan & Bai 2013). A more effective and interactive learning environment can be developed by employing IT, and learners can make full use of all kinds of resources to improve their learning performance. These IT learning resources have become a paramount medium for students to learn knowledge in modern universities, making it more convenient for them to search for information, collect information, obtain knowledge and improve their learning performance (Li 2007).
(2) Promoting the students' active self-learning

In an open and friendly cooperative learning environment, students are more willing to independently search for information and conduct in-depth analysis, which promotes the students' active self-learning behavior. Students can use IT software/tools to find a large amount of learning resources from the internet. Currently, more and more IT software and tools are being used to achieve a higher level of knowledge integration, which provides a basis for students to engage in the deep learning (Ruan & Bai 2013).
(3) Promoting the students' collaborative learning

Students can significantly enhance the effectiveness of communication and cooperation through the use of IT in the learning process, and make communication between students and teachers more smooth, without time and space constraints. Students can find and use teaching resources through the IT software and tools more easily, communicate more conveniently with classmates and teachers, and reflect learning experience (Ruan & Bai 2013). Flipping classroom is one of the applications of IT in teaching. The typical feature of this teaching approach is the improvement of learning interaction in the classroom. The communication and interaction between the teachers and students can be enhanced significantly. By learning as group, students can achieve their own individual learning goals and group learning goals (Ma 2016).

(4) Reforming the form of teaching implementation

Education using IT is different from the traditional teaching, such as blackboard and chalk & talk. It makes individualized learning as well as group learning become a reality. Each student can progress learning on his or her own merit. In addition, the teacher may use IT as the effective source to achieve a variety of learning outcomes, such as group learning, collaborative learning, and individual learning. Hence, the teacher should change the teaching approach, strategy, and organization when using IT in teaching to optimize the positive influences in improving the learning performance (Bo & Shen 2003).

(5) Accelerating the change of learning style

With the rapid development of IT, the learning style of university students changes significantly. Learners no longer focus solely on the learning at traditional classroom. They can learn at any time and anywhere through virtual resource sharing and network information storage. They can also use IT software and tools to settle the learning problems emerged in the learning process in a timely manner through effective interaction with the teachers (Xiong et al. 2012). IT has changed the way to receive and feedback information by the students in the learning process. Due to this type of learning style, the students have more opportunities of obtaining a variety of thinking and ideas, which may improve their critical thinking skill and problem solving capabilities (Ma 2016).

2.4.2 *Negative influence of IT on learning performance*

(1) Wechat information flood and distraction of online games

In a society with the rapid development of IT and the fast popularization of internet, the extracurricular life of college students is becoming more and more colorful. However, the flood of information makes students feel at a loss. They have to develop an excellent skill in handling huge amount of trivial information. They also need a more strict self control behavior to resist the temptation of online games. If they fail to do so, their learning performance will be negatively influenced (Wang et al. 2008). As an example, the frequent use of Wechat and online game play make it easy for university students distracted from learning, and their learning efficiency is decreased, which leads to the decline of their academic performance. If the university students immense in the virtual world of Wechat and online game for a long time, they will develop an illusion of virtual reality. In this situation, they may have difficulties in settling the real problem in life and continuing with their normal study (Lv 2013).

(2) Learning pace out of control

When IT is used in teaching, the learning pace is significantly accelerated, which may lead to some difficulties in effective learning for the students. For example, comparing with the traditional blackboard and chalk & talk teaching, most students feel difficult to follow the modern projector and presentation lecturing model. Hence, the learning performance may decrease due to the adoption of IT in some cases, if the teachers do not pay attention to this issue (Cai 2013).

(3) Restricting the cultivation of learners' creativity

IT provides great convenience for acquiring, processing and disseminating information. At the same time, it also become easier for anyone to copy information obtained from the internet, including the academic research findings. In this case, the university students may get used to rely heavily on IT resources, and may be unwilling to improve their creativity through continuous and regular practice of individual thinking. In addition, the combination of graphic and textual IT and virtual illustration has simplified a specific concept for easy understanding and catching the eyeball of the students, which may also make learners think in a stereotype form, and as a consequence their abstract thinking capability, creativity ability and imagination capability will not be well developed (Huang et al. 2009).

(4) Wasting learning resources significantly

Since the learning resources can be repeatedly used, they are much cheaper comparing with the traditional classroom learning resources. In addition, for a specific topic, the students may be very easy to find lots of learning materials. As a result, the university students may spend more time in collecting learning materials, instead of digesting them. In other cases, the teacher may also provide lots of teaching materials to students for learning. This may also lead to waste, since most of the time these materials are not well learned by the students. For example, for some online distance learning, the students' learning performance is evaluated partially based on the hits of the virtual material. In this case, the students may only complete the hits instead of real learning (Zhang 2006).

3 QUESTIONNAIRE SURVEY

In order to explore the influence of IT on the learning performance of university students, this research conducted a questionnaire survey to the university students majoring in construction management. A total of 120 questionnaires were distributed with 114 being collected. Among them, 111 samples were valid, which produced a valid rate of 97.37%. As to the valid samples, 89 questionnaires were collected through face-to-face method, and 22 were collected through the online tool of Questionnaire Star. The detail information of the respondents is shown in Table 1.

In terms of the gender distribution, 67.57% of the respondents were male. The gender distribution of the respondents is shown in Table 2.

Since the freshmen know less about the learning at the university, this study does not include them as the target investigation population.

4 DATA ANALYSIS RESULTS

4.1 Frequency of using IT in study by the university students

As to the frequency of IT use in the study, 87.39% of the respondents claimed that they frequently use IT when they study the disciplinary courses in construction management. The detail information is shown in Table 3.

4.2 Time spent in using the IT by the university students

As to the time spent in using IT (including study and other use), most (47.75%) of the respondents claimed that they spend 6-8 hours in each day. More than half (54.97%) of the

Table 1. Detail information of the respondents.

Data collection approach	Year of the students	Number of collected questionnaires	Valid questionnaires
	Senior	35	35
face-to-face	Junior	44	43
	Sophomore	11	11
On-line	Senior	24	22
Total		114	111

Table 2. Gender distribution of the respondents.

Gender	Number of respondents			Total	Ratio
	Sophomore	Junior	Senior		
Male	8	30	37	75	67.57%
Female	3	13	20	36	32.43%
Total	11	43	47	111	100%

Table 3. Frequency of using IT in study.

| Item | Number of respondents | | | Total | Ratio |
	Sophomore	Junior	Senior		
Frequently using	11	37	49	97	87.39%
Rarely using	0	6	8	14	12.61%
Total	11	43	57	111	100%

respondents spend over 6 hours a day on handling IT issue. The detail information is shown in Table 4.

4.3 Type of the IT used in learning by the university students

Based on literature review and interview to the university students, the type of IT used in the learning includes internet technology, digital learning, MOOC, intelligent technology, multimedia, digital library and virtual laboratory. According to the statistic results of questionnaire survey, it is found that intelligent technology (including the intelligent cell phone) was used by all the respondents in learning, followed by internet technology, which was used by 94.59% of the respondents. The detail information is shown in Table 5.

4.4 Influence of IT on the learning performance of university students

According to the literature review results, the influence of IT on the learning performance are divided into three categories: learner factors, teacher factors and environment factors. In order to investigate the influence level of IT on the learning performance, Likert scale instrument is used to collect the viewpoints of respondents, in which 1 represents "strongly disagree", 3 represents "neutrality", and 5 represents "strongly agree".

Table 4. Time spent in using the IT in each day by the respondents.

| Item | Number of respondents | | | Total | Ratio |
	Sophomore	Junior	Senior		
Less than 2 hours	0	0	3	3	2.70%
2-4 hours	1	2	6	9	8.10%
4-6 hours	1	18	19	38	34.23%
6-8 hours	7	21	25	53	47.75%
Over 8 hours	2	2	4	8	7.22%
Total	11	43	57	111	100%

Table 5. Type of the IT used in learning by the university students.

| Item | Number of respondents | | | Total | Ratio |
	Sophomore	Junior	Senior		
Internet technology	11	42	52	105	94.59%
Digital learning	4	18	23	45	40.54%
MOOC	11	41	48	100	90.09%
Intelligent technology	11	43	57	111	100%
Multimedia	9	37	36	82	73.87%
Digital library	1	8	9	18	16.22%
Virtual laboratory	0	0	3	3	2.70%
Others	0	2	2	4	3.60%

4.4.1 *Influence of IT on the learning performance of university students–from the perspective of learner factors*

Twelve questions were included in the Likert scale instrument to solicit the viewpoints of university students as to the influence of IT on the learning performance from the perspective of learner factors. The statistic results were shown in Table 6.

According to the statistic results in Table 6, it is found that:

(1) In terms of positive influence of IT on the learning performance of university students, the most significant aspects were accelerating the change of learning style (3.51), promoting the cooperative learning with classmates (3.27), and making the learning motivation clearer (3.26). The application of IT also slightly enhanced the learning ability (3.05) of the students.

(2) In terms of negative influence of IT on the learning performance of university students, the respondents slightly agreed that they can't resist the temptation of Wechat and online games. However, the respondents also slightly disagreed that the application of IT led to the waste of learning resources.

4.4.2 *Influence of IT on the learning performance of university students–from the perspective of teacher factors*

Four questions were included in the Likert scale instrument to solicit the viewpoints of university students as to the influence of IT on the learning performance from the perspective of teacher factors. The statistic results were shown in Table 7.

According to the statistic results in Table 4-4, it is found that in terms of influence of IT on the learning performance of university students, the most significant teacher's aspect was to

Table 6. Influence of IT on the learning performance of university students (from the perspective of learner factors).

Question	Average value			
	Sophomore	Junior	Senior	Total
The application of IT makes my learning motivation clearer.	3.09	3.30	3.26	3.26
The application of IT makes my interest in learning stronger.	3.09	3.14	3.21	3.17
The application of IT improves my involvement in learning.	3.27	3.23	3.16	3.20
The application of IT promotes my collaborative learning with my classmates.	3.27	3.33	3.23	3.27
The application of IT accelerates the change of learning style.	3.55	3.53	3.49	3.51
I can't resist the temptation of Wechat and online games.	3.27	3.23	3.16	3.20
The application of IT has led to a great waste of learning resources.	3.18	3.02	2.75	2.90
The application of IT has enhanced my learning ability.	3.00	3.14	3.00	3.05
The application of IT has changed my learning attitude.	2.73	3.00	2.74	2.84
The application of IT has improved my learning behavior.	3.00	2.86	2.98	2.94
The application of IT has increased my learning outcome.	3.00	2.95	2.84	2.90
The application of IT has improved my learning satisfaction level.	2.91	3.05	2.95	2.98

Table 7. Influence of IT on the learning performance of university students (from the perspective of teacher factors).

Question	Average value			
	Sophomore	Junior	Senior	Total
The application of IT strengthens teachers' supervision and guidance.	3.18	3.17	2.96	3.07
The application of IT enhances the communication between teachers and students.	3.18	3.23	3.18	3.20
The application of IT increases teachers' input.	3.18	3.12	2.95	3.03
The application of IT reforms the form of teaching implementation.	2.81	3.05	3.02	3.01

Table 8. Influence of IT on the learning performance of university students (from the perspective of environment factors).

Question	Average value			
	Sophomore	Junior	Senior	Total
The application of IT improves the search efficiency of learning resources.	2.91	3.25	3.11	3.14
The application of IT constructs a better learning environment.	2.82	3.05	2.93	2.96
The application of IT enhances an atmosphere facilitating a better learning experience.	2.72	3.02	2.96	2.96

enhance the communication between teachers and students (3.51). However, the remaining three factors were only slightly influenced by the application of IT in the learning process, considering the learning performance of university students.

4.4.3 *Influence of IT on the learning performance of university students–from the perspective of environment factors*

Three questions were included in the Likert scale instrument to solicit the viewpoints of university students as to the influence of IT on the learning performance from the perspective of environment factors. The statistic results were shown in Table 8.

According to the statistic results in Table 4-4, it is found that in terms of influence of IT on the learning performance of university students, the most significant environment aspect was improving the search efficiency of learning resources (3.14). However, it did not have positive influence on constructing a better learning environment and enhancing an atmosphere facilitating a better learning experience.

5 CONCLUSIONS

Based on literature review on the IT and learning performance, questionnaire survey method was employed to the university students majoring in construction management to explore the influence of IT on learning performance. It was found that 87.39% of the respondents frequently used IT when they study the disciplinary courses. More than half (54.97%) of the respondents spend over 6 hours a day on handling IT data. Intelligent technology (including the intelligent cell phone) was used by all the university student respondents in learning, followed by internet technology, which was used by 94.59% of the respondents. The IT has both positive and negative influences on the learning performance of university students. In terms of positive influence, the most significant aspects were accelerating the change of learning style (3.51), promoting the collaborative learning with the classmates (3.27), and making the learning motivation clearer (3.26).The employment of IT in learning also slightly enhanced the learning ability (3.05) of the students. The use of IT also enhanced the communication between teachers and students (3.51) and improved the search efficiency of learning resources (3.14). As to negative influence, the respondents slightly agreed that they can't resist the temptation of Wechat and online games. In the future, interview method may be adopted to collect more comprehensive information regarding the influence of IT on learning performance to derive more robust research findings.

REFERENCES

Bo, J.S. & Shen, C.J. 2003. Influences of education informatization on students' learning styles. *Jiangsu Education* 23: 21-23. (In Chinese)

Cai, H.J. 2013. Research and cause analysis of the separation between learning and teaching in the information technology environment. *Research on Audiovisual Education* 34(02): 93-99. (In Chinese)

Chen, M. Huang, Z.Y. Liang, S. Zhou, C.Y. & Liu, Y. 2016. The influence of network environment on university students' learning performance. *Cooperative Economy and Technology* 04: 128-129. (In Chinese)

Gao, S.R. 2017. Study on the change of university students' learning style under the education informatization environment. *Education and Teaching Forum* 51: 199-200. (In Chinese)

Gu, X.Q. Wang, L. & Wang, F. 2016. Did the influence of information technology exist: the impact of educational informatization. *Research on Audiovisual Education* 37(10): 5-13. (In Chinese)

Hao, J.C. 2005. *The relationship between life stress, negative emotion adjustment, learning satisfaction and learning performance.* Tianjin University, Tianjin. (In Chinese)

He, X.M. 2014. A case study on the impact of teachers' input on students' learning input. *Educational Academic Monthly* 07: 93-99. (In Chinese)

Huang, W.L. Xiong, C.P. Jia, J.L. & Zhu, A.Z. 2009. Analyzing the double-influence of information technology on education from the two sides of technology. *China Audiovisual Education* 09: 23-25. (In Chinese)

Huang, T.H. & Zheng, Q.H. 2011. A review of influencing factors of learners' digital learning performance. *China Distance Education* 07: 17-23+95. (In Chinese)

Li, F. Sun, X.H. & Zhao, Y.P. 2015. Analysis of the current situation and countermeasures of Chinese universities to meet the challenges of information technology. *Journal of National College of Educational Administration* 01: 15-20. (In Chinese)

Li, J. 2008. *Study on the correlation among learning motivation, learning strategies and academic performance effectiveness of senior high school students.* Jilin University, Jilin. (In Chinese)

Li, R. 2013. New trend of integration of information technology and curriculum. East China Normal University, Shanghai. (In Chinese)

Li, T.L. 2007. The influence of constructivist learning theory on learning in the information technology environment. *Journal of Xi'an University of Post and Telecommunication* 04: 126-128. (In Chinese)

Liu, C. Zhao, J.B. & Li, P. 2010. The influence of internet environment on university students' learning performance. *Journal of Hebei Radio and Television University* 15(06):26-29. (In Chinese)

Lv, W.J. 2013. The impact of Wechat on university students and countermeasure strategies. *Today's China Forum* 17: 432-433+435. (In Chinese)

Ma, X. 2016. Influence of information technology on students' learning style. *Asia-Pacific Education* 35: 286. (In Chinese)

Pan, L.J. 2015. *Research on the relationship between MOOC design, learner involvement and learning performance.* Zhejiang University, Hangzhou. (In Chinese)

Qiao, X.F. 2003. Functioning the role of "two factors" to improve the learning performance of university students. *China Metallurgical Education* 06: 36-38. (In Chinese)

Ren, L.L. 2006. Comparison of educational technology and information technology, and development of educational information technology. *Educational Informatization* 17: 13-14. (In Chinese)

Ruan, Q.Y. & Bai, Y. 2013. The positive impact of information technology on students' learning in teaching-Based on the analysis of constructivist learning Theory. *China Science and Education Innovation Guide* 04:134. (In Chinese)

Shi, N.Z. 2014. Future development of mathematics education. *Mathematics Teaching* 01: 1-3+18. (In Chinese)

Wang, B.R. Wu, H.L. & Cheng, J. 2008. The influence of modern information technology on university students' self learning. *Science and technology information* 33: 45+101. (In Chinese)

Wang, H.R. 2012. A study on the learning adaptability of middle-grade students in primary school under the influence of teacher factors. Tianjin Normal University, Tianjin. (In Chinese)

Wu, H. 2008. *Study on the evaluation index system of university students' learning performance.* Ningbo University, Ningbo. (In Chinese)

Xie, B.C. 2017. The relationship between mental health level and learning performance of university students in Nanjing. *Campus Psychology* 15(06): 438-441. (In Chinese)

Xiong, C.P. He, X.Y. & Wu, R.H. 2012. On the revolutionary impact of information technology on educational development. *Education Research* 33(06): 22-29. (In Chinese)

Zhang, G.Y. 2006. Causes and solutions of low performance in distance learning under internet environment. *China Audiovisual Education* 12: 35-38. (In Chinese)

Zhang, X. 2014. *Study on the evaluation index system of learning performance based on internet learning environment.* China Normal University, Wuhan. (In Chinese)

Sustainable Buildings and Structures: Building a Sustainable Tomorrow – Papadikis et al. (Eds)
© 2020 Taylor & Francis Group, London, ISBN 978-0-367-43019-1

Feasibility analysis of an automated construction progress management system based on indoor positioning technology

W. Shen & C. Zhang
Department of Civil Engineering, Xi'an Jiaotong-Liverpool University, Suzhou, China

ABSTRACT: Timely and accurate construction progress tracking is significant to the nowadays construction industry, considering that construction productivity has dropped by almost 20% over the past 50 years. Ultra-Wideband (UWB), as a kind of high precision, stable and real-time indoor positioning technology, provides the supports to the proposed automated construction progress management system. More importantly, it is necessary for the construction progress management to have the forward flow of design intent and feedback flow of real construction state information, in this way, the efficient and effective construction progress management could be realized. The proposed system is introduced in detail in this research, and it demonstrates the theoretical feasibility with six requirements but needs further verification by practical cases in the future.

1 INTRODUCTION

1.1 *Background*

The Architecture, Engineering & Construction (AEC) industry has a great impact on the economy, the environment, and the society. The productivity and efficiency of construction projects have the potential to improve via an advanced construction progress managements system (Heigermoser et al. 2019). By definition, a project progress measurement system should be used to measure the actual progress of a construction project periodically and then to compare it with the planned or expected progress (Howes 1984). The so-called periodicity, on the premise of ensuring that the overall project is completed as planned, it allows for a systematic evaluation and analysis of the construction planning in terms of productivity, manpower allocation, and quantification of a waste considering the short-term planning process, which promotes continuous improvement of future construction planning (Iyer & Jha 2005).

Nowadays, construction industry has increased demand for timely and accurate information of construction projects. On one hand, lack of clear understanding of current project progress would result in a series of mistakes and improper adjustments by the managing team. Generally, the construction period and cost will increase sharply if the abnormal events which are different from the original plan cannot be found and adjusted in a short time, which is amplified when that situation occurs to a complex integrated project (Kopsida et al. 2015). On the other hand, according to the report from World Economic Forum in 2018 (Philipp et al. 2016), over the past 50 years, the productivity of architecture, engineering and construction (AEC) industries has declined by almost 20%, while the productivity of non-agricultural enterprises has increased by more than 150%; therefore, it is urgent to seek an efficient low-cost construction progress management system to improve the productivity of the engineering industry.

Traditionally, construction progress management is performed by manually collecting data through visual inspections or the report from on-site construction workers. The required information includes but is not limited to actual construction progress, the transportation and consumption of construction materials (Memarzadeh et al. 2013) and workers' daily workload. The manual approach is time-consuming, error-prone and labor-intensive. If these shortcomings of manual construction progress managements cannot be well solved, they will bring a series of errors, such as inaccurate construction progress measurement, unreasonable resource allocation, wrong execution plan, cost overruns and so on. While an efficient progress monitoring system should help automate progress inspections, reduce the risks of error, facilitate proper and timely corrective actions, and prevent deviations in terms of cost and schedule (Turkan et al. 2012). Therefore, the present research aims to investigate the feasibility of applying UWB on construction site to enable an automated construction progress management system based on the literature review.

1.2 Non-manual approaches

Different from the manual approaches, numerous methods of non-manual approaches to obtain periodical construction progress have been investigated by researchers. They could be divided into two categories, vision-based approaches and non-vision-based approaches.

For vision based approaches, Pučko et al. (2018) proposed an automated continuous construction progress monitoring via using multiple workplace real time 3D scans. They installed a low-precision 3D scanner called Kinect V2 sensor in the worker's helmet to sense, build and update point cloud models through worker movement during construction. As shown in Figure 1, in this way, workers can capture all workplaces inside and outside the building in real time, and record the partial point cloud, location and time stamp. Each worker's individual point cloud would be integrated into the as-built point cloud of the building under construction. Finally, the automated continuous construction progress monitoring would be realized via a comparison between the 4D as-designed model and the 4D as-built model, which could achieve more efficient management. Compared with periodical scanning of the whole building under construction, it eliminated the shortcomings of high-costed and time-consuming (Fathi et al. 2015), but the laser scanners in workers' helmets are limited in battery life which means they cannot work for a long time unless it is equipped with a mobile battery. As to the other vision based approaches, Kopsida et al. (2015) compared the mobile AR, stationary AR, photogrammetry, vision based reconstruction and laser scanning in six aspects which are time efficiency, accuracy, level of automation, required preparation, training requirements, cost and mobility. The conclusion drawn by the comparison is the best approach would vary with the situation such as inside the building or outside. In addition, Siebert & Teizer (2014) utilized Unmanned Aerial Vehicles (UVA) to collect related data autonomously for construction progress management. The general conclusion is that vision-based approach is normally high-cost, even if some low-cost devices could be used, their portability and durability are poor, and sometimes the vision-based approaches may refer to the privacy information of the workers which encountered the workers' resistance.

For non-vision based approaches, it has two categories: wave propagation and inertial navigation (Ibrahim & Moselhi 2016). Wave propagation technologies are based on the physical

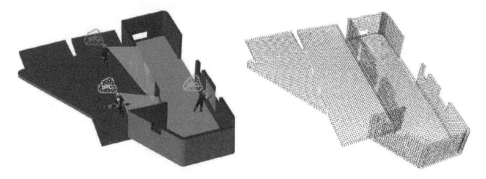

Figure 1.   Point cloud model built by multiple helmet-mounted scanners (Pučko et al. 2018).

345

propagation properties of radio, of which applications are Radio Frequency Identification System (RFID), UWB, infrared, WLAN and so on. Their common function is to automatically obtain the spatial information pertinent to the location of labor, construction materials and the core construction equipment. The construction progress, jobsite safety and even productivity could be estimated by localization and tracking of these resources on construction site. Montaser and Moselhi (2014) deployed the passive RFID tags which are treated as a reference point with a known location in the jobsite and let one worker carry the RFID reader, of which location can be identified by two approaches (triangulation and proximity) based on the signal measurements using Received Signal Strength Indicator (RSSI). Then the positioning method via RFID has been through the verification in the laboratory which shows the method has a mean error of less than 1.0 m. From a comprehensive perspective, the positioning methods based wave propagation have been summarized by Bose & Chuan (2007) into five categories: Cell Identity (Cell-ID), Time of Arrival (TOA), Time Difference of Arrival (TDOA), Angle of Arrival (AOA), and signal strength categories but their accuracy varies with the equipment and algorithms. As to inertial navigation, Ibrahim & Moselhi (2016) proposed an inertial navigation technique for an indoor position using a microcontroller equipped with an inertial measurement unit (IMU), which is independent of any infrastructure and avoids the trouble of laying too many readers on the construction site and improves the feasibility. The inertial navigation is composed of an accelerometer and a gyroscope. In practical application, it also has defects, the accelerometers are easily confused by non-human motion, and the gyroscopes are susceptible to electromagnetic equipment on the jobsite.

1.3 *Accuracy comparison*

In order to minimize the influence of uncertainties in the construction site and realize the stable automated construction progress management, the indoor position technology based on wave propagation is a better choice than inertial navigation. Li et al. (2016) summarized the utilization of different location technology in construction from 2005 to 2014 from 75 articles. The results showed that the Bluetooth and infrared (IR) have the highest accuracy in indoor environments, but neither of them deserves widespread application due to some unsuitable properties such as the Bluetooth can only obtain the two-dimensional positioning information, IR is limited by its short ranging distance, and RFID and UWB are the major indoor position technologies covered in the sample reviewed papers. The accuracy of RFID and UWB is of paramount importance. For RFID, it uses the principle of electromagnetic induction to excite the short-range wireless label by wireless, so as to realize the technology of information reading. Deployed in an indoor environment with wall and metallic objects, RFID only can realize 10.7 m of precision averagely (Pradhan et al. 2009), if it can be free from the obstacle, its accuracy cam improves to 1.9 m averagely (Luo et al. 2011). So, RFID is mainly used to identify whether a person exists in a certain area, and cannot be tracked in real time. For UWB, it has almost excellent penetration capacity and is based on a wireless technology under a very narrow pulse and no carrier. Through these advantages, UWB plays an incisive and vivid role and has displayed a very good effect. It usually uses TDOA for internal positioning, which generates, transmits, receives and processes the extremely narrow pulse signal in the difference of signal arrival time. UWB indoor positioning system includes UWB receiver, UWB reference tag and active UWB tag, of which signals transmitted by tags are received by UWB receivers in location determination. The signals containing effective information are obtained by filtering various noise interferences in the electromagnetic wave transmission. Then the location calculation and analysis are carried out by the central processing unit. In the indoor environments, UWB could realize 17 cm without obstacle tested by Cho et al. (2010) and 60 cm with obstacle tested by Shahi et al. (2012). As a conclusion, UWB is generally more accurate and has more utilization prospective than RFID.

2 METHODOLOGY

The proposed framework is presented in Figure 2. All the information in the design stage can be found in the models built by Building Information Modeling (BIM) technology. According to the schedule, different work spaces are divided on the construction site, and each work space should have a corresponding leader. The construction progress in the workspace is automatically estimated by the movement of equipment, the consumption of construction materials and the movement of workers. The movement of equipment and workers could be directly presented by the location data obtained from UWB. As to the consumption of construction materials, it needs to be assisted by the location data of equipment and workers which would be explained in detail in section 3.

By comparing the actual construction progress with the initial schedule, it could be judged whether the construction is carried out according to the expectation or not. If it is inconsistent with the initial schedule, the appropriate adjustment is needed. As UWB can reflect the location information in almost real time, the leader of each work space can adjust their construction

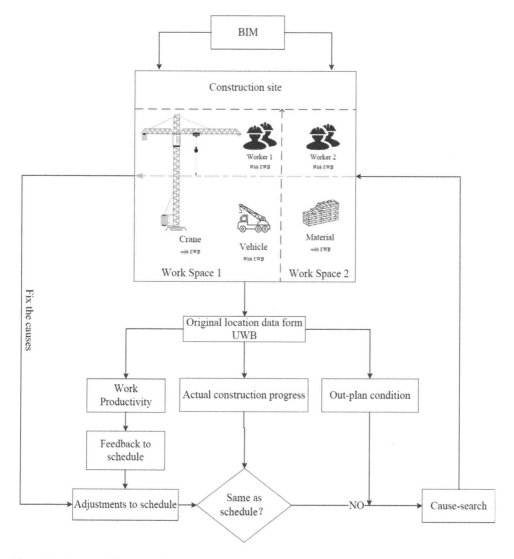

Figure 2.   Automated construction progress management system.

schedule in time according to current productivity reflected by UWB. The initial schedule is derived from the drawings which are unable to deal with unexpected issues in practice, such as the impact of weather conditions. Therefore, the actual construction productivity may be slower or faster than expected. Timely redistribution of labor resources and materials based on actual productivity will promote a more efficient construction. The real-time properties of UWB provide the possibility to deal with unexpected emergencies. The focus of the present research is to investigate the feasibility of estimating the construction progress based on location data collected from the UWB technology.

3 EXPERIMENTAL DESIGN

As shown in Figure 3, a partitioned work space is represented by a black quadrilateral, where on-site pouring of a concrete floor will be carried out. According to the designed schedule, the total demanded volume of concrete to be poured is 20 cubic meters, and the carrying capacity of each concrete truck is 8 cubic meters. Therefore, if only one concrete truck is rented, at least three trips to the working area could signify that the total volume of concrete had been met. The two elevations vertical coverage and four corners horizontal coverage of UWB Readers in work space could guarantee the accuracy of monitoring the trajectory and real-time position of workers and trucks, as once workers or trucks enter into the work space, their UWB tags would be read to transmit real-time location information. In addition, the productivity of workers can be calculated according to their working hours, and then more precise and reasonable schedule for similar tasks can be made, thus reducing the waste of labor and material resources.

4 FEASIBILITY ANALYSIS

After literature review and experimentation with UWB, six identified requirements have been summarized by Zhang et al. (2011). These are accuracy, visibility, scalability, real time, tag form factor, power, and networking requirements. Accuracy is of paramount importance to ensure access to valuable data. The relative high accuracy could be achieved by utilizing Angle of arrival (AOA) and time difference of arrival (TDOA) to locate UWB tags based on trilateration.

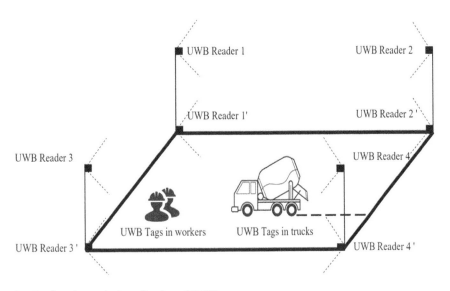

Figure 3. Designed practical application of UWB.

In addition, the calibration of the sensors and data filtering also is essential to obtain higher accuracy. Visibility requires the UWB sensors to be set in the way by using the field of view in both azimuth and altitude, and for one large range of monitored area, multiple cells should be formed by more sensors, for one crucial object, multiple tags also should be attached for better visibility. Scalability and real-time requirements need the balance of visibility and accuracy via choosing an applicable quantity of tags with suitable update rate, due to the more tags in one system, the less updated frequently, and vice versa. Tag form factor requires tags with different form factors to be applied in their specific way, and all the tags should be provided sustaining and stable power source for precision measurements. As to networking requirements, the choice of the type of the network (wired or wireless) should depend on the specific construction site condition. All these six requirements serve to the full realization of UWB's function of indoor real-time positioning function.

5 CONCLUSION

This paper presented the literature review of various technologies of construction progress management and proposed a new Automated Construction Progress Management System based on indoor positioning technology. In the context of continued decline in construction productivity (Philipp et al. 2016), Seeking a more efficient and effective construction progress management system is imminent. As the current manual construction management is time-consuming and error-prone, non-manual tracking methods found the best time to get on the stage of construction to replace the outdated and obsolete methods. After the sheer volume of comparisons, UWB became the best technology with its unparalleled accuracy and stability to make up the proposed automated construction progress management system. Much more critically, the UWB's property of real-time brings the possibility of timely adjustments to unexpected situations to the construction progress. As the construction progress necessitates the forward flow of design intent and feedback flow of real construction state information (Turkan et al. 2012), UWB has the ability to shorten the time of this kind of "flow". From the perspective of theoretical analysis, it is proved that the proposed system is practically feasible, while the downside is that this paper lacks the support of a real case, which is what needs to be done in the future.

ACKNOWLEDGEMENT

This research is funded by the Jiangsu University Natural Science Research Programme, 17KJB560011.

REFERENCES

Bose, A. & Chuan, H. F. 2007. A Practical Path Loss Model for Indoor WiFi Positioning Enhancement. *6th International Conference on Information, Communications and Signal Processing, ICICS.*

Cho, Y. K. Youn, J.H. & Martinez, D. 2010. Error Modeling for an Untethered Ultra-Wideband System for Construction Indoor Asset Tracking. *Automation in Construction* 19(1):43–54.

Fathi, H. Dai, F. & Lourakis, M. 2015. Automated as-built 3D reconstruction of civil infrastructure using computer vision: Achievements, opportunities, and challenges. *Advanced Engineering Informatics* 29(2):149–161.

Heigermoser, D. De Soto, B.G. Abbott, E.L.S. & Chua, D.K.H. 2019. BIM-based last planner system tool for improving construction project management. *Automation in Construction* 104:246–254.

Howes, N. R. 1984. Managing software development projects for maximum productivity. *IEEE Transactions on Software Engineering* SE-10(1):27–35.

Ibrahim, M. & Moselhi, O. 2016. Inertial measurement unit based indoor localization for construction applications. *Automation in Construction* 71:13–20.

Iyer, K. C. & Jha, K.N. 2005. Factors Affecting Cost Performance: Evidence from Indian Construction Projects. *International Journal of Project Management* 23(4):283–295.

Kopsida, M. Brilakis, I. & Vela, P.A. 2015. A Review of Automated Construction Progress Monitoring and Inspection Methods. *Proc. of the 32nd CIB W78 Conference* (June 2016): 421–431.

Li, H. Chan, G. Wong, J.K.W. & Skitmore, M. 2016. Real-time locating systems applications in construction. *Automation in Construction* 63:37–47.

Luo, X. O'Brien, W.J. & Julien, C.L. 2011. Comparative evaluation of received signal-strength index (RSSI) based indoor localization techniques for construction jobsites. *Advanced Engineering Informatics* 25(2):355–363.

Memarzadeh, M. Golparvar-Fard, M. & Niebles, J. C. 2013. Automated 2D detection of construction equipment and workers from site video streams using histograms of oriented gradients and colors. *Automation in Construction* 32:24–37.

Montaser, A. & Moselhi, O. 2014. RFID indoor location identification for construction projects. *Automation in Construction* 39:167–179.

Philipp, G. 2016. *Shaping the future of construction: A breakthrough in mindset and technology*. World Economic Forum (WEF) (May): 1–64.

Pradhan, A. Ergen, E. & Akinci, B. 2009. Technological assessment of radio frequency identification technology for indoor localization. *Journal of Computing in Civil Engineering* 23(4):230–238.

Pučko, Z. Šuman, N. & Rebolj, D. 2018. Automated continuous construction progress monitoring using multiple workplace real time 3D scans. *Advanced Engineering Informatics* 38:27–40.

Shahi, A. Aryan, A. West, J.S. Haas, C.T. & Haas, R.C.G. 2012. Deterioration of UWB positioning during construction. *Automation in Construction* 24:72–80.

Siebert, S. & Teizer, J. 2014. Mobile 3D mapping for surveying earthwork projects using an unmanned aerial vehicle (UAV) system. *Automation in Construction* 41:1–14.

Turkan, Y. Bosche, F. Haas, C.T. & Haas, R. 2012. Automated progress tracking using 4D schedule and 3D sensing technologies. *Automation in Construction* 22:414–421.

Zhang, C. Hammad, A. & Rodriguez, S. Crane pose estimation using UWB real-time location system. *Journal of Computing in Civil Engineering* 26(5):625–637.

SWOT analysis for using BIM for infrastructure across the whole lifecycle of transportation projects

Z. Li & F. Guo
Xi'an Jiaotong-Liverpool University, Suzhou, P.R. China

D. Schaefer
University of Liverpool, Liverpool, UK

ABSTRACT: The transportation infrastructure plays an important role in affecting the development of the national economy, which has developed rapidly in recent years. However, the cost of transportation projects is huge, and the traditional project process management causes a waste of time and resources. Building Information Modelling (BIM) is widely used in the building industry, but also can be used in the whole life cycle of transportation projects to improve the project efficiency. SWOT analysis is an important tool to assist the development of enterprises by analyzing their strengths, weaknesses, opportunities and challenges. In this paper the authors use SWOT analysis to investigate the BIM application across the whole lifecycle of transportation projects and to provide advice for their future development accordingly. The result shows that although BIM can save time, money and resources, facilitate management and information exchange, and has a wide application prospect, there are still some technical and normative problems. Interaction between different specialties, information aggregation and fusion, and new regulations to deal with new technologies are all the problems to be resolved.

1 INTRODUCTION

The transportation industry is an important prerequisite for the economic and social development of a country (Costin et al. 2018). Logistics and transportation promote economic development, but also bring challenges (Cleophas et al. 2019, Wann-ming 2019). The congestion problem brought by urbanization construction also poses challenges for transportation construction. China's population base is large and its growth rate is fast. Urbanization construction is also being rapidly promoted. The construction of transportation projects plays an auxiliary role in the national economy. In the process of developing transportation projects, the government should not only consider the economic benefits it brings, but also the requirements of its sustainable development (Hadi & Haghshenas 2018). Therefore, a more effective method is needed to plan and manage the whole life cycle of transportation projects (Costin et al. 2018). In recent years, the construction industry is gradually promoting the use of BIM (Olawumi & Chan 2018). BIM for infrastructure can help stakeholders better understand the design items, estimate and control costs, facilitate construction planning and monitoring, and improve project quality (Chan et al. 2019). Transportation project planning and management through BIM can optimize the traditional project operation mode.

Transportation projects are expensive and difficult to manage. Normally, they are constructed by private or state-owned companies and monitored by the government. These companies and government should not only serve the people, but also have profitable income. Therefore, it is necessary to adopt strategic methods to make important decisions as to whether a certain approach is desirable. SWOT is one such method (Nazarko et al. 2017). This method of analysis is derived from business management. It was first proposed by a professor of Management Economics at University of San Francisco, USA (Arslan & Er 2007).

As an analytical system summarized by Stanford Research (Jasiulewicz-kaczmarek 2016), it is suitable for profitable units such as enterprises (Arslan & Er 2007), because it is intuitive, flexible, and can produce practical outputs. SWOT is the abbreviation for strength, weakness, opportunities and threats, which can be defined as below (Nazarko et al. 2017, Jasiulewicz-kaczmarek 2016):

Strengths——favorable properties/features that can enhance its competition.
Weaknesses——unfavorable properties/features that can hinder its development.
Opportunities——external factors that can promote its development.
Threats——external factors that prevent from achieving the intended results.

In this paper the authors introduce the application of BIM for infrastructure across the whole life cycle of a project (design, construction, operation and maintenance phases) based on a comprehensive literature review. The reason for why BIM can reduce project cost, time and resource consumption is explained. SWOT is used to analyze the results of BIM application in the whole lifecycle transportation projects. The analysis provides suggestions for the development of transportation infrastructure.

2 METHODOLOGY

Three steps were adopted in conducting the literature reviews for this paper: 1) determining the database, 2) determining the keywords, 3) screening and classifying information.

Determine the database

In order to ensure the reliability of academic contents, relevant academic journal articles and international conference articles were reviewed. In order to ensure the timeliness of the academic content, literature from 2000 to the present was considered. Table 1 shows the database used for this article.

Table 1. Examined journals categorized by scope.

Category	Journal title	Publisher
Construction	Journal of Building Engineering	Elsevier
	Archives of Civil and Mechanical Engineering	Elsevier
	Journal of Building Engineering	Elsevier
	International Journal of Construction Education and Research	Elsevier
Built environment	Automation in Construction	Elsevier
	Construction and Building Materials	Elsevier
	Journal of Computing in Civil Engineering	ASCE
	Building and Environment	Elsevier
	Build Simulation	Tsinghua University Press and Springer-Verlag Berlin Heidelberg
	Building Research & Information	Taylor & Francis
Transportation	Journal of Traffic and Transportation Engineering	Elsevier
	Transportation Research Part A	Elsevier
	Transportation Research Part D	Elsevier
Energy	Procedia Engineering	Elsevier
	Energy and Buildings	Elsevier
	Procedia – Social and Behavioral Sciences	Elsevier

(Continued)

Table 1. (*Continued*)

Category	Journal title	Publisher
Civil and mechanical engineering	Archives of Civil and Mechanical Engineering	Elsevier
	Cities	Elsevier
	Archives of Civil & Mechanical Engineering	Elsevier
Informatics technology	Advanced Engineering Informatics	Elsevier
International Conference	International Conference on Computing in Civil and Building Engineering	ICCCBE
	International Workshop on Computing in Civil Engineering	ASCE
	International Construction Specialty Conference	ICSC
	IFAC-Papers On Line	Elsevier
Sustainable development	Sustainable Cities and Society.	Elsevier
Management	Process Safety Progress	Elsevier
	European Journal of Operational Research	Elsevier
	Journal of Management in Engineering	Elsevier
	Social and Behavioral Sciences	Elsevier

Determine the search keywords

The purpose of determining keywords was to find the relevant articles that can address the main questions of this study. Table 2 lists the questions to be discussed and the keywords corresponding to them.

Information screening and classification

Once the databases were determined, a keyword search could be performed in the database. Search results were filtered through abstracts and conclusions of articles. The relevant articles in this paper were classified and stored for information screening.

Table 2. Questions and keywords.

Questions	Keywords
What is the life cycle of infrastructure	Infrastructure; LCA
What are typical infrastructure projects	Road and highways; bridges; railways; port and harbor
What is the goal of infrastructure development	Infrastructure; development
What functions can BIM achieve	BIM
	Virtual design and construction
What functions of BIM are needed in infrastructure	BIM for infrastructure
	3D engineering models
	Civil information models
What does infrastructure management include	Infrastructure; management
How is BIM used in the whole life cycle of infrastructure	BIM for infrastructure; Life cycle
What are the advantages/benefits of using BIM in infrastructure	BIM for infrastructure; Advantages/benefits
What are the disadvantages/challenges of using BIM in infrastructure	BIM for infrastructure; Disadvantages/challenges
What are the application examples of BIM in infrastructure	BIM for infrastructure; Case study
What is SWOT analysis	SWOT analysis
What are the advantages of SWOT analysis method	SWOT analysis

BIM can build 3D models, 4D/5D models, even nd models, so the resulting information models can better visualize the details of the project, better control the time and cost of the project in the whole project cycle, and improve the quality of the project. It can help manage different project phases more effectively from planning and design to construction and maintenance (Costin et al. 2018, Olawumi & Chan 2018).

In the design stage

Planning in the design stage not only needs to meet the requirements of various standards and specifications, but also needs to reduce the cost as much as possible, avoid risks and reduce construction time to improve efficiency. BIM for infrastructure can be used to create, manage, and share design models, coordinate between different disciplines and reduce design-time spatial collisions (Costin et al. 2018, Ma & Ren 2017). Cheng et al. (2016) pointed out that BIM spatial visualization can be used in pipeline design of foundation projects to avoid pipeline overlap and to adjust the spatial relationship between different models, so as to avoid design errors. One of the most important benefits of BIM is clash detection, which can prevent possible major design errors at the beginning of the design and eliminate time and cost overruns caused by rework (Cheng et al. 2016). Visualization of design is also convenient for explaining design concepts and key points to owners, and for owners to make intuitive comparisons between various designs options. At this stage, it can be used in conjunction with Geographic Information System (GIS) technology to facilitate better integration of human and geographical environment of the project (Li et al. 2012).

BIM for infrastructure can be used to calculate the amount of work, such as the amount of materials used in construction, the amount of earth filled and excavated, and the use of machinery and equipment. It will also ensure the efficient use of resources by reducing the energy consumption. Moreover, it does not need a lot of time to recalculate the amount of work when the design changes, which can save designers' energy and time in the early design stage. It is also important to determine the construction process in the early design stage. Smooth and reasonable construction process can improve the utilization rate of mechanical equipment, improve work efficiency, reduce time and cost (Sarkar 2016). It can better cope with the current promotion of green sustainable development in China and prevent waste of resources.

In the construction stage

BIM for infrastructure provides better visualization operation to optimize the details of construction work, improve the accessibility and operability in the construction phase, enhance the coordination of various departments (Costin et al. 2018), and improve the quality of the project. It can greatly prevent the incorrect construction caused by inaccurate transmission of information from the designers to the constructors. When errors occur, BIM can also be used to remedy them. For example, if incorrect reinforcement is used in a beam, the original design can be replaced by the incorrect reinforcement method. The structure connected with the beam can be checked by BIM, and the joint mode of reinforcement can be adjusted to reduce material waste.

In the construction phase, the critical components that need to be managed effectively include process, material and goods, cost, machinery and equipment, risk and quality control. Although these factors have been considered in the design stage, there will be many uncertainties in the construction phase, so it is necessary to monitor the site in real time and adjust the construction plan according to the real conditions (Lin 2018). BIM for infrastructure can use the nd model to monitor the construction process in real time, judge whether the time and expenditure exceed the budget, adjust the budget and time as necessary, so that the project can be completed on time, and on budget. According to the site conditions at different construction stages, the possible risks a can be also assessment through the models. BIM for infrastructure can be used to plan the space of construction site and separate the construction scope of equipment from that of constructors, so as to avoid potential safety hazards to constructors caused by improper mechanical operation (Lin 2018).

In operating and maintenance stage

The data collected during the design phase and the data updated during the construction phase of the infrastructure can be stored by BIM and used in the operation and maintenance phase of the infrastructure (Cheng et al. 2016). BIM for infrastructure can use data organization and integration to carry out detailed geometric design (Bieē 2011). According to the geometric model data and data sensors, it can check whether the structural deformation exceeds the specified range and whether it needs maintenance or reconstruction. Thus, will maintain structural reliability, security and serviceability (Abdessemed et al. 2011). BIM for infrastructure can also carry out structural analysis and post-repair analysis of infrastructure with a certain service life according to the load conditions, so as to improve the feasibility of the repair scheme (Bieē 2011).

4 SWOT ANALYSIS OF USING BIM FOR INFRASTRUCTURE

SWOT analysis

In order to better utilize BIM in the whole life cycle of a transportation project, it is necessary to clearly identify its strengths, weaknesses, opportunities, and threats. The following SWOT analysis is conducted based on the existing literatures related to BIM for infrastructure.

Strengths

In early design, BIM for infrastructure can avoid collision problems that may occur when combining various design models, so as to avoid risks as early as possible. In the maintenance and renovation of transportation projects, BIM for infrastructure can also assist the spatial planning of construction sites to ensure normal traffic flow and maintain the operational efficiency of the transportation system (Sciara et al. 2017). Research by Kim et al. (2015) and Hu et al. (2011) confirms that BIM for infrastructure can combine geometric information, timetables and historical data to help plan and manage projects, to evaluate and analyze their security, and to identify and reduce risks quickly. According to the research of Golovina et al. (2016), the use of BIM for infrastructure can avoid the workspace conflict between staff and heavy equipment, thus reduce the risk of construction stage. BIM for infrastructure can also be used to estimate construction time and cost. By comparing the as-built conditions with the design model, it can also optimize the management during construction and operation and maintenance, and reduce the waste of time and cost. Based on the empirical study of Lu et al. (2014), BIM for infrastructure can reduce the construction cost by 8.61%. Visualization model can show the appearance and details of the project, which is convenient for comparison and selection between different schemes. It will also improve the overall comfort of the building, more in line with social and economic benefits (Gurevich et al. 2017). In the whole life cycle of the project, there is a need for efficient communication and collaboration among multiple parties. The visualization and information exchange produced by BIM can facilitate and help maintain effective communication platform (Takim et al. 2013). Using BIM and Life Cycle Assessment (LCA) to analyze the whole life cycle of the project, the design scheme with the least energy consumption and the least emission of greenhouse gases or harmful substances can be obtained (Basbagill et al. 2013).

Weaknesses

Usually a huge amount of investment funds are needed for the technology and the necessary training in the early stage (Costin et al. 2018). Using BIM for infrastructure at the design stage will result in an additional 45.93% effort (Lu et al. 2014). If different parties use different software to development models, there may be problems that the model cannot be interoperable and integrated when summarizing design schemes (Vilgertshofer & Borrmann 2017, Borrmann & Jubierre 2013). Eadie et al. (2013) also agreed that when different departments use new methods and software to model or organize information, problems such as incompatibility of models or lack of information will arise in the process of summarizing or communicating. It is hard for different disciplines to get very accurate information from the models produced by other incompatible software, and thus also difficult for them to check the clashes between different design components (Ku & Taiebat 2011). In addition, BIM for infrastructure needs huge amount of data to support different scenarios simulation when it is used for urban traffic simulation, which has certain economic and time limitations (Abou-senna et al. 2018). BIM for infrastructure can't store all the data of the whole project, for example, some information (e.g. material properties and functions) makes it necessary to set up other external database for storage (Jalaei 2013). In such case, it will be necessary to establish multiple databases to ensure information integrity, which makes it more difficult for information management and search. At the same time, large file storage systems require additional maintenance costs (Guo et al. 2017).

Opportunities

Currently, there is a trend of moving towards smart city. Maintenance and construction of transportation infrastructure often involves a large number of land resources, which will affect the local ecological environment (Amiril et al. 2014) and the daily life of residents. Therefore, in the direction of sustainable development, transportation departments should focus on the three areas of environment, society, and economy (Odeck & Kjerkreit 2019). BIM for infrastructure can collaborate with ALMSUN (Advanced Interactive Microsimulation of Urban and Non-Urban Networks) and Traffic Simulator System (TSS) to help realize the concept of sustainable transportation, and achieve the goal of reducing environmental damage, improving public satisfaction and reducing costs (Abou-senna et al. 2018). Various models produced by BIM for infrastructure can be combined and managed to help with integrated management of the city (Nazarko et al. 2017). With the assistance of big data, BIM for infrastructure can be used to store information and help built smart cities (Olawumi et al. 2018). The models can also be used as reference to improve the efficiency of maintenance or transformation programs, save resources and contribute to sustainable development. For road projects, it is necessary to carry out benefit-cost analyses (BCAs) after the completion of the project construction to check whether the project income and investment cost are balanced (Hadi & Haghshenas 2018). BIM for infrastructure simulates the traffic volume and traffic growth rate to predict the project income in advance, so as to prevent the situation that investment outweighs earnings. BIM and other surveying and mapping tools can be used to manage and analyze existing for infrastructure that have not be modeled before. If the facility has become old enough that affects its use or has potential safety hazards, BIM can be used to reconstruct a more energy-efficient and environmental-friendly facility (Bribián et al. 2011). Using BIM in the whole life cycle of infrastructure projects can help facilitate a more environmental-friendly city (Phillips et al. 2017).

Threats

When using BIM together with other technologies in the whole life cycle of projects, it will produce a huge amount of digital data(Khosrowpour 2015). This requires an integrated system to preserve, manage, search, and use the information, but such system is hard to establish. To build an integrated city, it is necessary to use BIM and other information collection and processing tools together. This also brings incompatibility issues. In order to achieve collaboration among different divisions, information models are shared with each other, which might bring challenges in the copyright and information loss (Al-Shalabi et al. 2015). In the process of project data exchange, there will be the problem of responsibility sharing. BIM for infrastructure provides a convenient information exchange method for the construction phase of transportation projects, and allows real-time modification according to the actual situation. However, the construction site of transportation projects is related to the local environment, so it is difficult to have data reference to determine whether the modification is feasible. Moreover, there is no uniform standard for determining who should be given the right to modify the model and who should be given the right to read the date only. After the modification of the model, how to judge the responsibility attribution is also not specified (Guo et al. 2017, Costin et al. 2018). New standards and regulations need to be established to resolve these challenges.

5 CONCLUSION

In this paper the authors use SWOT analysis to analyze the application of BIM in the whole cycle of transportation projects. Through the analysis of the advantages, disadvantages, opportunities and challenges of the current BIM application, the issues that need to be addressed in the development of transportation infrastructure are summarized. BIM plays an important role in reducing energy consumption, saving time and cost. It can also provide convenience in management (including construction, operation and maintenance), and provide technical support for future urban intellectualization. However, it will require a great amount of time and cost spent on the training of staff. Moreover, there may be incompatibility issues resulting from the information exchange between different specialties. New standards and regulations are also needed to better cope with the adoption of new technology.

ACKNOWLEDGEMENT

The materials presented in this paper are a part of RDF-17-02-60 project sponsored by Xi'an Jiaotong - Liverpool University.

REFERENCES

Abdessemed. M. Kenai, S. Bali, A. & Kibboua, A. 2011. Dynamic analysis of a bridge repaired by CFRP: Experimental and numerical modelling. *Construction and Building Materials* 25: 1270–1276.

Abou-senna, H. Radwan, E. & Navarro, A. 2018. ScienceDirect Integrating transportation systems management and operations into the project life cycle from planning to construction: A synthesis of best practices. *Journal of Traffic and Transportation Engineering (English Edition)* 5(1): 44–55.

Al-Shalabi, F. A. Turkan, Y. & Laflamme. S. 2015. BrIM implementation for documentation of bridge condition for inspection. *Icsc15-the Csce International Construction Specialty Conference* 262: 1–8.

Amiril, A. Nawawi, A.H. Takim, R. & Farhana Ab.Latif, S.N. 2014. Transportation Infrastructure Project Sustainability Factors and Performance. *Procedia - Social and Behavioral Sciences* 15: 90–98.

Arslan O, & Er. I. D. 2007. A SWOT analysis for successful bridge team organization and safer marine operations. *Process Safety Progress* 27: 21–28.

Basbagill, J. Flager, F. Lepech, F. & Fischer, M. 2013. Application of life-cycle assessment to early stage building design for reduced embodied environmental impacts. *Building and Environment* 60: 81–92.

Bieē, J. 2011. Modelling of structure geometry in Bridge Management Systems. *Archives of Civil & Mechanical Engineering* 11(3): 519-532.

Borrmann, A. & Jubierre. J. 2013. Computing in Civil Engineering - A Multi-Scale Tunnel Product Model Providing Coherent Geometry and Semantics. *ASCE International Workshop on Computing in Civil Engineering* 33: 291-298.

Bribián, I. Z. Capilla, A.V. & Usón, A. A. 2011. Life cycle assessment of building materials : Comparative analysis of energy and environmental impacts and evaluation of the eco-ef fi ciency improvement potential. *Building and Environment* 46(5): 1133–1140.

Chan, D. W. M. Olawumi, T.O. & Ho, A.M. L. 2019. Perceived benefits of and barriers to Building Information Modelling (BIM) implementation in construction : The case of Hong Kong. *Journal of Building Engineering* 25(4): 100764.

Cheng, J. C. P. Lu, Q. & Deng, Y. 2016. Automation in Construction Analytical review and evaluation of civil information modeling. *Automation in Construction* 67: 31–47.

Cleophas, C. Cottrill, C. Ehmke, J.F. & Tierney, K. 2019. Collaborative urban transportation : Recent advances in theory and practice. *European Journal of Operational Research* 273(3): 801–816.

Costin, A. Adibfar, A. Hu, H. & Chen, S.S. 2018. Automation in Construction Building Information Modeling (BIM) for transportation infrastructure – Literature review, applications, challenges, and recommendations. *Automation in Construction* 94(7): 257–281.

Eadie, R. Browne, M. Odeyinka, H. McKeown, C. & McNiff, S. 2013. Automation in Construction BIM implementation throughout the UK construction project lifecycle: An analysis. *Automation in Construction* 36: 145–151.

Golovina, O. Teizer, J. & Pradhananga, N. 2016. Automation in Construction Heat map generation for predictive safety planning: Preventing struck-by and near miss interactions between workers-on-foot and construction equipment. *Automation in Construction* 71: 99–115.

Guo, F. Jahren, C.T. Turkan, Y. & David Jeong, H. 2017. Civil Integrated Management: An Emerging Paradigm for Civil Infrastructure Project Delivery and Management. *Journal of Management in Engineering* 33(2): 1–10.

Gurevich, U. Sacks, R. & Shrestha, P. 2017. *BIM adoption by public facility agencies : Impacts on occupant value.* 3218.

Hadi, M. & Haghshenas, H. 2018. Micro-scale sustainability assessment of infrastructure projects on urban transportation systems : Case study of Azadi district, Isfahan, Iran. *Cities* 72: 149–159.

Hu. Z. Z. & Zhang. J.P. 2011. BIM-and 4D-based integrated solution of analysis and management for conflicts and structural safety problems during construction: 2. Development and site trials. *Automation in Construction* 20(2):167–180

Jalaei, F. 2013. Integrating building information modelling with sustainability to design building projects at the conceptual stage. *Building Simulation* 429–444.

Jasiulewicz-kaczmarek, M. 2016. ScienceDirect SWOT analysis for Planned Maintenance SWOT study SWOT analysis analysis for strategy – a case study. *IFAC-PapersOnLine* 49(12): 674–679.

Khosrowpour, M. 2015. Encyclopedia of Information Science and Technology, First Edition. *Information Science Reference*.

Kim, H. Lee, H.S. Park, M. & Chung, B. 2015. Information Retrieval Framework for Hazard Identification in Construction. *Journal of Computing in Civil Engineering* 29(3): 04014052.

Ku, K. & Taiebat. M. 2011. BIM experiences and expectations: the Constructors' perspective. *International Journal of Construction Education and Research* 7: 175–197.

Li, H. Chan. N.K.Y. Huang, T. Skitmore, M. & Yang, J. 2012. Virtual prototyping for planning bridge construction. *Automation in Construction* 27: 1–10.

Lin, J. 2018. Towards BIM-based model integration and safety analysis for bridge construction Towards BIM-based model integration and safety analysis for bridge construction.

Lu, W. Fung, A. Peng, Y. Liang, C. & Rowlinson, S. 2014. Cost-benefit analysis of Building Information Modeling implementation in building projects through demystification of time-effort distribution curves. *Building and Environment* 82: 317–327.

Ma, Z. & Ren, Y. 2017. Integrated Application of BIM and GIS : An Overview. *Procedia Engineering* 196: 1072–1079.

Nazarko, J. Ejdys, J. Halicka, K. Magruk, A. Nazarko, L. & Skorek, A. 2017. Application of Enhanced SWOT Analysis in the Future-oriented Public Management of Technology. *Procedia Engineering* 182: 482–490.

Odeck, J. & Kjerkreit, A. 2019. *The accuracy of benefit-cost analyses (BCAs) in transportation : An ex-post evaluation of road projects* 120: 277–294.

Olawumi, T. O. Chan, D.W.M. Wong, J.K.W. & Chan, A.P.C. 2018. Barriers to the integration of BIM and sustainability practices in construction projects : A Delphi survey of international experts. *Journal of Building Engineering* 20: 60–71

Olawumi, T. O. & Chan, D. W. M. 2018. Identifying and prioritizing the benefits of integrating BIM and sustainability practices in construction projects : A Delphi survey of international experts. *Sustainable Cities and Society* 40: 16–27.

Phillips, R. Troup, L. Fannon, D. & Eckelman, M.J. 2017. Do resilient and sustainable design strategies conflict in commercial buildings? A critical analysis of existing resilient building frameworks and their sustainability implications. *Energy & Buildings* 146: 295–311.

Sarkar. D. 2016. Risk based building information modeling (BIM) for urban infrastructure transport project;, Advanced Engineering Informatics. *International Scholarly and Scientific Research & Innovation* 10(8): 1022–1026.

Sciara, G.C. Bjorkman, J. Stryjewski, E. & Thorne, J.H. 2017. Mitigating environmental impacts in advance : Evidence of cost and time savings for transportation projects. *Transportation Research Part D* 50: 316–326.

Takim, R. Harris, M. & Nawawi, A.H. 2013. Building Information Modeling (BIM): A new paradigm for quality of life within Architectural, Engineering and Construction (AEC) Industry. *Procedia - Social and Behavioral Sciences* 101: 23–32.

Vilgertshofer, S. & Borrmann, A. 2017. Advanced Engineering Informatics Using graph rewriting methods for the semi-automatic generation of parametric infrastructure models. *Advanced Engineering Informatics* 33: 502–515.

Wann-ming, W. 2019. Constructing urban dynamic transportation planning strategies for improving quality of life and urban sustainability under emerging growth management principles. *Sustainable Cities and Society* 44: 275–290.

Sustainable Buildings and Structures: Building a Sustainable Tomorrow – Papadikis et al. (Eds)
© 2020 Taylor & Francis Group, London, ISBN 978-0-367-43019-1

Financial subsidies to further construction waste resources utilization through a game theory model

S.H. Chen, J.L. Hao, B.Q. Cheng & C.M. Herr
Xi'an Jiaotong-Liverpool University, Suzhou, Jiangsu, China

J.J. Wang & J. Xu
Anhui Jianzhu University, Hefei, Anhui, China

ABSTRACT: Resource utilization is a significant approach to address the problem of construction waste and the government plays a key role in this process through financial policies. This paper proposes a model by using game theory to discuss how the government can use financial subsidies to motivate the contractor to utilize the construction waste. The best subsidy level of the government is worked out through finding the Nash equilibrium point between the government and the contractor. The research result can provide a reference for policy makers to design and optimize the subsidy policy on resource utilization of construction waste.

1 INTRODUCTION

Construction waste, accounting for over 40% of the total amount of solid waste in China, has put severe problems on the "Smart Growth" of china's cities (Golej 2014). Actually, wastes are misplaced resources which means that resource utilization is a significant approach to reduce negative impacts of construction waste. However, the resource utilization rate of construction waste in China is only about 5%, which is obviously less than the average level (about 90%) of developed countries (Lu et al. 2011). The inconsistencies between the objectives of the government and the contractor, two main agents of construction waste management, are the main reason influencing the resource utilization of construction waste. The government aims to protect the public interest and the contractor aims to make high profits (Peng & Liu 2018). How to balance the interests of both sides and achieve a win-win situation is an important problem to be solved in the promotion of resource utilization of construction waste. The traditional view thought the best policy choice was that the government could subsidize the contractors (Luo 2018). However, this is not always useful (Zhang 2019). Successful subsidy policy hinges on getting this right: if the subsidy is too small, the contractor will tend to pay landfill for these wastes; if it is too high, the government will lose a large amount of money (Wang et al. 2016). Most previous policies were developed based on the experience of decision-makers without consideration of the gaming relationship between the two sides (Sun et al. 2017). Therefore, research on the decision-making process of the treatment of construction waste between the government and the contractor is significant. Then game theory, which can possess a best strategy for a competitive two-person game, can be used to analyze the decision-making process (Kapliński & Tamošaitienė 2010).

This paper presents a game model to design and optimize the subsidy policy on resource utilization of construction waste based on the game theory. It attempts to find the best subsidy level through finding the Nash equilibrium point. The result can provide reference for policymakers when designing the subsidy policy on resource utilization of construction waste.

2 MODEL DEVELOPMENT

The model is built by using game matrix. In order to use the game matrix, basic assumptions are made which are as follows:

Assumption 1: There are only two game parties in the game process: the government and the contractor.

Assumption 2: The government will make timely response and give corresponding financial subsidies to the contractor when the contractor utilizes the construction waste.

Assumption 3: Both the government and the contractor are rational agents, which means that the game parties will both take the action to maximize the profit.

Assumption 4: The game process between the government and the contractor is a game of perfect information in which the game parties know the full information of the game.

In the static game with perfect information, the government can always know whether the contractor resource utilizes the construction waste and if the contractor utilizes the waste, the government will give corresponding subsidies to the contractor. The contractor will decide how to dispose the construction waste in consideration of whether the government will provide corresponding subsidies.

The strategies of the government are subsidy (β_1) and no subsidy $(1 - \beta_1)$ and the strategy of the contractor are landfill (α_1) and resource utilization (α_1). S is the subsidy on recourse utilization of construction wasted the government paid for the contractor. C_1 is the landfill fees that the contractor needs to pay to the refuse landfill site. C_2 is the resource utilization fees that the contractor needs to pay for utilizing the construction waste. R_1 is the environmental benefit that the contractor brings to the government when the construction waste is utilized. R_2 is the resident satisfaction acquired by the government when the government successfully uses subsidy to let the contractor utilize the construction waste (Liu 2015). In this way, this paper established the game matrix between the government and the contractor shown in Table 1.

Therefore, a static game model with perfect information used to analyze the reasons why the contractor chooses landfill or resource utilization in construction waste management is developed. Based on the condition that the government will give financial subsidy to the contractor if the construction waste is utilized, when the contractor uses landfill to dispose the construction waste, the benefit that the government will get is R_2, which is the resident satisfaction acquired by the government when the government successfully uses subsidy to let the contractor utilize the construction waste. When the contractor utilizes the construction waste, the government will gain the benefit $R_1 + R_2 - S$, which is the environmental benefit plus the resident satisfaction, minus the subsidy that the government pays to the contractor.

Based on the condition that the government won't provide subsidy for the contractor if the construction waste is utilized, when the contractor uses landfill to dispose the construction waste, the benefit that the government can acquire is 0. When the contractor utilizes the construction waste, the government will gain the environmental benefit R_1.

When the contractor uses landfill to treat the construction waste, no matter the government will provide subsidy for the contractor or not, the benefit that the contractor requires is $-C_1$ because the contractor cannot get financial subsidies from the government if the waste is not utilized and the benefit turns out to be the landfill fees that the contractor needs to pay to the refuse landfill site, which is negative.

Table 1. Game matrix between two parties.

Contractor Government	landfill (α_1)	resource utilization $(1 - \alpha_1)$
subsidy (β_1)	$R_2, -C_1$	$R_1+R_2-S, S-C_2$
no subsidy $(1 - \beta_1)$	$0, -C_1$	$R_1, -C_2$

Based on the condition that the contractor utilizes the construction waste, when the government provides the contractor corresponding subsidies, the benefit that the contractor can acquire is $S - C_2$, which is the subsidies that the government provides minus the resource utilization fees that the contractor needs to pay to the resource utilization plant. When the government won't provide subsidies for the contractor, the benefit that the contractor gains is $-C_2$, which is the resource utilization fees that the contractor needs to pay to the resource utilization plant.

3 DECISION MAKING ANALYSIS

Since the game matrix analyzed above doesn't exist the Nash Equilibrium of pure strategy, this paper has to solve the game matrix of mixed strategy Nash equilibrium.

3.1 Decision making analysis of the government

When the probability of the contractor utilizing the construction waste is $(1 - \alpha_1)$, for the government, the strategic expected benefit of providing financial subsidies ($\beta_1 = 1$) and not providing financial subsidies ($\beta_1 = 0$) respectively are:

$$\pi_g(1, 1 - \alpha_1) = (R_1 - R_2 + S)(1 - \alpha_1) + R_2\alpha_1 \tag{1}$$

$$\pi_g(0, 1 - \alpha_1) = R_1(1 - \alpha_1) + 0\alpha_1 \tag{2}$$

The Nash equilibrium is reached when $\pi_g(1, 1 - \alpha_1) = \pi_g(0, 1 - \alpha_1)$, from which the result can be calculated that the optimal probability of the contractor utilizing the construction waste is:

$$(1 - \alpha_1) * \frac{R_2}{2R_2 - S} \tag{3}$$

Then it can be found that, in the static game model with perfect information, the probability of the contractor utilizing the construction waste $(1 - \alpha_1)$ is proportional to the resident satisfaction acquired by the government R_2, which means that if the government uses the strategy that provides subsidies for the contractor to utilize the construction waste to enhance the government reputation and gain the resident satisfaction, the motivation of the contractor to utilize the construction waste will be largely increased and then the probability of the contractor utilizing the construction waste will increase. Moreover, the probability of the contractor utilizing the construction waste $(1 - \alpha_1)$ is in an inverse ratio to the government's strategic subsidies $2R_2 - S$. The result indicates that since excessive financial subsidies may exceed the government yield, the government, as a rational agent, should adjust the subsidies according to the difference of income and expenditure when the probability of the contractor utilizing the construction waste largely increases, so as to the get the maximum benefits.

When $\pi_g(1, 1 - \alpha_1) < \pi_g(0, 1 - \alpha_1)$, the probability of the contractor utilizing the construction waste is $(1 - \alpha_1) < \frac{R_2}{2R_2 - S}$. At this time, the prospective earnings of the government providing subsidies is smaller than the prospective earnings of the government not providing subsidies and the optimal strategy of the government is not to provide subsidies.

When $\pi_g(1, 1 - \alpha_1) > \pi_g(0, 1 - \alpha_1)$, the probability of the contractor utilizing the construction waste is $(1 - \alpha_1) > \frac{R_2}{2R_2 - S}$. At this time, the prospective earnings of the government providing subsidies is larger than the prospective earnings of the government not providing subsidies and the optimal strategy of the government is to provide subsidies.

When $\pi_g(1, 1 - \alpha_1) = \pi_g(0, 1 - \alpha_1)$, the probability of the contractor utilizing the construction waste is $(1 - \alpha_1) = \frac{R_2}{2R_2 - S}$. At this time, the prospective earnings of the government providing subsidies is equal to the prospective earnings of the government not providing subsidies and the optimal strategy of the government is to provide subsidies or not to provide subsidies.

3.2 Decision making analysis of the contractor

When the probability of the government providing financial subsidies is β_1, for the contractor, the strategic expected benefit of using landfill to dispose the construction waste ($\alpha_1 = 1$) and using resource utilization to dispose the construction waste ($\alpha_1 = 0$) are respectively:

$$\pi_g(\beta_1, 1) = (-C_1)\beta_1 + (-C_1)(1 - \beta_1) \tag{4}$$

$$\pi_g(\beta_1, 0) = (S - C_2)\beta_1 + (-C_2)(1 - \beta_1) \tag{5}$$

The Nash equilibrium is reached when $\pi_g(\beta_1, 1) = \pi_g(\beta_1, 0)$, from which the result can be calculated that the optimal probability of the government providing financial subsidies is:

$$\beta_1 * \frac{C_2 - C_1}{S} \tag{6}$$

Then it can be found that, in the static game model with perfect information, the probability of the government providing financial subsidies β_1 is proportional to the difference of the resource utilization fees and the landfill fees $C_2 - C_1$, which means that the probability of the government providing subsidies will become larger when the difference of the resource utilization fees C_2 and the landfill fees C_1 gets larger. Otherwise, the probability of the government providing subsidies will become smaller. The reason is that when the resource utilization fees is much higher than the landfill fees, the possibility of the contractor using resource utilization to dispose the construction waste will be much lower. Then the government needs to provide financial subsidies for the contractor to compensate the partial expected revenue that the contractor has lost from using resource utilization to dispose the construction waste, so as to increase the probability of the contractor utilizing the construction waste. Moreover, the probability of the government providing financial subsidies β_1 is in an inverse ratio to the subsidy that the government will pay to the contractor S. The result indicates that if the government has provided the contractor with corresponding financial subsidies, the probability of the government increasing financial subsidies will greatly decrease afterwards, which means that the government will not endlessly provide the financial subsidies in order to reach the strategic expected revenue.

When $\pi_g(\beta_1, 1) > \pi_g(\beta_1, 0)$, the probability of the contractor utilizing the construction waste is $\beta_1 < \frac{C_2 - C_1}{S}$. At this time, the prospective earnings of the contractor using landfill to dispose the construction waste is larger than the prospective earnings of the contractor using resource utilization to dispose the construction waste and the optimal strategy for the contractor is to use landfill to dispose the construction waste.

When $\pi_g(\beta_1, 1) < \pi_g(\beta_1, 0)$, the probability of the contractor utilizing the construction waste is $\beta_1 > \frac{C_2 - C_1}{S}$. At this time, the prospective earnings of the contractor using landfill to dispose the construction waste is smaller than the prospective earnings of the contractor using resource utilization to dispose the construction waste and the optimal strategy for the contractor is to use resource utilization to dispose the construction waste.

When $\pi_g(\beta_1, 1) = \pi_g(\beta_1, 0)$, the probability of the contractor utilizing the construction waste is $\beta_1 = \frac{C_2 - C_1}{S}$. At this time, the prospective earnings of the contractor using landfill to dispose the construction waste is equal to the prospective earnings of the contractor using resource utilization to dispose the construction waste and the optimal strategy for the contractor is to use landfill or resource utilization to dispose the construction waste.

Therefore, conclusions can be made that the contractor will utilize the construction waste with a probability of $\frac{R_2}{2R_2 - S}$ and the government will subsidize the contractor with a probability of $\frac{C_2 - C_1}{S}$ to meet the Nash equilibrium point of the static game model with perfect information.

4 CONCLUSIONS AND RECOMMENDATIONS

Through the analysis of the static game with perfect information, scientific prediction can be made for under what circumstances will the government adjust the financial subsidies to motivate the contractor to utilize the construction waste instead of using landfill to dispose the construction waste. The factors that influence the probability of the contractor utilizing the construction waste are the subsidy that the government will pay to the contractor utilizing the construction waste and the resident satisfaction. For these two factors, the probability of the contractor utilizing the construction waste is in an inverse ratio to the government's strategic subsidies and is proportional to the resident satisfaction acquired by the government.

There are two administrative means that can promote the contractor utilizing the construction waste, which are developing the environmental technology and setting financial subsidies reasonably. By developing the environmental technology, the cost of utilizing the construction waste can be reduced, which will increase the probability of the contractor using resource utilization to dispose the construction waste. Furthermore, the environmental monitoring technology will be improved, which can help the government to clearly recognize the condition of the construction waste utilization and make an effective regulation. By setting reasonable financial subsidies, the contractor can be motivated to utilize the construction waste. Although the motivation will become greater when the subsidies get larger, excessive subsidies will cause the financial loss of the government. Therefore, the upper and lower limit of the subsidies should be taken into consideration to set the reasonable amount of the subsidy. For the upper limit, the subsidies should be lower than the resource utilization fees, otherwise the subsidies will be profitable for the contractor. Moreover, the subsidies should be lower than the environmental benefit, or the subsidies cannot be renewable. For the lower limit, the subsidies should be higher than the landfill fees, otherwise the low subsidies may fail to motivate the contractor to utilize the construction waste, which will stagnate the environmental improvement.

ACKNOWLEDGEMENTS

The authors wish to acknowledge the financial support from Suzhou Soft Science Fund (2019), Summer Undergraduate Research Fellowships, Xi'an Jiaotong-Liverpool University (201932) and National Innovation Training Program for College Students Fund (2019).

REFERENCES

Golej, J. 2014. Smart growth and sustainable development: New trends in land development. *Proceedings of the International Scientific Conference People, Buildings and Environment, Kroměříž, Czech Republic* 15-17.

Kapliński, O. & Tamošaitienė, J. 2010. Game theory applications in construction engineering and management. *Technological and Economic Development of Economy* 16(2): 348-363.

Liu, Q. 2015. Design of government subsidy policy and optimization of manufacturer decision under the background of low carbon economy. (Doctoral Dissertation, Southeast University). *Southeast University*.

Lu, W. Yuan, H.P. Li, J. Hao, J. L. Mi, X. & Ding, Z. K. 2011. An empirical investigation of construction and demolition waste generation rates in Shenzhen city, South China. *Waste management* 31(4): 680-687.

Luo, Y. F. 2018. Game analysis of government and manufacturers under energy conservation and emission reduction. *Business Economy* (3): 116-119.

Peng, H. & Liu, Y. 2018. How government subsidies promote the growth of entrepreneurial companies in clean energy industry: An empirical study in China. *Journal of Cleaner Production* 188: 508-520.

Sun, Y. Y. Song, Y.T. & Zhu, J. H. 2017. Research on enterprise illegal emission from the perspective of game theory. *Theory Monthly* 52(10): 115-119.

Wang, C. Zhang, Y. Zhang, L. & Pang, M. 2016. Alternative policies to subsidize rural household biogas digesters. *Energy Policy* 93: 187-195.

Zhang, P. H. 2019. Research on financial subsidy of resource enterprise low-carbon innovation based on signal game. *Journal of Zhengzhou Institute of Light Industry (Social Science)* 20(2): 73-79.

Sustainable Buildings and Structures: Building a Sustainable Tomorrow – Papadikis et al. (Eds)
© 2020 Taylor & Francis Group, London, ISBN 978-0-367-43019-1

Lean construction application: A case study in Suzhou, China

W. Xing & J. Hao
Xi'an Jiaotong-Liverpool University, Suzhou, Jiangsu, China

D. Wu & L. Qian
BOSCH, Shanghai, China

K.S. Sikora
University of Wollongong in Dubai, Dubai, UAE

ABSTRACT: Chinese Architecture, Engineering, Construction (AEC) industry is under the reform resulting from the large amount of waste production, huge energy consumption and severe environmental pollution. Aiming to maximize the project value while reducing the waste and cost, lean thinking is referred from manufacturing industry to construction since two industries have several similarities. This paper reviewed the change of lean thinking from production to construction, the differences between traditional and lean construction (LC), and the current situation of Chinese LC situation. Besides, a case study was assessed to examine how the LC management works in a practical project in China. Through the research, project schedule is divided into several hundred sub-items to assist the site supervisors to manage the construction progress. In addition, utilizing prefabrication parts and digital technology are two effective approaches to conduct the LC in the project.

1   INTRODUCTION

Chinese AEC industry has been a pillar industry since 1978, till now its scale expends to at least 20 times and its impact on the national economy is significant (He et al. 2013, Liao 2018). In 2018, for every 10,000 RMB output value increasing in the AEC industry, it brings 24,045 RMB benefits to other industries as both direct and indirect productivities (Wang 2019).

Nevertheless, some issues are exposed to its rapid development, which contributes to a large amount of waste production, huge energy consumption, and severe environmental pollution. Considering the whole life-cycle of the building, the AEC industry consumes 46.7% of total energy consumption in China annually. The construction wastes, dust, and noise account for 30% to 40%, 22% and 8% of the relevant pollution sources in the city (Yang 2016). Therefore, the public and government are aware that such the extensive but ineffective development way in the AEC industry is urgent to be reformed, to sustain the national economy and protect the environment. Accordingly, some scholars suggest that lean thinking could be applied to Chinese AEC industry to activate its sustainable development.

This paper focused on how LC was applied in a construction project in Suzhou from reviewing the transform of lean thinking to the construction industry and its current situation in China, and analyzing the implementation of it to the case study.

2   LITERATURE REVIEW

2.1   *From lean production to lean construction*

Lean production/manufacturing is derived from the automotive industry, which is primarily introduced from the principles of Toyota Production System (TPS) that obtain higher quality with few resources and respect for humanity (Salem et al. 2006, Wang & Ma 2014). The main elements of TPS are defined as just-in-time, autonomation, workforce flexibility, and creative thinking (Salem et al. 2006). A key point of lean production is to eliminate any waste from production by identifying correct and standardized processes, therefore it is necessary to optimize the workflow to add production value and improve the performance (Liu 2015). From vast numbers of research and comparisons, such the production mode, or TPS, is the most proper management approach for the modern manufacturing industry.

The success of lean production arises the interest of the public to extend its ideas to other industries. The concept of LC is raised by Lauri Koskela at International Group for Lean Construction in August 1993 first (Ballard & Howell 2003). Having put forward the LC concept and done the analysis on the application of LC, Koskela then discusses in 2000 that transformation, flow, and value (TFV) are three fundamental elements that need to be encompassed in the relevant theory to support the lean applications in construction industry (Jorgensen & Emmitt 2008, Yang, 2016). In addition, Ballard et al. (2001) define that maximizing value and minimizing waste is a universal goal and framed a new project planning system, which promotes the lean thinking to delivery from manufacturing to construction. Ballard & Howell (2003) investigate how to implement the LC to the project, and ways to apply Lean Project Delivery System (LPDS) to achieve higher productivity.

Through the continuous explorations on its field, LC has been successfully utilized in the AEC industry. For example, Tommelein (1988) introduces "pull-scheduling" to the construction project to increase the output of the material installation. Moreover, "Just-in-time" principle is used to several projects in terms of site management and construction material supply (Grasso 2005, Salem et al. 2006). Furthermore, Salem et al. (2006) conduct a case study which implemented more complete LC system to the project, including last planner, visualization, huddle meeting, first-run studies, five S's (Sort, Straighten, Standardize, Shine, and Sustain) and fail-safe for quality, and they obtain the active responses from the construction personnel.

2.2 *From traditional construction to lean construction*

In general, traditional construction only concentrates on the organizational structuring and division of the work to be finished, which cause a series of disadvantages such as blocked information sharing, high rework rate, low quality of the project and uncontrolled construction schedule. Moreover, traditional construction is blamed to a huge amount of construction wastes from steel and cement used, extravagant energy usage and water pollution, and inefficient labor arrangement (Tam & Le 2019). From the research conducted by Wang & Ma (2014), the construction sites exist a lot of invalid or repeated works that increase the labor and time costs, which could be avoided with scientific management method. As a result, the client and the manager are no longer satisfied with the traditional construction method since the unnecessary waste and additional cost are harmful to the project and society.

LC differs from the traditional construction in regards to the goals, the structure of the phases, the relationship between phases, and the participants (Ballard & Howell 2003). LC management carries out dynamic management of complex construction processes by quantitative analysis. Table 1 summarizes the differences between traditional and lean construction from extensive literature review (Yang 2016, Wang & Ma 2014, Liao 2018, He et al. 2013, Su et al. 2018, Tam & Le 2019, Yang 2017, Salem et al. 2006).

Further, according to Su et al. (2018), their project benefits from the LC method, presenting as 112 days completion ahead of schedule, 18.6% saving of labor number, and 2% material waste reduction to traditional construction approach. Compared to the traditional ways, the advantages of the application of LC are tangible.

2.3 *Lean construction in China*

LC is introduced in the world in 1993, while Chinese construction enterprises start to apply it to their projects in 2005. So far, only a few companies would like to implement LC, mainly

Table 1. Differences between traditional and lean construction.

	Traditional construction	Lean construction
Goal	Defined in contract	Maximizing value and minimizing waste
Management method	Static	Dynamic
Integration level	Low integration	Integrated project delivery
Project quality	Low	High
Project organization	Massy	Precise
Construction material	High waste	Low storage and waste
Group cooperation	Low	High
Information sharing	Low	High

due to lack of comprehensive ability to carry out new mode, the willing to change and insufficient theory support (Liao 2018). In addition, it is difficult to follow the experiences concluded by developed countries, as the applications of LC in China is not large-scale and under the initial period.

However, Chinese construction projects which implement LC method gain considerable benefits. Liao (2018) evaluates four construction enterprises, all of them show the positive results in several aspects, which are the strengthened sense of responsibility for the job, the increased profit, the earlier schedule completion, the reduced waste, and the increased customer satisfaction. Other scholars are investigating the supplementation of other advanced technology and methods to LC. For instance, Zhu & Tan (2015) try to define a cost control system for LC in China based using WSR analysis approach. Several researchers combine the ideas of green building and/or industrialized building with LC, to reduce the waste and cost in its original (Yang 2016, Yang 2017, Jiang et al. 2018).

Now the Chinese government is adjusting the structure of the AEC industry and its development pattern, the construction quality becomes the main target and LC is concentrated on. The government of Jiangsu Province also puts forward that LC, digital technology, green building and prefabrication construction are main directions to promote the sustainable construction in China (Jiang et al. 2018, Yang 2017). With the support of the government, the gap between China and developed countries in the field of LC will be filled soon.

3 METHODOLOGY

The research objectives are achieved through the investigation of a construction project in Suzhou (China) with the applications of LC and corresponding management, prefabrication, BIM and digital inspection approach. Case study is carried out from mid-June of 2019 to mid-July 2019. During the period, the authors are allowed to visit the construction site, participate in the meetings, have access to the design drawings and BIM model, and assist to organize the relevant document files. The observations are recorded and the typical information related to LC are summarized. In addition, a brief interview to the stakeholders is conducted to inquire their experiences and learnings from the LC.

4 CASE STUDY

4.1 Project information

A case that BOSCH Suzhou parking house project is researched and evaluated as it applies to LC management throughout the project. The parking house is six-storey occupying 31,349.47 m², which is planned to complete in 400 calendar days. To carry out the LC, a temporary group is set to manage and solve the issues occurring on the construction site. The group includes different stakeholders, which are representative from the client, the design institute, the consultant company, the site supervision company, the contractor and subcontractors.

4.2 Lean construction approach

Before the project starts, a master schedule has been determined from the tendering, design and construction to handover. The detailed schedule is updated weekly based on the design alteration and construction progress. With the aid of the Gantt chart, the master schedule is refined as more than 600 project sub-items. In that case, last planner system is implemented to pull the schedule, and the project managers can track the updated tasks to be completed by look-ahead schedule. Combined with the daily huddle meetings, weekly work plan and post-it notes on walls, all construction stakeholders and personnel are familiar with the project progress, so that participate in controlling the construction schedule and improving the construction quality. Figure 1 shows examples of the applications of LC management in the project.

The prefabrication rate in this project is approximately 50%, which is beneficial to maximize the project value and minimize the waste. Compared to cast in situ, prefabrication is superior to time consumption, quality control, waste reduction, and on-site material storage. According to the feedback from the site manager and supervisor, the standardized prefabricated parts used in the project diminish the rework, enlarge the available interior space, and save the construction time. The rest cast in situ parts follow five S's principle of LC. Moreover, the application of the modular construction method promotes the efficiency of the construction process, to achieve the optimized resource distribution. Such an approach requires that the staff focus particularly on the tasks assigned, accordingly to cut down the personnel waste and add the project value.

In addition, Building Information Modelling (BIM) is implemented to the construction management in this project. Through 3D visualization, construction process simulation and clash detection, see Figure 2, it is effective to find out unreasonable issues during design and construction period so that to guide the construction activities. Meanwhile, the construction simulation is referred to, assisting to master the detailed schedule and adjust the next assignments.

Regarding the inspection and acceptance of the construction tasks, a new checking procedure is applied, see Figure 3. Once site supervisors detect the undesirable on-site issues, they record it in Aconex Field platform to inform the contractors and track the subsequent

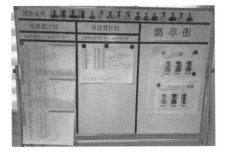

Figure 1.   Examples of lean construction management application.

Figure 2.   BIM applications in lean construction management.

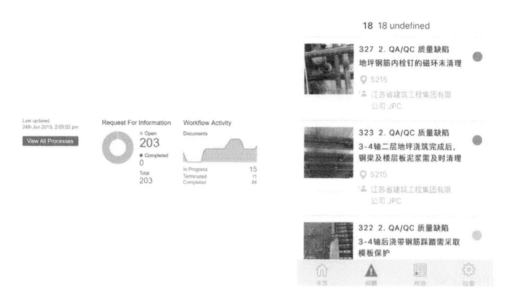

Figure 3.   Inspection and acceptance by Aconex Field.

solution. The contractors can follow the details given such as the issue type, the location, the issue description, and due date. They will respond to the supervisors through the mobile phone anytime when they rectify the issues on site, instead of submitting a paper report, which avoids the checking omission, shortens the response time and raises the management level.

5   FINDINGS

It is observed that with the application of LC management, the schedule is followed fundamentally, when the quality of the project and safety of the staff are provided as well. Prefabricated elements used in the project reduce the on-site waste and risks while accelerates the construction progress. Further, implementations of digital technology/tool that BIM and Aconex Field are advanced for a construction project since the construction staff are accustomed to use paper-based drawing and report in the traditional projects. The 3D visualization, the construction process simulation, the clash detection, the in-time inspection and acceptance by digital technology/tool allow the project to be in control of the management group well.

Based on the responses from the management group, not all the construction sites use the same management method as that applied to the BOSCH parking house. In China, LC management is still at the preliminary stage, therefore it is difficult to follow the specific management approach to conduct the LC. However, this construction project combines LC and corresponding management approach, prefabrication, BIM, and digital inspection tool in it, giving a good example to Chinese AEC industry for LC development. Due to the concept of 'lean' comes from the manufacturing industry, the current LC management refers to its main elements and principles to guide the on-site construction activities. Such a management approach is improved according to the benefits and lessons obtained from the previous implementations, to form a standard method gradually.

To accelerate the project value while reducing the waste and cost, it is a suitable way to combine LC with the prefabricated construction method and digital technology such as BIM. Both the case study and previous experiences from managers have indicated that they have positive effects on the project progress and performance, and wastes in terms of the personnel, the material, and the equipment. China is now developing the building industrialization and informalization, hence, the LC with implementation of prefabricated construction method and digital technology will be a tendency in the Chinese AEC industry.

6 CONCLUSIONS

Aiming to maximize the project value while reducing the waste and cost, lean thinking is referred from the manufacturing industry to construction since two industries have several similarities. LC brings advantages to the project value by reduced wastes, controlled construction progress, proper labor number, and just-in-time material preparation. Through the investigation of a construction project using LC method, several findings are summarized. All the construction staff is involved and responsible to improve the project performance, when supervisors are capable to manage the construction progress efficiently. Utilizing prefabrication elements and digital technology are effective to conduct LC in the project.

ACKNOWLEDGEMENT

The authors wish to acknowledge the financial support from the fund provided by Xi'an Jiaotong-Liverpool University (XJTLU RDF 2016-32) and KSF (KSF-E-29).

REFERENCES

Ballard, G. & Howell, G.A. 2003. Lean Project Management. *Building Research & Information* 31(1): 1-15.

Ballard, G. Koskela, L. Howell, G. & Zabelle, T. 2001. Production System Design in Construction. *Proceedings of the 9th Annual Conference of the International Group for Lean Construction* 1-15.

Grasso, L.P. 2005. Are ABC and RCA Accounting Systems Compatible with Lean Management? *Management Accounting Quarterly* 7(1): 12.

He, Q. Zhao, J. & Dong, S. 2013. Research on lean construction theory and application obstacles. *Journal of Engineering Management* 27(3): 13-17.

Jiang, L. Chen, J. Cheng, J. & Su, Z. 2018. Research on lean construction technology system building and appropriate technology adoption. *Construction Economy* 39(6): 100-103.

Jorgensen, B. & Emmitt, S. 2008. Lost in Transition: The Transfer of Lean Manufacturing to Construction. *Engineering, Construction and Architectural Management* 15(4): 383-398.

Liao, Y. 2018. Research on implementation strategy of lean construction management in Chinese construction enterprises. *Advances in Social Science, Education and Humanities Research* 205: 984-987.

Liu, Y. 2015. Construction of process management system based on the lean thinking. *Value Engineering* 20: 219-220.

Salem, O. Solomon, J. Genaidy, A. & Minkarah, I. 2006. Lean construction: From theory to implementation. *Journal of Management in Engineering* 22(4): 168-175.

Su, K. Huang, S. & Zhang, J. 2018. Research on housing project implementation based on lean construction – A case study of the Jiuzhou Garden 58# Project. *Journal of Engineering Management* 32(2): 131-135.

Tam, V.W.Y. & Le, K.N. 2019. *Sustainable Construction Technologies: Life Cycle Assessment*. Oxford: Kidlington.

Tommelein, I.D. 1998. Pull-driven scheduling for pipe-spool installation: Simulation of a lean construction technique. *Journal of Construction Engineering and Management* 124(4): 279-288.

Wang, J. & Ma, R. 2015. Research on the core method of lean construction mode and its implementation. *Advanced Materials Research* 853:500-505.

Wang, S. 2019. Analysis on the development prospects of Chinese construction industry in 2019. *Construction and Architecture* 873(1): 15-20.

Yang, H. 2016. Study on construction quality green construction management based on lean construction. *Jilin Jianzhu University*. Jilin.

Yang, J. 2017. The Research on green construction management model based on lean construction. *Journal of Changchun Institute Technology* 18(1): 77-79.

Zhu, J. & Tan, F. 2015. Establishment of the lean construction cost control system based on WSR. *Construction Economy* 36(10): 63-67.

Sustainable Buildings and Structures: Building a Sustainable Tomorrow – Papadikis et al. (Eds)
© 2020 Taylor & Francis Group, London, ISBN 978-0-367-43019-1

Application of an intelligent safety helmet to construction site safety management

X. Zhang, M. Zhou & L. Dai
Institute of Civil Engineering, Nantong Institute of Technology, Nantong, China

K. Zhou
Nantong Construction Engineering Group, Nantong, China

ABSTRACT: On-site safety is the most crucial issue that the construction industry focuses on, and safety helmet is one piece to on-site personnel protection and accident reduction. Construction safety managers expect to communicate and locate the on-site personnel in-time, thus an upgrade of traditional safety helmet is established. This paper proposes an intelligent safety helmet that works more than a personal protection equipment. It effectively takes the dynamic nature of site operatives and the needs of safety management into account, using a platform to deal with safety management issues on construction sites. Through the combinations of ergonomics, radio frequency identification technology and wireless communication, the intelligent safety helmet can be used to record the attendance, locate the on-site personnel, alarm the security and communicate in-time.

1 INTRODUCTION

With the continuous development of China's economy and improving living standards, the construction industry has seen an unprecedented expansion (Li 2017). However, the increased volume of construction work has brought about an increase in construction-related accidents, exposing China's poor safety record in this regard. Construction accidents now rank third behind road traffic and coal mines accidents (Feng & Sai 2014). The question is how to reduce the occurrence of construction site accidents and mitigate their impact when they do occur. Studies have shown that safety helmets play a significant role in protecting against injury. Not only can helmets withstand the impact of most falling objects, but they also reduce the degree of damage when a construction worker falls to the ground; and can even save lives. With that being said, traditional safety helmets have some unavoidable drawbacks with regard to safety management on construction sites.

2 ANALYSIS OF ON-SITE SAFETY MANAGEMENT

2.1 *Poor communication of information*

Construction site personnel is the most scattered part throughout the site and communication among them is often inconvenient and insufficient. This means that the dissemination of real-time information becomes difficult, which impedes construction progress, affects quality and leads to safety hazards. The immediate and smooth spreading of information on a construction site is therefore an important issue.

2.2 *Unable to locate personnel effectively*

Construction site personnel working in an organized and effective way not only improve work efficiency but also guarantee the safety and reduce accidents. However, due to a large amount of construction site personnel and the complicated nature of the work, daily safety inspection is a time-consuming activity (Hinze et al. 2013). For many large-scale construction projects, it is now almost a necessity to know the location of workers in real-time.

2.3 *Poor awareness to wear a safety helmet*

Some workers have poor awareness of safety issues. For example, they remove their helmet on the construction site or simply put the helmet on their head without strapping it. Studies have shown that more than 30% of on-site safety incidents are related to unproper wear safety helmets.

3 INTELLIGENT HELMET TECHNOLOGY

With the continuous advancement of information technology in China, the traditional construction model has also been upgraded, and intelligent management has become popular. In the latest national standards for helmets, requirements such as color, classification, and structural form have been removed (Park & Kim 2013). This is an opportunity for intelligent helmets to make a breakthrough in term of their usage on construction sites as part of intelligent site management. In addition to the basic protective function of traditional helmets, the intelligent helmet can realize functions of one-key intercom, personnel positioning, and anti-cap warning through the technology of the Internet of Things to develop an efficient, convenient, dynamic and intelligent security management system.

4 FUNCTION DESIGN

4.1 *Product design purpose*

Safety is a crucial issue of the daily work on construction sites. In order to reduce injury or death by accidents during the construction process, site personnel is often required to wear personal protective equipment according to the work type and the actual working environment. However, safety helmet is one piece of personal protective equipment (PPE) that always needs to be worn under all working conditions. Related studies have shown that using PPE can prevent up to 37% of occupational injuries. Moreover, it can provide both physical protection and an increased sense of psychological safety in workers, thereby promoting work quality and work efficiency.

However, the design of traditional construction helmets focuses only on the basic safety protection function, while ignoring the intelligent aspect that the helmet platform can offer. Since all personnel is required to wear the helmet in the construction site, it is ideal to design secondary intelligent functions that would allow for real-time monitoring and communication between personnel. This would be effective for informed decision making by communication between management and field operators to understand the actual site conditions, but would also help to identify safety issues. Based on the safety helmet and the theories of ergonomics, radio frequency identification technology (RFIT) and wireless communication, a new intelligent safety helmet technology is proposed to remedy the shortcomings of current construction site safety helmets.

4.2 *Principle and structure of the product*

The proposed intelligent helmet is designed to realize intelligent functions, including job attendance monitoring, voice communication, personnel positioning, and automatic capping alarm. This new type of helmet is more treated as a platform connecting rear management and frontline operators than only a safety device. The platform is divided into two parts: a hard hat terminal and an intelligent monitoring application platform. Based on RFIT, combined with computer technology and wireless communication, it provides a practical, cost-effective, safe and reliable management solution for site safety management. The terminal in the helmet's platform is designed to realize on-site data collection and communication, and integrates attendance punching, uncap automatic alarm, personal positioning and voice communication functions. Whereas, the platform's intelligent monitoring application is designed to accomplish background data processing and transmission, then realize the unified scheduling and command of different operation sites and operation personnel. The components of an intelligent helmet are presented in Figure 1.

The overall functional design of the safety helmet terminal and intelligent monitoring platform includes several items. These items are: (1) A wireless voice transceiver module for receiving and transmitting voice signals to realize remote command and field two-way communication; (2) A label (transponder) that stores basic personal identity information, automatically responds to the reader at the access control, verifies identity information, realizes attendance checking, and actively sends position signals to the receiver, so as to realize real-time positioning of operators; (3) A miniature pressure transducer (WYBS) determines whether the wearer's helmet has been taken off and, according to changes in pressure, whether their head has been struck and whether or not to issue an alarm prompt; (4) A single chip which coordinates communication and positioning functions to achieve stable and effective operation of various functions; and (5) Computer software processing various background data.

In terms of hardware structure, the intelligent helmet terminal adopts ergonomic design, label, integrated circuit, and internal wiring for collaborative consideration and unified design to ensure maximum safety and comfort. To facilitate identification, the label is located on the inner left side of the safety helmet (above the wearer's left ear). In addition, the integrated circuit is designed at the back of the safety helmet, and the helmet shape of the installation integrated circuit is redesigned to adapt to the installation of the integrated circuit (Ardalan et al. 2014).

It should be noted that although transponder, headset, integrated circuit module, and other related accessories are added, the overall weight of the safety helmet does not increase significantly enough to compromise comfortable wearing. In addition, the reasonable layout and partition isolation design of earphones and integrated circuits can ensure the safety protection of wearers. The integrated circuit (IC) board is installed inside the safety helmet, which

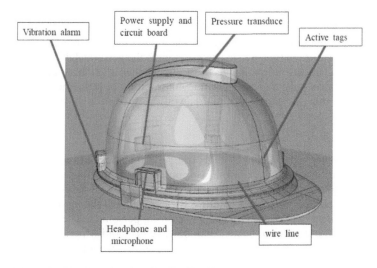

Figure 1.   Product function distribution diagram.

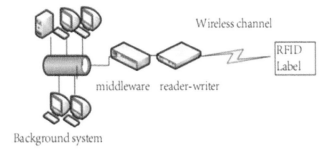

Wireless channel

middleware   reader-writer

RFID
Label

Background system

Figure 2.   Product system composition diagram.

includes power supply module, wireless voice transceiver module, main chip, and other components. The core component of the intelligent helmet terminal is an IC, which plays the role of location, communication, and alarm. The intelligent monitoring application platform is a remote system matched to the intelligent safety helmet terminal (Ning 2014). The platform displays all on-site construction personnel who are online and wearing intelligent helmets, so as to facilitate decision-makers to locate the position of construction operatives and realize automatic real-time communication between office and on-site staff. The product system composition is shown in Figure 2.

5   FUNCTION REALIZATION

5.1   *Attendance (punch in)*

Each employee needs a safety helmet (the active tag is embedded in the left position of the safety helmet) with a tag to identify their identity and serve as a data carrier for entering and leaving the site. The information of an employee is first stored into the database and matched with the corresponding safety helmet label. A system installed at the gateway to the site allows employees to pass through without having to line up and can automatically input the attendance information according to the chip content and transmit it to the background system (Zhang & Fu 2019). In addition, employees and outsiders who do not wear custom safety helmets are forbidden from entering to avoid the unnecessary problem for site safety management.

The application of RFIT can realize the identification and positioning of the staff entering the construction site from the time they pass through the site gate to the time they leave the site, and the internal label of the safety helmet is their unique identification (Han & Lee 2013). A label reader is installed in the field to read the labels within the range of frequency band and save the records of access in real-time (this whole process realizes the personnel displacement information record). The automatic registration management of personnel attendance can be realized by analyzing the real-time personnel displacement flow information of personnel and using the corresponding attendance rule calculation formula. The process module is shown in Figure 3.

5.2   *Personal locator*

The active label in the safety helmets sends out position information, and the reader in the field sends information to the background computer for calculation when the position signal is received. Through the display of the current position and real-time dynamics of personnel in the physical plane, the personnel entering the site can be distinguished according to the management personnel, technical workers, and visiting personnel. The different identities can then be divided into four types of helmet: red, blue, yellow, and white. It can also be displayed by a plane diagram combined with a table to show for any given period of time, the identity, number and distribution of personnel in a certain area, as well as the actual location of one or

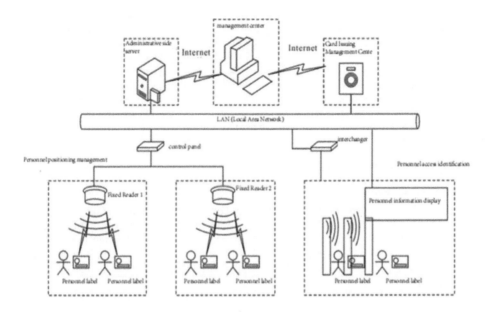

Figure 3. Flow module diagram.

more personnel and items of equipment. It can also show a series of information, such as the arrival and departure time and total working time of relevant personnel and equipment at any location, which effectively checks that safety management personnel are carrying out on-site inspections at various places on time.

5.3 Security alarm

The micro pressure sensor inside the helmet detects changes in pressure and sends a signal to the chip at the side of the intelligent safety helmet. The chip then sends the signal to the background to confirm the current location of the label, which will determine whether removal of the helmet has occurred in the operating area and whether it is necessary to emit a sound to prompt the wearer (Zhang et al. 2019). The system can also automatically send advance warning to the background when a heavy object strikes a wearer's helmet so as to function as a safety alarm and reduce the possibility of an accident by pinpointing a potential danger. Records of each staff wearing a safety helmet are displayed in the background statistics and analysis function. When an employee wears the helmet correctly, the record will be green; if not, the record will be red. The management department can effectively adjust the intensity and strength of management by regularly reviewing the violations of all staff, so as to minimize potential safety hazards on site. In addition, this feature can provide convincing data for accident liability tracing when an accident does occur.

5.4 Real-time communication

The communication environment on a construction site is almost always complicated and often phone signals do not adequately cover the site. Therefore, to ensure dependable communications on-site, a wireless voice transceiver chip is installed in the intelligent helmet, and thanks to its strong anti-interference ability, stable and reliable working frequency, low power consumption, suitability for use with portable products, and low transmission power and high receiver sensitivity, the chip can meet the requirements of wireless control. It requires no license, it inputs and outputs data directly and constitutes a complete radio frequency

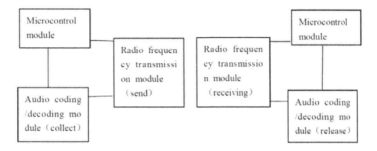

Figure 4. Speech interaction diagram.

transceiver with only a small number of external components such as a single chip microcomputer, which is especially suitable for the communication of complex situations on the site (Zhang et al. 2015). The voice is first converted into radio signals through the micro-control module and audio editing module, which are sent out through the radio frequency transmission module, received by the other end and then released to realize the exchange of voice information, as shown in Figure 4.

6 CONCLUSIONS

The use of intelligent safety helmet technology effectively reduces the probability of accidents happening and plays an important role in protecting the personal safety of front-line workers. It also allows for the dynamic and continuous supervision of operatives and for managers to make fully-informed decisions based on real-time situations. To gain full value from the intelligent helmet, it needs to be used in conjunction with best engineering and safety practices and there needs to be continuing research on ways to improve and expand its use.

REFERENCES

Ardalan, K. Juan, C. & Mani, G.F. 2014. Vison-based workface assessment using depth images for activity analysis of interior construction operations. Automation in Construction 48: 74-87.
Feng, P. & Sai, Y. 2014. Common safety hazards and corresponding countermeasures on construction site. Construction Safety 6:33-35.
Hinze, J. Thurman, S. & Wehle, A. 2013. Leading indicators of construction safety performance. Safety Science 51: 23-28.
Han, S. & Lee, S. 2013. A vison-based motion capture and recognition framework for behavior-based safety management. Automation in Construction 35(11):131-141.
Li, J. 2017. Research on application of intelligent safety management in construction site. Construction Technology 46(20): 139-141.
Ning, X. 2014. Application of internet of things technology in safety management of construction engineering. Construction Economy 12:30-33.
Park, C.S. & Kim, H.J. 2013. A framework for construction safety management and visualization system. Automation in Construction 44:95-103.
Zhang, H. & Fu H. 2019. Performance appraisal and incentive mechanism for safety behavior of construction workers combined with intelligent safety helmet. Journal of Safety Science and Technology 15 (3): 180-186.
Zhang, M. Cao, Z. Zhao, X. & Yang, Z. 2019. Research on safety helmet wearing recognition of construction workers based on deep learning. Journal of Safety and Environment 02.
Zhang, J. Han, Y. & Ma, G. 2015. Research of intelligent early warning system for falling accidents based on BIM and RFID for construction workers. Journal of Engineering Management 29(6):17-21.

Developing sustainability performance evaluation indicators of PPP projects: A case study

Q. Sun, S. Zhang & W. Lu
School of Civil Engineering, Suzhou University of Science and Technology, Suzhou, China

ABSTRACT: The traditional performance evaluation indicators (PEIs) of PPP projects are insufficient, and specifically lack of attention to the project sustainability. Based on critical literature review of relevant research, this paper identifies sustainability PEIs of PPP projects, in terms of three dimensions including economy, society and environment. Nine critical PEIs were elaborated in detail. Based on this result, the identified critical PEIs are further analyzed through three different types of cases. The findings are valuable for researchers embarking on relevant study in the future.

1 INTRODUCTION

Public-private partnership (PPP) refers to the benefit sharing, risk sharing and long-term cooperative relationship between the public and private established by the government to enhance the supply of public products and services, improve supply efficiency, and establish franchise, purchase services, and equity cooperation (China National Development and Reform Commission 2014).

The performance evaluation of the PPP project includes a comprehensive evaluation of the technical, economic, social and environmental factors that emerged during the project implementation process from the perspective of project input, process, outcomes (Wang et al. 2014). Comprehensive and effective performance evaluation is one of the critical success factors for PPP projects (Yuan et al. 2009). Due to the complexity of PPP projects, it is generally necessary to evaluate their performance through a series of indicators systematically and comprehensively. Therefore, it is necessary to conduct the study on the PEIs of PPP projects. Since PPP projects are generally large-scale public infrastructure projects, their construction and operation have more significant impact on the economy, society and environment. Currently, the entire construction community has paid increasing attention to promote sustainable development of construction projects. If the PPP project performance evaluation does not consider sustainability, it is obviously incomplete and does not meet the requirements of sustainable development.

2 THE IMPORTANCE OF PEIS FOR CONSTRUCTING PPP PROJECT FROM THE PERSPECTIVE OF SUSTAINABILITY

2.1 *Policy support for sustainability performance evaluation of PPP projects*

Sustainability is defined as a condition and measure through a range of indicators (Salvado et al. 2015), and sustainability performance indicators are critical to setting goals, monitoring progress, and determining relative performance (Hueskes 2013). The sustainability of PPP projects means that PPP projects can maintain a dynamic balance between various costs and benefits throughout their life cycle. The development speed and development quality of the project are balanced and adopted to achieve the desired objectives of the project (Zhou 2016). Sustainability performance evaluation of PPP projects can avoid mistakes in investment decisions, choose the best investment solutions, and improve investment efficiency (Gan et al. 2009). The Chinese government attaches great importance to the sustainability of PPP project performance evaluation and has issued a series of policy documents. The macro-strategies developed by the government need to be transferred to concrete actions (Ugwu et al. 2006), which will drive the industry to pay more attention to the sustainability of PPP project performance.

2.2 *Researchers' attention to the sustainability performance evaluation of PPP projects*

Improving the sustainability performance of PPP projects can make an important contribution to the life-cycle value of the building (Shen et al. 2016). More and more researchers are also calling for sustainable performance evaluation of PPP projects to improve project sustainability (Ugwu et al. 2006, Harkness & Bourne 2015).

The concession period of PPP projects is generally 20-30 years, and the private entity normally paid excessive attention to short-term return on investments. The sustainability performance of PPP projects can only be obtained from long-term development (Koppenjan & Enserink 2009). Governments may evaluate a particular project with more comprehensive methods through sustainable performance evaluations, which will motivate private entity to focus more on project sustainability. From this perspective, sustainability performance evaluation indirectly regulates the distribution of benefits between government and social capital.

3 DEVELOPMENT OF SUSTAINABILITY PEIS FOR PPP PROJECTS

3.1 *Literature review on PEIs from the perspective of sustainability for PPP projects*

Since the application of PPP in developed countries started earlier, more research of sustainability PEIs of PPP projects were conducted based on the scenario of these countries. For example, Haavaldsen et al. (2014) argued that when assessing the sustainability of infrastructure projects, the evaluation indicators should be established from the economic, social and environmental dimensions, and the impact of each dimension on the future of the project cannot be overlooked. Hueskes et al. (2017) used the framework of Devold & Block (2015) as a basis for the measurement of sustainability level and developed an evaluation framework that includes 54 sustainability indicators. Zhou et al. (2013) stated that the achievement of sustainable development for PFI (Private Finance Initiative) project in UK needs to balance the economic, social and environmental dimensions, and needs to incorporate sustainable technological factor.

In recent years, PPP has been widely employed in China, and it is a hot research topic in the field of construction management. For example, Li & Lu (2015) pointed out that sustainability evaluation is critical to maximize the comprehensive benefits of economy, society and environment in rural safety drinking water projects, and achieve comprehensive, balanced and sustainable development. Zhou (2016) claimed that the post-sustainability evaluation index system of PPP projects should include operational, socio-economic and environmental resources. In addition to focus on project operational and socio-economic impacts, it should also consider the analysis of program design, risk sharing and licensing. The rationality of the operating period reflects the critical aspects of the effectiveness of PPP. Li & Lang (2017) developed an evaluation index system with consideration of economy, society, ecology and management, through analyzing the main influencing factors of sustainable development of public rental housing PPP projects.

In summary, researchers in China and abroad usually construct sustainability PEIs for PPP projects from three dimensions: economy, society and environment. The economy, society and the environment are recognized as principles of sustainable development and widely described as the triple bottom line (Griffith & Bhutto 2008). Therefore, this paper searches for papers in CNKI (China National Knowledge Infrastructure) and SCI (Science Citation Index) with sustainability and performance evaluation indicators as keywords, then selectes 20 papers with high citation rates and identifies the sustainability PEIs of PPP projects from the three dimensions of economy, society and environment. The literature review and statistical results are shown in Table 1.

Table 1. Literature review and statistical results of sustainability PEIs for PPP projects.*

	Evaluation index	1	2	3	4	5	6	7	8	9	10	11	12	13	14	15	16	17	18	19	20	Total
Economy	Overall life cycle cost	√		√	√		√		√	√			√			√	√		√	√	√	13
	Project financing source	√					√	√	√	√	√		√			√	√	√				10
	Internal rate of return						√	√	√				√	√	√	√	√	√				9
	Payback period						√		√				√	√		√	√	√				7
	Project investment plan			√				√	√							√	√					5
	Life cycle benefit/profit	√			√					√						√				√		5
	Risk sharing			√				√	√	√	√											5
	Economic advantages of technical methods						√	√								√	√			√		5
Society	Impact on local development	√	√			√	√	√	√		√	√	√		√	√	√	√	√	√	√	16
	Provide employment opportunity	√				√	√		√	√	√		√	√	√	√	√	√	√	√	√	15
	Safety and health			√	√	√	√	√	√		√		√			√	√		√	√	√	13
	Stakeholder satisfaction		√			√			√		√	√	√			√	√			√	√	10
	Provision of local facilities				√		√	√				√			√	√					√	7
	Protection of cultural heritage						√					√				√		√	√	√	√	6
Environment	Impact on soil pollution	√	√			√	√	√	√	√	√	√	√			√		√	√		√	14
	Impact on air quality	√	√				√	√		√			√	√	√	√	√	√	√	√	√	14
	Impact on water quality	√	√			√		√		√			√	√	√	√	√	√		√	√	13
	Noise impact		√	√	√			√	√					√	√	√	√	√		√	√	12
	Energy consumption	√	√	√		√		√		√			√				√		√	√	√	11
	Ecological effect	√						√	√		√		√	√	√	√				√		9
	Environmental protection measures in project design							√	√				√			√	√	√	√		√	8
	Project waste treatment						√						√	√	√		√	√	√	√		8
	Impact on local landscape				√	√							√		√		√			√	√	7

* The literature 1-20 are as the following: 1. Haavaldsen et al. (2014); 2. Hueskes et al. (2017); 3. Zhou et al. (2013); 4. Boz & El-adaway (2015); 5. Cong & Ma (2018); 6. Ugwu et al. (2006); 7. Shen et al. (2011); 8. Gan et al. (2009); 9. Zhou (2016); 10. Chen (2018); 11. Ye & Deng (2014);12. Gan & Li (2013); 13. Li & Lu (2015); 14. Sun & Li (2013); 15 Zhang (2018); 16. Gan (2011); 17. Yi (2014); 18. Gan (2014); 19. Huang (2017); 20. Liang (2016).

3.2 Analysis on the critical PEIs from the perspective of sustainability for PPP projects

3.2.1 Sustainability PEIs based on economical dimension

(1) Overall life cycle cost

For PPP projects, the analysis from the overall life cycle cost is necessary due to the longer period of PPP project implementation, which can avoid the problem of short-term effectiveness instead of the long-term effectiveness (Du et al. 2018).

(2) Project financing source

The development, construction and operation of PPP projects require a large amount of financial support. Attracting the project finance funding from various sources is an important precondition for the smooth development of the PPP project (Gan 2011). The financing models of China's PPP projects mainly include government-dominated financing, trust financing and insurance funding.

(3) Internal rate of return

The internal rate of return (IRR) is the discount rate when the current value of the net cash flow of the project investment plan is equal to zero during the calculation period. The economic meaning is the profitability of the unrecovered funds in the investment plan (Liu 2006). Ensuring the viability of the project from IRR is an important factor that investors must consider in the project decision-making stage. The value of IRR will affect the amount of investment shared between the government and private entity in the PPP project.

3.2.2 Sustainability PEIs based on social dimension

(1) Impact on local development

PPP is widely used in the infrastructure sector, and infrastructure has always been considered a prerequisite for social development (Gan 2011). PPP projects have a direct impact on local development. For example, the PPP project of underground integrated pipe gallery can overcome the drawbacks of traditional direct burial municipal pipelines which utilize more land. PPP projects also have an indirect impact on the local economic development. For example, the development of metro PPP projects usually results in the increase of housing price near the station (Wang & Cao 2017).

(2) Provision of employment opportunity

The construction of major projects can provide more employment opportunities (Gan 2011). During the overall project life cycle, the planning and construction of PPP projects require a large number of participants and various types of professionals. Labor employment is a social issue closely related to economic development. Providing adequate employment opportunities not only promotes the economic development, but also enhances the sustainability of PPP projects and promotes sustainable social development (Pan 2002).

(3) Safety and health

Due to the large scale of PPP projects, it is necessary to focus on safety production and management during their overall life cycle. For example, safety supervision departments should establish long-term effective measures to control and reduce the occurrence of safety accidents in construction companies to ensure safety (Chen & Zhang 2015).

3.2.3 Sustainability PEIs based on environmental dimension

(1) Impact on soil pollution

Soil pollution is one of the important indicators affecting the sustainable use of land (Xie et al. 2015). The sustainable use of land refers to the continuous increase of the total amount of land capital (including human capital, i.e. knowledge technology, man-made capital, and natural environmental capital) in the land use process and the sustainable use of land resources. The impact of PPP projects on soil pollution affects the sustainability of the land, which in turn affects the sustainability of the PPP project.

(2) Impact on air quality

Air quality can have a direct impact on ecosystems and people's health (Qi & Zhang 2015). Due to the large scale of PPP projects, more dust and gas emission will be generated during construction, which will adversely affect the local ecosystem and personnel health. To protect the health of participants and various types of personnel is an important component to improve the sustainability of construction projects (Wu 2004). The impact of PPP projects on air quality affects the health of local ecosystems and people, which in turn affects the sustainability of the PPP project.

(3) Impact on water quality

The wastewater generated during the construction of PPP projects will penetrate into the underground or flow into rivers, destroy the ecological balance of water resources, and adversely affect the local ecosystem and personnel health (Zhao 2015). Health is an important part to improve the sustainability of engineering projects (Wu 2004). The impact of PPP projects on water quality affects the health of local ecosystems and people, which in turn affects the sustainability of the PPP project.

This paper first expounds the importance of developing PEIs for PPP project from the perspective of sustainability, and then identifies PPP project sustainability PEIs based on literature review method, and elaborates on the meaning of critical PEIs, followed by case study method.

4 CASE STUDY

4.1 Background of the cases

Case study is an effective research method to explore lessons learned from case projects and to summarize management implications for future infrastructure investors (Yin 2014), hence, the application of sustainability PEIs of PPP projects is validated in detail through three different types of PPP project cases.

In the literature review stage, this paper critically reviewed 20 articles which include a total of 17 cases. Three cases were finally selected as the cases because they meet the following criteria: (1) three cases include all the necessary information of specific details of the project and entire performance evaluation process; (2) three cases are different types of projects; (3) three cases are located in different cities and operated in different social, economic, political and legal environments. All the above criteria for the case selection are to produce a more robust research finding.

In Case 1, it is a PPP project for the development of a pleasant and healthy town near the Sihe River, which is a new project comprising part of the urban comprehensive development scheme. The construction site is located in Quanlin National Forest Park of southen Sishui County, Jining City, Shandong Province. The aim of the project construction is to develop a theme town with "great health facility" and "typical tourism park", which adopts the healthy thinking in modern construction industry in China. The focus of the PPP project implementation includes four aspects: healthy pension, ecological construction, accurate poverty alleviation and cultural tourism. The project construction period is planned to be 6 years, and the project will be executed in three phases (i.e. 2015-2017, 2018-2019, 2020) with a operation period of 24 years.

The Case 2 is a centralized heating project, which is a community project initiated by the government for investment and construction. After completion, the PPP project will be responsible for the supply of central heating in the urban area. The concession period is 25 years. Beijing Yuantong Thermal Power Co., Ltd., as the private entity, and the government authorized the investment representative Zhangjiakou City State-owned Assets Operation and Management Center to establish a project company as the Special Purpose Vehicle (SPV) with an equity ratio of 10% to 90%. The project company recovers its initial investment by collecting heating fees and heating facilities operation fees. The project was implemented in the form of BOT model.

The Case 3 is the construction of a sewage treatment project, which includes sewage treatment plant, sewage piping systems and four pumping stations. After completion, the project will treat the wastewater to discharge standards, improve water environment quality and reduce environmental pollution. The project concession period is 17 years. After the concession period, the project assets will be transferred to the local government without compensation. At the same time, the SPV needs to upgrade the main equipment during the operation period to ensure normal and long-term operation after the project transfer. The project contract stipulated that the SPV should provide the government with sewage treatment services, which have to meet the wastewater quality standards stipulated by the environmental protection department. The government will pay the service fee to the operation company.

The general background information of the three PPP cases is summarized in Table 2.

4.2 *Sustainability performance of the three case projects*

This section describes the sustainability performance of the three cases through content analysis, combined with the critical PEIs identified through literature review method in section 4.2, as shown in Table 3.

The following conclusions can be obtained from Table 3:

(1) PPP projects are focusing on sustainability performance evaluation

In Case 1, when the government department provided finance of 50% of the total investment, the project's optimal sustainable benefits were obtained. When the government did not provide any investment, the sustainable benefit of the project was higher than the project which is invested totally by the government (Zhang 2018). Case 1 shows that in PPP projects, the appropriate investment ratio between government and social capital will make PPP projects more sustainable (Shen et al. 2016). The instability of financing sources will affect the sustainability of PPP projects. Case 1 adopts the PPP approach and uses the industry integrated model to operate the project, which solve the project's funding problem scientifically and reasonably (Gan 2011). In Case 2, the actual operation efficiency of the boiler was increased to 83% due to increased environmental protection investment. Energy saving and emission reduction are able to increase the project's income by 20% (Chen 2018), and also enhance environmental sustainability. In Case 3, the project used advanced technology to effectively reduce costs, and government regulation effectively increased the profitability of social capital, thereby effectively reducing the project investment (Zhou 2016).

(2) The performance of current PPP projects should be further improved to meet the sustainability standard

In case 1, before adopting the PPP approach, the factors such as the planning and design of the town have been determined by the government and cannot be changed. These problems can only be settled partially in the construction process and subsequent operation stage (Zhang 2018). In Case 2, because Beijing Yuantong Thermal Power Co., Ltd. still has established negative image in providing the public services, some minor problems happen, it will have cascading effects on the public sentiments. However, with the joint efforts of government and private entity, the project results got positive feedback from the general public and end users (Chen 2018). In Case 3, the cost and benefits of retrofitting the project need to be improved (Zhou 2016).

(3) Proper PEIs are critical to improve the sustainability performance for PPP project

As mentioned above, Haavaldsen et al. (2014) claimed that the evaluation indicators should achieve a balance between the three dimensions of economy, society and environment, but the analysis results of case 1 show that the performance of economic and social sustainability of the project is high, whereas the performance of the environmental sustainability is average. Therefore, the project needs to pay more attention to environmental sustainability in the subsequent construction and operation phase (Zhang 2018). A further analysis on the three cases showed that different types of PPP projects have differences in their construction objectives, investment backgrounds, and operational approaches. Specific sustainability performance evaluation criteria should also be established and applied (Zhou 2016).

Table 2. General background information of the three PPP cases.

Case No.	Project name/location	Project type	Total investment	Construction period	Concession period	Concessionaire	Public sector	Source
(1)	A pleasant and healthy town near the Sihe River, Jining, China	Comprehensive urban development	9300 million RMB	6 years	24 years	Shandong Wanziyuan Tourism Development Co., Ltd.	Sishui County Urban Construction Investment Development Co., Ltd.	Zhang (2018)
(2)	Central Heating Project in Qiaoxi District, Zhangjiakou, China	Heating	415 Million RMB	1 year	25 years	Beijing Yuantong Thermal Power Co., Ltd.	Zhangjiakou City State-owned Assets Operation and Management Center	Chen (2018)
(3)	Guangzhou Wastewater Treatment Plant, Guangzhou, China	Sewage treatment	985 million RMB	3 years	17 years	American Tyco Asia Investment Company	Guangzhou Municipal Government	Zhou (2016)

Table 3. Sustainability performance analysis based on critical PEIs.

Sustainability PEI		Case No.	Project performance
Economy	Overall life cycle cost	1	Reasonably arrange the use of investment funds.
		2	The design, construction, operation and maintenance of the project are integrated in three phases, and reasonable operating cost and maintenance cost plans are established to reduce the total life cycle cost.
		3	The project construction and operation investments are reasonable, which is able to generate sufficient cash flow to achieve the predetermined financial goals, and to use advanced technology to effectively reduce the costs.
	Project financing source	1	Using the PPP model, the private entity will bring experience capital to operate the system.
		2	The economic-related risks are reasonably shared between government departments and social capital, such as the fluctuation of exchange rate.
		3	Guangzhou Sewage Treatment Co., Ltd. invested 110 million Yuan, and Tyco Asia Investment Co., Ltd. invested 223 million Yuan, and obtained about 667 million Yuan of project mortgage from commercial banks.
	Internal rate of return	1	When the proportion of government investment is 50%, the project has the highest sustainability value.
		2	Estimating the cost of future changes to specific services.
		3	According to the construction and operation costs of the sewage treatment plant, the two parties will determine the price of the sewage treatment service fee, based on factors such as reasonable profit rate.
Society	Impact on local development	1	The project is built following the health thinking as an emerging industry in China, introducing other industries such as rehabilitation, life science, technology development, and tourism.
		2	The service of the project during the handover of the project and after the handover meets the public requirements.
		3	Driving the economic growth in the surrounding area.
	Provision of employment opportunity	1	Improve the income of farmers by building tea gardens, organic farms, ecological orchards, etc.
		2	Promoting employment for the local people.
		3	Experienced project leaders provide reliable support for the follow-up of the project.
	Safety and health	1	Creating a pleasant and healthy town is in line with the region's high-level strategic planning.
		2	Distributing health and safety manuals to the public and regularly report to the public on health and safety in operational services.
		3	Taking safety measures to prevent accidents.
Environment	Impact on soil pollution	1	Focusing on planning the overall layout of the town and environmental green development.
		2	Increasing public space and green space.
		3	Municipal Environmental Protection Bureau conducts environmental supervision on the environmental impact of project construction and operation.

(*Continued*)

Table 3. (*Continued*)

Sustainability PEI	Case No.	Project performance
Impact on air quality	1	The green ecological concept will be integrated into the whole process of project construction, with higher standards and stricter measures to control water quality, and use clean sky slogan to improve the attractiveness of example projects.
	2	Using energy-saving systems and advanced management methods such as lean construction thinking to minimize carbon emissions during the construction, operation and maintenance phases.
	3	Project implementation has no negative impact on the air quality.
Impact on water quality	1	Improving the regional environment by integrating the comprehensive management of the source of the Lishui River Basin.
	2	Using green building technology solutions and innovative solutions to reduce pollution to the water environment.
	3	Sewage treatment significantly improving the quality of local environment.

5 CONCLUSION

The systematic study on the sustainability performance evaluation of PPP projects is not only critical to improve the performance of PPP projects, but also fulfill the requirements of Chinese government of developing PPP projects. Based on literature review, the critical PEIs were identified with elaboration on their meanings. Three cases were employed to further explore the sustainability performance evaluation issue of PPP projects in practice. Based on above analysis, this paper concludes with the following remarks: (1) 24 PEIs for PPP projects were identified in the literature review process from economic, social and environmental dimensions, which will be a theoretical basis for researchers to embark on PPP performance research in the future. (2) Based on analysis of three different types of PPP cases, it was found that part of the sustainability PEIs for PPP projects was implemented in practice. However, some of the PEIs were overlooked which resulted in the imbalance of the performance evaluation indicators and as a consequence the low performance level of the PPP projects in sustainability. Some limitations exist in the current study, which will become the future research directions of the authors. Firstly, only qualitative research method was adopted in this study. The authors will conduct relevant research employing quantitative method to derive more comprehensive findings. Secondly, only three cases were analyzed in this study. More cases will be analyzed to achieve a more broad research goal. These research efforts will contribute to the development PPP knowledge in China.

ACKNOWLEDGEMENT

The work described in this paper was supported by the Postgraduate Research & Practice Innovation Program of Jiangsu Province (SJCX18-0876).

REFERENCES

Boz, M.A. & El-adaway, I.H. 2015. Creating a holistic systems framework for sustainability assessment of civil infrastructure projects. *Journal of Construction Engineering and Management* 141: 1-11.
Chen, X.Y. 2018. *Research on Value-for-Value Evaluation System of PPP Project Based on Sustainable Development Goals*. Shandong: Shandong University.

Chen, W.K. & Zhang, K. 2015. Research on Safety Management of Subway Construction Project Based on Game Theory. *Project Management Technology* 13(3): 37-40.

China National Development and Reform Commission. 2014. *Guiding Opinions of the National Development and Reform Commission of China on Launching Government and Social Capital Cooperation*, No. 2724.

Cong, X.H. & Ma, L. 2018. Performance Evaluation of Public-Private Partnership Projects from the Perspective of Efficiency, Economic, Effectiveness, and Equity: A Study of Residential Renovation Projects in China. *Sustainability* 10(6): 1-21.

Devolder, S. & Block, T. 2015. Transition thinking incorporated: towards a new discussion framework on sustainable urban projects. *Sustainability* 7(3): 3269-3289.

Du, Y.Q. Wu, N.W. Ding, D. & Liu, P.Y. 2018. Life Cycle Cost Study of Rural Environmental Governance PPP Model. *China Population Resources and Environment* 28(11): 162-170.

Gan, L. 2011. *Application Research of PPP Mode in Rural Water Conservancy Infrastructure Construction Project*. Chongqing: Chongqing University.

Gan, L. Shen, L.Y. & Fu, H.Y. 2009. Research on Evaluation Index System of Infrastructure Project Based on Sustainable Development. *China Civil Engineering Journal* 42(11): 133-138.

Gan, X.L. 2014. *Research on Decision Model of Sustainable Construction Scheme for Infrastructure Project Based on Stakeholder Theory*. Chongqing: Chongqing University. (In Chinese)

Gan, X.L. & Li, S.R. 2013. Research on Construction Method of Sustainability Evaluation Index of Infrastructure Project Based on Risk Management. *Construction Economics* 3: 89-92.

Griffith, A. & Bhutto, K. 2008. Improving environmental performance through integrated management systems (IMS) in the UK. *Quality* 19(5): 565-578.

Harkness, M. & Bourne, M. 2015. *Is complexity a barrier to effective performance measurement?* Auckland: Proceedings of the PMAA Conference.

Huang, J. 2017. *Dynamic Evaluation of Sustainability of Urbanization Infrastructure Projects Based on ANP Modified Model*. Tianjing: Tianjing University.

Hueskes, M. 2014. *Improving sustainability: Inevitable trade-offs in civil regulation. A case study of the IDH fruits and vegetables program*. Utrecht: Utrecht University.

Haavaldsen, T. Lædre, O. Volden, G.H. & Lohne, J. 2014. On the concept of sustainability-assessing the sustainability of large public infrastructure investment projects. *International Journal of Sustainable Engineering* 7(1): 2-12.

Hueskes, M. Verhoest, K. & Block, T. 2017. Governing public-private partnerships for sustainability, An analysis of procurement and governance practices of PPP infrastructure projects. *International Journal of Project Management* 35: 1184-1195.

Koppenjan, J.F.M. & Enserink, B. 2009. Public-Private partnerships in urban infrastructures: reconciling private sector participation and sustainability. *Public Administration Review* 69(2): 284-296.

Li, Y.R. & Lu, X.G. 2015. Research on Sustainability Evaluation of Rural Drinking Water Safety Engineering Based on PPP Financing Model. *Ecological Economy* 31(4): 137-140.

Li, L.X. & Lang, Q.G. 2017. Research on Sustainability Evaluation of Public Rental Housing PPP Project. *China Real Estate* 8: 64-71.

Liang, Y. 2016. *Research on Sustainability Evaluation Model of Metro Project Based on Stakeholder Satisfaction*. Tianjing: Tianjing University.

Liu, X.J. 2006. *Engineering Economics*. Beijing: China Architecture and Building Press.

Pan, W.Q. 2002. An Industrial Structure Optimization Model Based on Sustainable Development. *System Engineering Theory and Practice* 7: 23-29.

Qi, Y. & Zhang, J.S. 2015. Review of Ecological Governance and Global Environmental Sustainability Indicators. *Social Science Abroad* 21-34.

Salvado, M.F. Azevedo, S.G. Matias, J.C.O. & Ferreira, L.M. 2015. Proposal of a sustainability index for the automotive industry. *Sustainability* 7(2): 2113-2144.

Shen, L.Y. Tam, V.W.Y. Gan, L. Ye, K.H. & Zhao, Z.N. 2016. Improving sustainability performance for Public-Private-Partnership (PPP) projects. *Sustainability* 8(3): 289-304.

Shen, L.Y. Wu, Y.Z. & Zhang, X.L. 2011. Key assessment indicators for the sustainability of infrastructure projects. *Journal of Construction Engineering and Management* 137(6): 441-451.

Sun, H.L. & Li, X.W. 2013. Construction of Evaluation System for Sustainable Development of Large-scale Infrastructure. *Statistics and Decision* 16: 28-31.

Ugwu, O.O. Kumaraswamy, M.M. Wong, A. & Ng, S.T. 2006. Sustainability appraisal in infrastructure projects (SUSAIP) Part 1. Development of indicators and computational methods. *Automation in Construction* 2: 239-251.

Wang, C. Zhao, X.B. & Wang, S.Q. 2014. Research on Performance Evaluation Index of PPP Project Based on CSF and KPI. *Project Management Technology* 12(8): 18-24.

Wang, B. & Cao, J.M. 2017. Impact of Rail Transit on Residential Prices. *Journal of Civil Engineering and Management* 34(3): 81-85.

Wu, Z.Z. 2004. Safe Production is an Important Part of China's Sustainable Development. *China Safety Science Journal* 14(9): 3-6.

Xie, H.L. Liu, Q. Yao, G.R. & Tan, M.H. 2015. Measurement of Regional Land Use Sustainability Level Based on PSR Model: A Case Study of Poyang Lake Ecological Economic Zone. *Resource Science* 37(3): 449-457.

Ye, X.S. & Deng, Y. 2014. Research on the Ways to Realize the Sustainability of PPP Infrastructure Projects from the Perspective of Partnership. *Science and Technology Management Research* 12: 189-193.

Yi, H.L. 2014. *Evaluation and Analysis of Sustainability of Large Public Projects*. Guangzhou: South China University of Technology.

Yin, R.K. 2014. *Case Study Research: Design and Methods*. Los Angeles: Sage Publishing.

Yuan, J.F. Zeng, A.Y. Skibniewski, M.J. & Li, Q.M. 2009. Selection of performance objectives and key performance indicators in public-private partnership projects to achieve value for money. *Construction Management and Economics* 27: 253-270.

Zhang, F. 2018. *Research on Sustainability Evaluation of PPP Projects in Characteristic Towns*. Beijing: Beijing University of Civil Engineering and Architecture.

Zhao, Y. 2015. Failure Case Analysis and Risk Prevention of PPP Projects in China. *Sub National Fiscal Research* 6: 52-56.

Zhou, L. Keivani, R. & Kurul, E. 2013. Sustainability performance measurement framework for PFI projects in the UK. *Journal of Financial Management of Property and Construction* 18: 232-250.

Zhou, M.L. 2016. *Post-sustainability Evaluation of PPP Projects*. Dalian: Dongbei University of Finance and Economics.

Sustainable Buildings and Structures: Building a Sustainable Tomorrow – Papadikis et al. (Eds)
© 2020 Taylor & Francis Group, London, ISBN 978-0-367-43019-1

Enlightenment of the construction of zero waste city in Singapore to China

J.F. Zhu, T.B. Yang, S.H. Chen, Z.H.L. Huang, J.L. Hao, B.P. Jose & B.Q. Cheng
Xi'an Jiaotong-Liverpool University, Suzhou, Jiangsu, China

ABSTRACT: With the acceleration of economic development and urban expansion, urban solid waste has become an urgent problem that needs to be solved in environmental development. By analyzing the present situation of the treatment and resource of solid waste in China and combining the construction of "zero waste city" in Singapore. This paper views the national conditions and methods of China, analyzes the issue and elements restricting reuse of solid waste in China, and pushes forward some reasonable countermeasures for promoting the construction of "zero waste city" in China.

1   INTRODUCTION

Cities are the main places of human activities in today's society. A good urban environment is of great significance to human health and living standards (Min et al. 2018). However, with the rapid economic development, urban expansion is accelerating, which leads to the population explosion, and the amount of solid waste produced in cities getting larger and larger. According to incomplete statistics, the annual output of urban waste in China reached as much as 1.55 billion tons (Jiao & Du 2019). According to the data of the ministry of environmental protection, with the annual increase rate of 5%~8%, two-thirds of the more than 600 large and medium-sized cities in China are facing the "garbage siege" crisis (Jiao & Du 2019). On the other hand, taking construction waste as an example, which accounts for 70% of solid waste, the utilization rate of China's construction waste is less than 10%, far lower than the average level of developed countries in Europe and America (Hao et al. 2018). In order to solve the problem that China's waste production is large but the utilization rate of resources is low, and to realize the sustainable development of cities, it is necessary to continuously promote the reduction of solid waste and the construction of "zero waste city" through resource utilization. In the opinions of the CPC central committee and the state council on comprehensively strengthening ecological and environmental protection and resolutely fighting pollution prevention and control, it is clearly proposed to promote the battle of pure land preservation, strengthen the prevention and control of solid waste pollution, and carry out the pilot project of "zero waste city".

"Zero waste city" neither means not producing solid waste, nor does it mean that solid waste can be completely utilized, but it is an advanced urban management concept aiming at the ultimate goal of minimum production, full utilization and safe disposal of solid waste of the whole city. In January 2019, the general office of the state council issued the "'zero waste city' construction pilot work plan", pointing out that the "zero waste city" is an urban development mode that can minimize the environmental impact of solid waste by promoting green and green-life development and reducing the amount of solid waste at its source and utilizing the solid waste continuously, guided by the newly developed concept of innovation, coordination, green, open, and sharing.

At present, the construction of "zero waste cities" in China has just been laid out and is still in the initial exploration stage. Learning the advanced concepts and methods of "zero waste construction" from other countries is of far-reaching significance to the construction of "waste free city" in China. As a "garden city", Singapore has formed a complete and relatively scientific construction plan for "zero waste city" after many years of exploration and has achieved initial results. Currently, the comprehensive utilization rate of solid waste in Singapore has reached 61 % and is expected to reach 70 % by 2030 (Wang & Li 2018). This proves that solid waste has high utilization value and if the solid waste can be used properly, it will bring significant environmental, economic and social benefits, and that it is a feasible way to build "zero waste city".

This paper aims to discuss the construction of "zero waste city" in Singapore from the perspective of the basic national conditions and the construction plan of "zero waste city", so as to provide reference for the better development of "zero waste city" pilot work in China.

2 SINGAPORE "ZERO WASTE CITY" DEVELOPMENT

2.1 *Singapore*

Singapore is with a population of 5.80626 million and a large number of tourists coming to visit Singapore every year (World Population Review 2019). Therefore, Singapore generates about 21,000 tons of waste every day (Wang & Li 2018). However, territorial resources of Singapore are limited, only about 719.1 square kilometers (World Population Review 2019). In this case, if the problem of waste disposal is not solved properly, it may occupy a large area of land resources. This affects economic development of Singapore and quality of life of the residents. As a result, this particular 'zero waste city' development has been become the focus of Singapore government.

2.2 *Supporting policies*

At present, Singapore has a complete environmental legislation and waste management policy system. Singapore has enforced several pollution control laws, which set clear limits on the treatment and discharge of various types of waste, so as to provide laws for all kinds of projects, industrial and commercial activities and daily life. The *green plan, national recycling plan, zero waste action* and other economic recycling policies launched by the Singapore government have all provided policy support for promoting the utilization of waste and improving the utilization rate of waste resources. In addition, the *Singapore sustainability blueprint* (2015) proposes five initiatives as well, based on Singapore's national conditions, aiming to improve waste utilization and to advocate active citizen participation (Wang & Li 2018).

1) Packaging Agreement: Organizations from all walks of life are appealed to participate this agreement, and the signatories should try their best to reduce the production of packaging waste. The Singapore government has set up an annual award as well to encourage signatories to pursue sustainable development;
2) Mandatory Waste Report in Large Commercial Places: Major business sites in Singapore are required to report waste status on time, so as to urge the companies to strengthen waste management, to timely dispose the waste and to let the companies be environmentally conscious and reduce waste production;
3) 3R fund: The government established the fund to recognize the efforts of businesses to reduce waste;
4) Food Waste Recycling Strategy: The government advocates reasonable purchasing of food to avoid food waste caused by expired storage and to encourage citizens to donate surplus food;
5) National Voluntary e-waste Recycling Partnership: Repeal e-retailers and manufacturers to develop cooperation plans for recycling of e-waste and the e-waste management system of Singapore is taking shape.

2.3 Adopted methods

Referring to the management method of water resources, the solid waste management method can be extracted, and the solid waste treatment and management can be conducted in six directions, including consumption, collection, landfill, production, recycle and power generation.

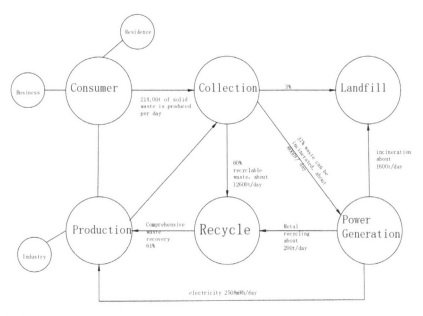

Figure 1. Singapore waste management process chart.

Based on the four measures for waste closed-loop management:

1) Control waste from the source: As mentioned above, control from the source will help reduce waste generation. For example, for construction waste, prefabricated buildings refine and quantify the construction process and achieve the purpose that most of the building materials can be used through advance survey and careful design, thus reducing the generation of construction waste (Jaillon et al. 2009);
2) Recycling: Use certain collection and reuse system, complete the recycling of waste, exhaust the use value of goods, increase the service life of goods;
3) Incineration for Power Generation: For non-recyclable items, the current common treatment method is to use incineration to generate electricity. Nowadays, with the continuous development of science and technology in Singapore, more and more new technologies, such as microbial degradation, are being applied into the conversion and recycling of articles;
4) Reduce Landfill: Landfill is not a good way to degrade all the waste, and it is easy to cause secondary pollution. For Singapore, providing land to build landfills is not an option that can be used once and for all. The shortage of land requires Singapore to reduce landfills as much as possible and increase the service life of landfills.

3 SUCCESSFUL FACTORS FROM SINGAPORE "ZERO WASTE CITY" DEVELOPMENT

At present, Singapore has certain waste conversion capacity, and with the support of policies and systems, it can recycle 61% of waste, 37% of waste for burning to get the power generation. It is also able to bury burnt residue and other combustion products and 2% of waste that cannot be recycled. Thus, Singapore's waste management system has achieved some success. In November 2014, the Singapore government released the Singapore sustainable blueprint 2015, aiming to achieve 70% waste recycling rate by 2030 (Wang & Li 2018).

From the "zero waste development" in Singapore, eight successful factors can be analyzed.

1) Estimate the current daily waste production capacity and formulate a reasonable plan for the construction of waste treatment plant, considering the waste treatment capacity;
2) Formulate reasonable goals and relevant implementation policies for improvement according to the current national waste treatment situation;
3) Draw lessons from the field of water resources management and apply the reasonable management methods to the management of solid waste;
4) Unite industry associations, companies, non-governmental organizations and waste management companies to conduct targeted treatment of the waste accounting for a large proportion in the whole amount of waste, like packaging waste;
5) Carry out mandatory waste report in large shopping malls to improve the management capacity of the managers in shopping malls and to develop better waste management plans;
6) Establish funds to motivate Singapore's companies, non-profit organizations, non-government organizations, city councils, schools and other regulatory agencies to increase the utilization of solid waste or reduce the production of solid waste;
7) Strengthen publicity, develop education, issue guidelines on food waste disposal for households to reduce kitchen waste and encourage the citizens to reduce the amount of waste in the form of competition;
8) Establish partners for electronic waste utilization to set out to establishing a standardized electronic waste management system, developing a reasonable collection and recycling system for the electronic waste.

4 TOWARDS "ZERO WASTE CITY" DEVELOPMENT IN CHINA

4.1 Environmental vision in China

To begin with, the environmental protection problem in China is still serious. The situation of excessive waste, which once occurred in super cities such as Beijing, Shanghai and major provincial capitals, affects people's quality of life (Chen et al. 2010). However, in recent years, this problem has been increasingly valued by the public, and some exploration has been carried out under the effective planning of the leadership, the call of "clean water and green mountains are gold and silver mountains" has been issued as well. China maintains an attitude of learning from the new technologies and systems of environmental protection, which is not only for the purpose of environmental protection in China, but also for deepening the partnership between China and other countries. Thus, it can be seen that China actively dealing with environmental issues and holds a positive attitude towards Singapore's new concept of "zero waste city" as a responsible country. As a result, China will take measures to address the growing trend of waste.

On April 30, 2019, the ministry of ecology and environment announced 11 pilot cities for "zero waste" construction, including Xuzhou in Jiangsu province, Shenzhen city in Guangdong province, Baotou city in Inner Mongolia autonomous region and Tongling city in Anhui province. Meanwhile, Xiongan new area in Hebei province, Beijing economic and technological development zone, china-Singapore Tianjin eco-city, Guangze prefecture in Fujian province, and Ruijin city in Jiangxi province are the special cases to be promoted with reference to the "zero waste city" construction pilot.

4.2 Comparison of national conditions in China and Singapore

Singapore has a national territorial area of 719.2 square kilometers and a population of 5.80626 million (World Population Review 2019). China has a national territorial area of 9.6 million square kilometers and a population of 1395.38 million (National Bureau of Statistics 2018). Although China has a vast territory, large population leads to small per capita land area. Therefore, in the case of Singapore, the construction of large landfill sites is not the best way to deal with waste, and the amount of landfill should be reduced as much as possible. Waste recycling is a major problem in environmental protection sector in China. According to the data, the recycling rate of waste in South Korea can reach 90% (Su & Wang 2018), many developed countries are close to that figure. However, recycling rate in China is less than 5%. The source of this problem is the unreasonable private disposal of waste (such as straw burning in the open air, the failure to implement waste classification, the backward waste recycling technology and insufficient publicity.

Table 1. Comparison and contrast between China and Singapore in land area, population, and urbanization rate and population density.

	Land area	Population (2019)	Urbanization rate	Population density
China	9.6 million square km	1.39538 billion	59.58%	147.67persons/square km
Singapore	719.1 square km	5.806269 million	100%	7915.73perons/square km

After the comparison, the national conditions of China and Singapore are similar to some extent. Both countries have large population in a unit area. However, the urbanization rate in Singapore was almost twice as much as the urbanization rate in China. Therefore, the reference to the construction of waste free city in Singapore is feasible in China.

5 CONCLUSIONS AND RECOMMENDATIONS

In conclusion, as a developing country, China needs to actively deal with environmental protection while committing itself to economic development. As for the increasingly serious waste disposal problems, "zero waste city" has been put on the agenda of China's socialist construction. As a pioneer of "zero waste city" and a friendly partner of China, Singapore's waste management experience is worthy of China's reference. But the specific implementation measures need to be adapted to China's national conditions. China has the ability to try controlling the waste at its source and cooperating with manufacturers and retailers to recycle it. But China still needs scientific research to improve recycling rates and the methods to utilize the waste.

From the previous analysis, it can be found that four approaches and five initiatives for waste disposal in Singapore are connected and mutually supportive. Combined with the development of Chinese characteristics, it is feasible to control the waste from the source. As stated above, the government should increase publicity efforts and combine the actual situation of each production department and manufacturers to estimate and quantify the potential waste, so as to minimize the waste production. At the same time, reports, agreements and awards should be used to encourage production departments to recycle and reuse the waste and raise the awareness of waste management. In addition, some research can be carried out on the generated waste that cannot be processed at present, so as to develop new technologies for waste utilization and transformation.

Based on the results of the lessons learned from Singapore, four detailed recommendations are suggested for the future evaluation of waste management systems:

1) To establish a completely regulation and supervision system:
   China's current laws and regulations on waste disposal are not complete, so relevant laws and regulations should be formulated as soon as possible to clarify the responsibilities of personnel at all levels and coordinate the work of all departments. In addition, relevant departments need to be established to supervise waste disposal matters, and relevant rules and regulations should be used as tools and guidelines, such as "mandatory waste reporting in large commercial premises" in the Singapore five initiative, to refine responsibilities;
2) Strengthen the publicity of waste recycling management:
   Cause by China's large population, propaganda should be targeted. Raise the awareness of

waste recycling among builders, designers, governments and managers. Such publicity is conducive to the implementation of laws and regulations, the smooth progress of supervision and the promotion of resource-based products;

3) Develop waste recycling technology and contact relevant industries:

At present, there is still a lot of waste in China that cannot be recycled and treated. In addition to introducing new technologies and equipment from abroad, Chinese technical developers also need to develop solutions suitable for China's waste situation. In addition, manufacturers and sellers located in the upstream and downstream of the same product chain can try to connect them together to jointly develop waste recycling strategies, reduce manufacturers' reproduction costs and make profits for sellers;

4) Improve the reward and punishment system:

Singapore's 3R fund and "packaging agreements", which recognize and commend companies to reduce the most waste each year, have helped to encourage the development and maintenance of Singapore's waste management system. In addition, certain policies can be set up to encourage relevant industries to use environment-friendly materials for production, such as subsidies for the cost of environment-friendly materials.

ACKNOWLEDGEMENT

The authors wish to acknowledge the financial support from Summer Undergraduate Research Fund, Xi'an Jiaotong-Liverpool University (XJTLU) and Suzhou Soft Science Fund (2019).

REFERENCES

Chen, X. Geng, Y. & Fujita, T. 2010. An Overview of Municipal Solid Waste Management in China. *Waste Management* 30(4): 716-724.

Hao, J.L. Chen, Z.K. Bao, Z.K. Guo, F.Y. & Xing, W.Q. 2018. Quantifying construction waste reduction through the application of prefabrication in China. In *NAXOS 6th International Conference on Sustainable Solid Waste Management.*

Jaillon, L.C. Poon, C.S. & Chiang, Y.H. 2009. Quantifying the waste reduction potential of using prefabrication in building construction in Hong Kong. *Waste Management* 29(1): 309-320.

Jiao, X.D. & Du, H.Z. 2019. Research on the status quo, problems and management countermeasures of urban waste treatment in China. *Renewable Resources and Circular Economy* 12(01): 23-27.

Min, J. Cheng, B. Gao, S. Luo, Z. Zhu, K. & Liu, B. 2018. Quantitative Assessment of City's Smart Growth Level in China Using Fuzzy Comprehensive Evaluation. In *MATEC Web of Conferences (Vol. 175, p. 04023).*

Singapore Population 2019. Retrieved July 9, 2019, from: World Population Review.

Su, Y. B. & Wang, L. 2018. Enlightenment of Resource Utilization of Korea's Construction Waste to China. *Construction Economy* 39(12): 22-26.

Total Population. 2018. National Bureau of Statistics.

Wang, Y. Y. & LI, P. W.2018. Make "Zero Waste City" Cooperation between China and Singapore to a Green 'One Belt, One Way' Cooperation Paradigm, Zero Waste, How Did Singapore Do? *Ecological Civilization of China* 26(4): 86-88.

Author index